T0271920

An Advanced Course in Probability and Stochastic Processes

An Advanced Course in Probability and Stochastic Processes provides a modern and rigorous treatment of probability theory and stochastic processes at an upper undergraduate and graduate level. Starting with the foundations of measure theory, this book introduces the key concepts of probability theory in an accessible way, providing full proofs and extensive examples and illustrations. Fundamental stochastic processes such as Gaussian processes, Poisson random measures, Lévy processes, Markov processes, and Itô processes are presented and explored in considerable depth, showcasing their many interconnections. Special attention is paid to martingales and the Wiener process and their central role in the treatment of stochastic integrals and stochastic calculus. This book includes many exercises, designed to test and challenge the reader and expand their skillset. *An Advanced Course in Probability and Stochastic Processes* is meant for students and researchers who have a solid mathematical background and who have had prior exposure to elementary probability and stochastic processes.

Key Features:
- Focus on mathematical understanding
- Rigorous and self-contained
- Accessible and comprehensive
- High-quality illustrations
- Includes essential simulation algorithms
- Extensive list of exercises and worked-out examples
- Elegant and consistent notation

Dirk P. Kroese, PhD, is a Professor of Mathematics and Statistics at The University of Queensland. He has published over 140 articles and seven books in a wide range of areas in applied probability, mathematical statistics, data science, machine learning, and Monte Carlo methods. He is a pioneer of the well-known Cross-Entropy method—an adaptive Monte Carlo technique, which is being used around the world to help solve difficult estimation and optimization problems in science, engineering, and finance.

Zdravko Botev, PhD, teaches Computational Statistics and Applied Probability at the University of New South Wales in Sydney, Australia. He is the recipient of the 2018 Christopher Heyde Medal of the Australian Academy of Science for distinguished research in the Mathematical Sciences and the 2019 Gavin Brown prize for his work on kernel density estimation, for which he is the author of one of the most widely used Matlab software scripts.

An Advanced Course in Probability and Stochastic Processes

Dirk P. Kroese and Zdravko I. Botev

CRC Press

Taylor & Francis Group

Boca Raton London New York

CRC Press is an imprint of the
Taylor & Francis Group, an **informa** business

A CHAPMAN & HALL BOOK

First edition published 2024
by CRC Press
2385 NW Executive Center Drive, Suite 320, Boca Raton FL 33431

and by CRC Press
4 Park Square, Milton Park, Abingdon, Oxon, OX14 4RN

CRC Press is an imprint of Taylor & Francis Group, LLC

Library of Congress Cataloging-in-Publication Data
Names: Kroese, Dirk P., author. | Botev, Zdravko I., 1982- author.
Title: An advanced course in probability and stochastic processes / Dirk P. Kroese and Zdravko I. Botev.
Description: First edition. | Boca Raton, FL : CRC, 2024. | Includes bibliographical references and index.
Identifiers: LCCN 2023034568 (print) | LCCN 2023034569 (ebook) | ISBN 9781032320465 (hbk) | ISBN 9781032324401 (pbk) | ISBN 9781003315018 (ebk)
Subjects: LCSH: Stochastic processes.
Classification: LCC QA274 .K76 2024 (print) | LCC QA274 (ebook) | DDC 519.2--dc23/eng/20231016
LC record available at https://lccn.loc.gov/2023034568
LC ebook record available at https://lccn.loc.gov/2023034569

ISBN: 978-1-032-32046-5 (hbk)
ISBN: 978-1-032-32440-1 (pbk)
ISBN: 978-1-003-31501-8 (ebk)

DOI: 10.1201/9781003315018

Typeset in TeXGyreTermesX-Regular font
by KnowledgeWorks Global Ltd.

Publisher's note: This book has been prepared from camera-ready copy provided by the authors.

In memory of Janny and Gerrit

— *DPK*

To Sarah, Sofia, George, and my parents

— *ZIB*

CONTENTS

PREFACE

This book resulted from various probability and statistics courses that we have taught over the years. Of course, there are many good books available that teach probability from a very elementary to a highly advanced level, so why the need for another one? We wanted to create a probability course that was (1) mathematically rigorous at an upper-undergraduate/lower-graduate level, and (2) would cover most of the important topics in advanced probability and stochastic processes, while (3) still being concise enough to fit into a one-semester curriculum. Another reason why we wrote the book is that it is enjoyable to try to explain mathematical "truth" in as simple a way as possible. We hope that we have succeeded in this. To paraphrase Richard Feynman: "if you cannot explain something in simple terms, you don't really understand it".

Naturally, in writing this book, we have been influenced by our own teachers. In particular, Dirk has been fortunate to have experienced the probability lectures of Erhan Çinlar in person at Princeton, and several results in this book have been inspired by his lecture notes, which later appeared in Çinlar (2011). Other main sources used were Billingsley (1995), Chung and Williams (1990), Grimmett and Stirzaker (2001), Kallenberg (2021), Karatzas and Shreve (1998), Klebaner (2012), Kreyszig (1978), Kroese et al. (2019), Kroese et al. (2011), and Mörters and Peres (2010). The online lecture notes of Lalley (2012) were also very helpful.

Ideally, the reader should be well-versed in mathematical foundations (calculus, linear algebra, real- and complex-analysis, etc.), and should have had some exposure to elementary probability and stochastic processes, including topics such as Markov chains and Poisson processes, although the latter is not essential.

An outline of the rest of the book is as follows. We start by emphasizing the crucial role of mathematical notation and introduce the "house-style" of the book.

An advanced course in probability should have measure-theoretic foundations. For this reason, Chapter 1 reviews the fundamentals of measure theory. We introduce σ-algebras, measurable spaces, measurable functions, measures, and integrals.

Probability starts in earnest in Chapter 2. We explain how concepts such as probability spaces, random variables, stochastic processes, probability distributions, and expectations, can be elegantly introduced and analysed through measure theory. We also discuss L^p spaces, moment and characteristic functions, the role of independence in probability, and how filtrations of σ-algebras can model information flow. Important stochastic processes such as Gaussian processes, Poisson random measures, and Lévy processes make their first appearance at the end of the chapter.

Chapter 3 deals with convergence concepts in probability. We discuss almost sure convergence, convergence in probability, convergence in distribution, and L^p convergence. The notion of uniform integrability connects various modes of convergence. Main applications are the Law of Large Numbers and the Central Limit Theorem.

Chapter 4 describes how the concept of conditioning can be used to process additional knowledge about a random experiment. We first introduce conditional expectations and then conditional distributions. Two existence results for probability spaces are also presented. At the end of the chapter we introduce Markov chains and Markov jump processes, and show how they are related to Lévy processes and Poisson random measures.

Martingales form an important class of stochastic processes. They are introduced and discussed in Chapter 5. Stopping times, filtrations, and uniform integrability play important roles in their analysis. The key results are Doob's stopping theorem and the martingale convergence theorem. Example applications include proofs for the Law of Large Numbers and the Radon–Nikodym theorem.

Chapter 6 deals with the Wiener process and Brownian motion. We prove the existence of the Wiener process and put some of its many properties on display, including its path properties, the strong Markov property, and the reflection principle. Various martingales associated with the Wiener process are also discussed, as well as its relation to the Laplace operator. We show that the maximum and hitting time processes have close connections to Lévy processes and Poisson random measures.

Chapter 7 concludes with a detailed introduction to stochastic integrals with respect to the Wiener process, Itô diffusions, stochastic differential equations, and stochastic calculus.

There are many definitions, theorems, and equations in this book, but not all of them are equally important. The most important definitions and theorems are displayed in blue and yellow boxes, respectively:

Definition 1: Complex i	**Theorem 3: Euler's Identity**
The number i is given by	It holds that
(2) $\quad i = \sqrt{-1}.$	(4) $\quad e^{i\pi} + 1 = 0.$

It is important to note that definitions, theorems, equations, figures, etc., are numbered consecutively using the *same* counter. This facilitates searches in the text. Less important definitions may appear within the text, and are often *stressed* by the use of italics. Less fundamental theorems are stated as propositions or lemmas and are displayed in boxes with a lighter color. Sections, subsections, and exercises are numbered separately.

We have included many exercises throughout the book and encourage the reader to attempt these, as actions speak louder than (reading) words. Similarly, the inclusion of various algorithms, as well as pseudo and actual MATLAB code, will help with a better understanding of stochastic processes, when actually simulating these on a computer. Solutions to selected exercises (indicated by a ∗) are given in Appendix A. Appendix B summarizes some important results on function spaces. Appendix C gives a complete proof of the existence of the Lebesgue measure.

Acknowledgments

The front cover shows a beautiful random process from nature: the path created by a larva of the Scribbly Gum Moth (*ogmograptis scribula*) from southern Queensland, Australia. Many thanks to Shanna Bignell for providing this photo. We thank Nhat Pham, Thomas Taimre, Ross McVinish, Chris van der Heide, Phil Pollett, Joe Menesch, Pantea Konn, Kazu Yamazaki, Joshua Chan, and Christian Hirsch for their very helpful comments. Of course, this book would not have been possible without the pioneering work of the giants of probability: Kolmogorov, Lévy, Doob, and Itô. This work was supported by the Australian Research Council, under grant number DP200101049.

Dirk Kroese and Zdravko Botev
Brisbane and Sydney, 2023

NOTATION

Notation is the heart of mathematics. Its function is to keep the myriad of ideas and concepts of the mathematical body connected and alive. A poorly performing notation system confuses the brain.

We have tried to keep the notation as simple as possible, descriptive, easy to remember, and consistent. We hope that this will help the reader to quickly recognize the mathematical objects of interest (functions, measures, vectors, matrices, random variables, etc.) and understand intricate ideas. The choice of fonts plays a vital role in the notation. In this book we use the following conventions:

- Boldface font is used to indicate composite objects, such as column vectors $x = [x_1, \ldots, x_n]^\top$ and matrices $\mathbf{X} = [x_{ij}]$. Note also the difference between the upright bold font for matrices and the slanted bold font for vectors.

- Random variables are generally specified with upper case roman letters X, Y, Z and their values with lower case letters x, y, z. Random vectors are thus denoted in upper case slanted bold font: $X = [X_1, \ldots, X_n]^\top$.

- Sets are generally written in upper case roman font D, E, F and their σ-algebras are in calligraphic font $\mathcal{D}, \mathcal{E}, \mathcal{F}$. The set of real numbers uses the common blackboard bold font $\mathbb{R}$. Expectation $\mathbb{E}$ and probability $\mathbb{P}$ also use the latter font.

- Probability distributions use a sans serif font, such as Bin and Gamma. Exceptions to this rule are the standard notations $\mathcal{N}$ and $\mathcal{U}$ for the normal and uniform distributions.

A careful use of delimiters (parentheses, brackets, braces, etc.) in formulas is important. As a rule, we omit delimiters when it is clear what the argument is of a function or operator. For example, we prefer $\mathbb{E}X^2$ to $\mathbb{E}[X^2]$ and $\ln x$ to $\ln(x)$. Summation and other indexes will also be suppressed when possible. Hence, we write (x_n) rather than $(x_n, n \in \mathbb{N})$, $\lim x_n$ instead of $\lim_{n \to \infty} x_n$, and $\sum_n x_n$ for $\sum_{n=0}^{\infty} x_n$. When defining sets and other *unordered* objects, we will use curly braces

---

{ }. In contrast, parentheses ( ) and square brackets [ ] signify *ordered* objects, such as sequences and vectors.

Measures are generally denoted by Greek letters λ, μ, ν, and π. An exception is the occasional use of the notation Leb for the Lebesgue measure. Whether π plays the role of a measure or of the fundamental constant $3.14159\ldots$ will be clear from the context.

For (Lebesgue) integrals we favor the notations $\int \mu(dx) f(x)$, $\int f\, d\mu$ or just μf over $\int f(x)\mu(dx)$; and when μ is the Lebesgue measure, we simply write $\int dx\, f(x)$. Putting the integrating measure directly after the integral sign avoids unnecessary brackets in repeated integrals; for example,

$$\int \mu(dx) \int \nu(dy) f(x, y) \quad \text{instead of} \quad \int \left(\int f(x, y)\nu(dy) \right) \mu(dx).$$

We make an exception for stochastic and Stieltjes integrals, which have the integrator at the end, such as in $\int F_t\, dW_t$.

The qualifiers *positive* and *increasing* are used in a "lenient" sense. Thus, a positive function can take values ≥ 0 and an increasing function can remain constant in certain intervals. We add the adjective *strict* to indicate the stringent sense. Thus, a strictly positive function can only take values > 0.

A function f is a mapping from one set to another set. Its function *value* at x is $f(x)$. The two should not be confused.

Reserved letters

$\mathcal{B}$	Borel σ-algebra on $\mathbb{R}$
$\mathbb{C}$	set of complex numbers
d	differential symbol
$\mathbb{E}$	expectation
e	the number $2.71828\ldots$
$\mathbb{1}\{A\}$ or $\mathbb{1}_A$	indicator function of set A
i	the square root of -1
ln	(natural) logarithm
$\mathbb{N}$	set of natural numbers $\{0, 1, \ldots\}$
O	big-O order symbol: $f(x) = O(g(x))$ if $|f(x)| \leq \alpha g(x)$ for some constant α as $x \to a$
o	little-o order symbol: $f(x) = o(g(x))$ if $f(x)/g(x) \to 0$ as $x \to a$
$\mathbb{P}$	probability measure
π	the number $3.14159\ldots$; also used for measures
$\mathbb{Q}$	set of rational numbers

$\mathbb{R}$	set of real numbers (one-dimensional Euclidean space)
$\mathbb{T}$	index (time) set of a stochastic process
Ω	sample space
$\mathbb{Z}$	set of integers $\{\ldots, -1, 0, 1, \ldots\}$

General font/notation rules

x	scalar
$\boldsymbol{x}$	vector
X	random vector
$\mathbf{X}$	matrix

Matrix/vector notation

$\mathbf{A}^{\top}, \boldsymbol{x}^{\top}$	transpose of matrix $\mathbf{A}$ or vector $\boldsymbol{x}$
$\mathbf{A}^{-1}$	inverse of matrix $\mathbf{A}$
$\det(\mathbf{A})$	determinant of matrix $\mathbf{A}$
$|\mathbf{A}|$	absolute value of the determinant of matrix $\mathbf{A}$

Specific notation

$\cap$	$A \cap B$ is the intersection of sets A and B
$\cup$	$A \cup B$ is the union of sets A and B
$\backslash$	$A \backslash B$ is the set difference between sets A and B
$\subseteq$	$A \subseteq B$ means that A is a subset of, or is equal to, B
$\supseteq$	$A \supseteq B$ means that A contains or is equal to B
$\vee$	$x \vee y$ is the maximum of x and y. Also, $\mathcal{F} \vee \mathcal{G}$ is the smallest σ-algebra generated by the sets in $\mathcal{F} \cup \mathcal{G}$
$\wedge$	$x \wedge y$ is the minimum of x and y
$\circ$	$f \circ g$ is the composition of functions f and g. Also used to define the Itô–Stratonovich integral $\int Y_s \circ dX_s$
$\otimes$	$\mathcal{E} \otimes \mathcal{F}$ is the product σ-algebra of $\mathcal{E}$ and $\mathcal{F}$. Also, $\mu \otimes \nu$ is the product measure of measures μ and ν.
$\in$	is an element of, belongs to
$\forall$	for all
$\sim$	is distributed as
$\overset{\text{iid}}{\sim}, \sim_{\text{iid}}$	are independent and identically distributed as

$\overset{\text{approx.}}{\sim}$	is approximately distributed as
$C[0,1]$	space of continuous functions on $[0,1]$
$\mathbb{E}_{\mathcal{F}_s}$, $\mathbb{E}_s$	conditional expectation given $\mathcal{F}_s$
$\mathbb{E}^x$	expectation operator under which a process starts at x
$\mathcal{F}_+$	set of positive $\mathcal{F}$-measurable numerical functions
$L^2[0,1]$	space of square-integrable functions on $[0,1]$
$\overline{\mathbb{N}}$	extended natural numbers: $\mathbb{N} \cup \{\infty\}$
$\mathbb{P}^x$	probability measure under which a process starts at x
$\overline{\mathbb{R}}$	extended real line: $\mathbb{R} \cup \{-\infty, \infty\}$
$\mathbb{R}^n$	n-dimensional Euclidean space
$\mathbb{R}_+$	positive real line: $[0,\infty)$
$\partial_i f$	partial derivative of f with respect to component i
$\partial_{ij} f$	second partial derivative of f with respect to components i and j
$\boldsymbol{\partial} f$	gradient of f
$\boldsymbol{\partial}^2 f$	Hessian matrix of f
Δf	Laplace operator acting on f
f^+	$\max\{f, 0\}$
f^-	$\max\{-f, 0\}$
$\approx$	is approximately
$\ll$	is absolutely continuous with respect to
$:=, =:$	is defined as, is denoted by
$\Rightarrow$	implies
$\rightarrow$	converges/tends to
$\uparrow$	increases to
$\downarrow$	decreases to
$\mapsto$	maps to
$\overset{\text{a.s.}}{\rightarrow}$	converges almost surely to
$\overset{\text{cpl.}}{\rightarrow}$	converges completely to
$\overset{\text{d}}{\rightarrow}$	converges in distribution to
$\overset{\mathbb{P}}{\rightarrow}$	converges in probability to
$\overset{L^p}{\rightarrow}$	converges in L^p norm to
$|\cdot|$	absolute value

$\|\cdot\|$	norm; e.g., Euclidean norm or supremum norm. Also, $\|\Pi\|$ is the mesh of a segmentation
$\|\cdot\|_p$	L^p norm
$\lceil x \rceil$	smallest integer larger than x
$\lfloor x \rfloor$	largest integer smaller than x
$\langle X \rangle$	quadratic variation process of X
$\langle \cdot, \cdot \rangle$	$\langle f, g \rangle$ is the inner product of functions f and g. Also, $\langle X, Y \rangle$ is the covariation process of stochastic processes X and Y

Probability distributions

Ber	Bernoulli
Beta	beta
Bin	binomial
Exp	exponential
F	Fisher–Snedecor F
Geom	geometric
Gamma	gamma
Itô	Itô process
$\mathcal{N}$	normal or Gaussian
Pareto	Pareto
Poi	Poisson
t	Student's t
$\mathcal{U}$	uniform

Abbreviations

a.s.	almost surely
cdf	cumulative distribution function
iid	independent and identically distributed
MGF	moment generating function
Leb	Lebesgue measure
ODE	ordinary differential equation
pdf	probability density function (discrete or continuous)
SDE	stochastic differential equation
UI	uniformly integrable

CHAPTER 1

MEASURE THEORY

The purpose of this chapter is to introduce the main ingredients of *measure theory*: measurable spaces, measurable functions, measures, and integrals. These objects will form the basis for a rigorous treatment of probability, to be discussed in subsequent chapters.

1.1 Measurable Spaces

Measure theory is a branch of mathematics that studies measures and integrals on general spaces. A measure can be thought of as a generalization of a length, area, or volume function to an arbitrary one- or multi-dimensional space. We will see that the concept is, in fact, much more general. Why do we need a study of length? Is it not obvious that, for example, the length of an interval (a, b) with $a < b$ is equal to $b - a$? The following two examples indicate that there are potential complications with the "length" concept.

■ **Example 1.1 (Cantor Set)** Consider the set that is constructed in the following way, illustrated in Figure 1.2:

Figure 1.2: Construction of the Cantor set.

Take the interval $[0, 1]$. Divide it into three parts: $[0, \frac{1}{3}]$, $(\frac{1}{3}, \frac{2}{3})$, and $[\frac{2}{3}, 1]$. Cut out the middle interval $D_{0,1} := (\frac{1}{3}, \frac{2}{3})$. Next, divide the remaining two closed intervals in the same way by removing the middle parts (i.e., removing the open intervals $D_{1,1} := (\frac{1}{9}, \frac{2}{9})$ and $D_{1,2} := (\frac{7}{9}, \frac{8}{9})$), and continue this procedure recursively, *ad infinitum*. The resulting set is called the *Cantor set*. Figure 1.2 only shows a few steps of the construction, starting with the interval $[0, 1]$ of length 1 at the bottom and ending with a set that is the union of 32 intervals with a combined length of $(2/3)^5 \approx 0.13$ at the top. If we keep going, we end up with a "dust" of points, invisible to the eye, with a combined length of 0. However, the *cardinality* of this Cantor set (i.e., the number of elements in this set) is equal to the cardinality of the set of real numbers, $\mathbb{R}$; see Exercise 19. ∎

Although the above example may seem artificial, similar sets appear quite naturally in the study of random processes. For example, we will see in Chapter 6 that the set of times when a Wiener process is zero behaves very much like the Cantor set. The Cantor set shows that the "length" of a set is fundamentally different from its size or cardinality. Even more bewildering is that some sets may not even have a length, as shown in the following example. Here, we make use of the *axiom of choice*, which states that, given an arbitrary collection of sets, we may construct a new set by taking a single element from each set in the collection. The axiom of choice cannot be proved or disproved, but states a self-evident fact.

∎ **Example 1.3 (A Set Without Length)** We wish to construct a set C on the unit circle S that cannot have a length. We can denote the points on the circle by $e^{ix} = (\cos x, \sin x), x \in [0, 2\pi)$, and we know that the circle has length 2π.

Let us divide the circle into equivalence classes: each point e^{ix} is grouped into a class with all points of the form $e^{i(n+x)}$ for all $n \in \mathbb{Z}$, where $\mathbb{Z}$ is the set of integers. Each equivalence class has a countably-infinite number of points (one for each $n \in \mathbb{Z}$) and there are uncountably many of these equivalence classes.

Applying the axiom of choice, we can make a subset C of the circle by electing *one* member from each equivalent class as the representative of that equivalence class. The collection of representatives, C, can thus be viewed as a "congress" of the points on the circle.

The set C has uncountably many elements. What else can we say about C? Let $C_n := e^{in}C$ be a rotated copy of C, for each $n \in \mathbb{Z}$. Then, by construction of C, the union $\cup_n C_n = S$ and, moreover, all the copies are disjoint (non-overlapping).

Since the union of the C_n is the whole circle, i.e.,

$$S = C_0 \cup C_1 \cup C_{-1} \cup \cdots$$

and all copies C_n must have the *same* length as C (as they are simply obtained by a rotation) — if they indeed have a length —, we must have

(1.4) $2\pi = \text{length}(S) = \text{length}(C) + \text{length}(C) + \cdots$

Suppose that C has some length $c \geq 0$. Then, (1.4) leads to a contradiction, as we either get $2\pi = 0$ or $2\pi = \infty$. Hence, C cannot have a length. ∎

The above examples illustrate that, when dealing with large sets such as $\mathbb{R}$ or the interval $(0, 1]$, we cannot expect every subset therein to have a length. Instead, the best we can do is to define the length only for certain subsets.

In what follows, we let E be an arbitrary set; e.g., the interval $(0, 1]$ or the square $(0, 1] \times (0, 1]$. We wish to assign a "measure" (think of "length" or "area") to certain subsets of E. We can apply the usual set operations to subsets of E, as illustrated in Figure 1.5.

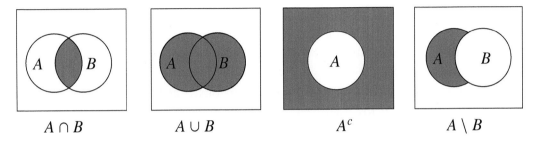

$A \cap B$ $\qquad\qquad A \cup B$ $\qquad\qquad\qquad A^c$ $\qquad\qquad\qquad A \setminus B$

Figure 1.5: Venn diagrams of set operations: intersection, union, complement, and set difference.

The collection of all subsets to which we wish to assign a measure is usually a σ-algebra:

Definition 1.6: σ-Algebra

A σ-*algebra* $\mathcal{E}$ on E is a collection of subsets of E that contains E itself, and that is closed under complements and *countable* unions; that is:
1. $E \in \mathcal{E}$.
2. If $A \in \mathcal{E}$, then also $A^c \in \mathcal{E}$.
3. If $A_1, A_2, \dots \in \mathcal{E}$, then also $\cup_n A_n \in \mathcal{E}$.

A σ-algebra is also closed under countable intersections: If $A_1, A_2, \dots \in \mathcal{E}$, then also $\cap_n A_n \in \mathcal{E}$; see Exercise 1. If instead in Definition 1.6 the collection $\mathcal{E}$ is closed only under *finite* unions (while Items 1. and 2. remain the same), then $\mathcal{E}$ is said to be an *algebra*. Convince yourself that the conditions of a σ-algebra are natural and are minimal requirements to have, for a collection of sets to which we wish to assign a measure.

■ **Example 1.7 (Algebra and σ-Algebra)** Let $E := (0, 1]$ and let $\mathcal{E}_0$ be the collection of finite unions of non-overlapping intervals of the form $(a, b]$, where $0 \le a < b \le 1$; we also add $\emptyset$ to $\mathcal{E}_0$. For example, the set $(0, 1/4] \cup (1/3, 1/2]$ lies in $\mathcal{E}_0$. Note that $\mathcal{E}_0$ is an algebra. However, it is *not* a σ-algebra; see Exercise 6.

For any set E, the collection of *all* subsets of E, the so-called *power set* of E, written as 2^E, is of course a σ-algebra. Unfortunately, as Example 1.3 indicates, it will often be too large to allow a proper measure to be defined thereon.   ■

Let E be a set with σ-algebra $\mathcal{E}$. The pair $(E, \mathcal{E})$ is called a *measurable space*. The usual way to construct a σ-algebra on a set E is to start with a smaller collection C of sets in E and then take the *intersection* of all the σ-algebras that contain C. There is at least one σ-algebra that contains C: the power set of E, so this intersection is not empty. That the intersection of σ-algebras is again a σ-algebra can be easily checked; see Exercise 7. We write $\sigma(C)$ (or also σC) for this σ-algebra and call it the σ-algebra that is *generated* by C.

■ **Example 1.8 (Borel σ-algebra on $\mathbb{R}^n$)** Consider $E := \mathbb{R}$, and let C be the collection of all intervals of the form $(-\infty, x]$ with $x \in \mathbb{R}$. The σ-algebra that is generated by C is called the *Borel σ-algebra* on $\mathbb{R}$; we denote it by $\mathcal{B}$. This $\mathcal{B}$ contains all sets of interest to us: intervals, countable unions of intervals, and sets of the form $\{x\}$. We can do the same in n dimensions. Let $E := \mathbb{R}^n$ and let C be the collection of subsets of the form $(-\infty, x_1] \times \cdots \times (-\infty, x_n]$, $x_1, \ldots, x_n \in \mathbb{R}$. The σ-algebra $\sigma(C)$ is called the Borel σ-algebra on $\mathbb{R}^n$; we denote it by $\mathcal{B}^n$.   ■

In the above example, we could have taken different collections C to generate the Borel σ-algebra on $\mathbb{R}$. For instance, we could have used the collection of *open sets* on $\mathbb{R}$. More generally, for any *topological space* (see Definition B.4), the σ-algebra generated by the open sets is called the *Borel σ-algebra* on this set.

Note that the collections C in Example 1.8 are closed under finite intersections; e.g., $(-\infty, x] \cap (-\infty, y] = (-\infty, x]$ if $x \le y$. Such a collection of sets is called a p-system (where p is a mnemonic for "product").

> **Definition 1.9: p-System**
>
> A collection C of subsets of E is called a *p-system* (or π-system) if it is closed under finite intersections.

One final collection of sets that will be important for us, mainly to simplify proofs, is the *Dynkin-* or d-system (an equivalent definition is given in Exercise 10):

> **Definition 1.10: d-System**
>
> A collection $\mathcal{D}$ of subsets of E is called a *d-system* (or λ-system) if it satisfies the following three conditions:
> 1. $E \in \mathcal{D}$.
> 2. If $A, B \in \mathcal{D}$, with $A \supseteq B$, then $A \setminus B \in \mathcal{D}$.
> 3. If $(A_n) \in \mathcal{D}$, with $A_1 \subseteq A_2 \subseteq A_3 \subseteq \cdots \subseteq \cup_n A_n =: A$, then $A \in \mathcal{D}$.

Similar to a σ-algebra, the intersection of an arbitrary number of d-systems is again a d-system, and the intersection of all d-systems that contains a collection C of sets is called the d-system that is *generated* by C; we write $d(C)$ for it. The two main theorems that connect p-systems, d-systems, and σ-algebras are given next.

Theorem 1.11: σ-Algebra as a d- and p-System

A collection of subsets of E is a σ-algebra if and only if it is both a p-system and a d-system on E.

Proof. Obviously, a σ-algebra is both a d- and p-system. We now show the converse. Suppose $\mathcal{E}$ is both a d- and p-system. We check the three conditions in Definition 1.6:

1. $E \in \mathcal{E}$, because $\mathcal{E}$ is a d-system.

2. $\mathcal{E}$ is closed under complements, as $E \in \mathcal{E}$ and for any $B \in \mathcal{E}$ we have $B^c = E \setminus B \in \mathcal{E}$ by the second condition of a d-system.

3. $\mathcal{E}$ is closed under finite unions: $A \cup B = (A^c \cap B^c)^c \in \mathcal{E}$, because $\mathcal{E}$ is closed under complements (as just shown) and is a p-system. It remains to show that it is also closed under countable unions. Let $A_1, A_2, \ldots \in \mathcal{E}$. Define $B_1 := A_1$ and $B_n := B_{n-1} \cup A_n$, $n = 2, 3, \ldots$. As $\mathcal{E}$ is closed under finite unions (just shown), each $B_n \in \mathcal{E}$. Moreover, by condition 3 of a d-system, $\cup_n B_n \in \mathcal{E}$. But this union is also $\cup_n A_n$. This completes condition 3 for a σ-algebra.

$\square$

Theorem 1.12: Monotone Class Theorem

If a d-system contains a p-system, then it also contains the σ-algebra generated by that p-system.

Proof. The proof requires the following result, which is easily checked: If $\mathcal{D}$ is a d-system on E and $B \in \mathcal{D}$, then

$$(1.13) \qquad \mathcal{D}^B := \{A \in \mathcal{D} : A \cap B \in \mathcal{D}\} \quad \text{is a d-system.}$$

Consider a d-system that contains a p-system C. We do not need to give it a name, as all we need to appreciate is that it contains the d-system $\mathcal{D} := d(C)$ that is generated by C, as $d(C)$ is the *smallest* d-system that contains C. We want to show that $\mathcal{D}$ contains $\sigma(C)$, and for this it suffices to show that $\mathcal{D}$ is a σ-algebra itself. By Theorem 1.11 we just need to show that $\mathcal{D}$ is a p-system (i.e., is closed under finite intersections).

First take a set B in C and consider the set $\mathcal{D}^B$ defined in (1.13). We know it is a d-system, as $B \in C \subseteq \mathcal{D}$. It also contains C, as C is a p-system. Hence, $\mathcal{D}^B = \mathcal{D}$. In other words: for any $B \in C$ the intersection of B with any set in $\mathcal{D}$ lies in $\mathcal{D}$. Now take any set $A \in \mathcal{D}$ and consider the set

$$\mathcal{D}^A := \{B \in \mathcal{D} : A \cap B \in \mathcal{D}\}.$$

It is again a d-system (by (1.13)), and we have just shown that it contains every set B of C. Hence, $\mathcal{D}^A \supseteq \mathcal{D}$. That is, $A, B \in \mathcal{D} \Rightarrow A \cap B \in \mathcal{D}$, as had to be shown. □

A final important definition in this section is that of a *product space* of two measurable spaces.

Definition 1.14: Product Space

The *product space* of two measurable spaces $(E, \mathcal{E})$ and $(F, \mathcal{F})$ is the measurable space $(E \times F, \mathcal{E} \otimes \mathcal{F})$, where $E \times F := \{(x, y) : x \in E, y \in F\}$ is the Cartesian product of E and F, and $\mathcal{E} \otimes \mathcal{F}$ is the *product σ-algebra* on $E \times F$ that is generated by the rectangle sets

$$A \times B, \quad A \in \mathcal{E}, B \in \mathcal{F}.$$

■ **Example 1.15 (Borel σ-algebra on $\mathbb{R}^2$)** The product space of $(\mathbb{R}, \mathcal{B})$ and $(\mathbb{R}, \mathcal{B})$ is $(\mathbb{R}^2, \mathcal{B}^2)$, with $\mathcal{B}^2 = \mathcal{B} \otimes \mathcal{B}$. We can think of the product σ-algebra $\mathcal{B} \otimes \mathcal{B}$ as the collection of sets in $\mathbb{R}^2$ that can be obtained by taking complements and countable unions of rectangles in $\mathbb{R}^2$. In particular, we can obtain all the usual geometric shapes: triangles, polygons, disks, etc. from this procedure.

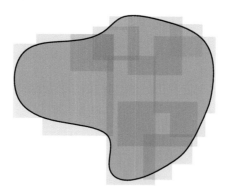

Figure 1.16: A shape being approximated by a countable union of rectangles in the product set $\mathcal{B}^2$.

■

1.2 Measurable Functions

Now we have established the basics of measurable spaces, we are going to define a class of "well-behaved" functions between such spaces.

Definition 1.17: Measurable Function

Let $(E, \mathcal{E})$ and $(F, \mathcal{F})$ be two measurable spaces, and let f be a function from E to F. The function f is called $\mathcal{E}/\mathcal{F}$-*measurable* if for all $B \in \mathcal{F}$ it holds that

(1.18) $$\{x \in E : f(x) \in B\} \in \mathcal{E}.$$

The set in (1.18) is called the *inverse image* of B under f, and is also written as

$$f^{-1}(B), \quad f^{-1}B, \quad \text{or} \quad \{f \in B\}.$$

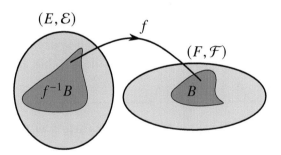

Figure 1.19: All elements in $f^{-1}B$ are mapped to somewhere in B.

Note that here f^{-1} does not mean the functional inverse of f. Notwithstanding, if f has an inverse $f^{\dagger}$, then $f^{-1}\{y\} = \{f^{\dagger}(y)\}$, so the notation f^{-1} is natural.

As most σ-algebras in practice are generated from smaller sets, the following theorem is useful in checking if a function is measurable:

Theorem 1.20: Measurable Function on a Generated σ-Algebra

A function $f : E \to F$ is $\mathcal{E}/\mathcal{F}$-measurable if and only if (1.18) holds for all $B \in \mathcal{F}_0$, with $\mathcal{F}_0$ generating $\mathcal{F}$.

Proof. Necessity is obvious. To prove sufficiency, we start from the fact that $\{f \in B\} \in \mathcal{E}$ for all $B \in \mathcal{F}_0$. Now consider the collection of sets $\mathcal{G} := \{B \in \mathcal{F} : f^{-1}(B) \in \mathcal{E}\}$. By assumption, this collection contains the sets in $\mathcal{F}_0$. Is $\mathcal{G}$ a σ-algebra? Let us check this.

1. Obviously, $F \in G$, as $E = f^{-1}(F)$.

2. Take $B \in G$. Is its complement also in G? Yes, because $f^{-1}(B^c)$ is the complement of $f^{-1}(B)$ (Hint: draw a picture).

3. Finally, take a sequence (B_n) of sets in G. Then, the inverse image of their union is

$$f^{-1}(\cup_n B_n) = \cup_n f^{-1}(B_n),$$

and so the union lies in G.

Consequently, G is a σ-algebra. Also, it contains every element of $\mathcal{F}_0$, and so it must be at least as large as (i.e., contain all elements of) $\sigma(\mathcal{F}_0) = \mathcal{F}$. That is, $G \supseteq \mathcal{F}$. But, by the definition of G, we also have $G \subseteq \mathcal{F}$. So the two must be equal. And hence $f^{-1}(B) \in \mathcal{E}$ for all $B \in \mathcal{F}$. $\qquad\square$

The next theorem shows that measurable functions of measurable functions are again measurable. Recall that the composition $g \circ f$ of functions g and f is the function $x \mapsto g(f(x))$.

Theorem 1.21: Composition of Measurable Functions

Let $(E,\mathcal{E})$, $(F,\mathcal{F})$, and $(G,\mathcal{G})$ be measurable spaces. If $f : E \to F$ and $g : F \to G$ are respectively $\mathcal{E}/\mathcal{F}$- and $\mathcal{F}/\mathcal{G}$-measurable, then $g \circ f : E \to G$ is $\mathcal{E}/\mathcal{G}$-measurable.

Proof. Take any set $B \in \mathcal{G}$ and let $C := \{g \in B\} = g^{-1}(B)$. By the measurability of g, $C \in \mathcal{F}$. Moreover, since f is $\mathcal{E}/\mathcal{F}$-measurable, the set $\{f \in C\}$ belongs to $\mathcal{E}$. But this set is the *same* as $\{g \circ f \in B\}$ (make a diagram to verify). Thus, the latter set is a member of $\mathcal{E}$ for any $B \in \mathcal{G}$. In other words, $g \circ f$ is $\mathcal{E}/\mathcal{G}$-measurable. $\quad\square$

Let $(E,\mathcal{E})$ be a measurable space. We will often be interested, especially when defining integrals, in real-valued functions that are $\mathcal{E}/\mathcal{B}$-measurable, where $\mathcal{B}$ is the Borel σ-algebra in $\mathbb{R}$. In fact, it will be convenient to consider the *extended real line* $\overline{\mathbb{R}} := [-\infty, +\infty]$, adding two "infinity" points to the set $\mathbb{R}$, and extending the arithmetic in a natural way, but leaving operations such as $\infty - \infty$ and ∞/∞ undefined. The corresponding Borel σ-algebra, denoted $\overline{\mathcal{B}}$, is generated by the p-system consisting of intervals $[-\infty, r]$, $r \in \mathbb{R}$.

Definition 1.22: Numerical Function

Let $(E,\mathcal{E})$ be a measurable space.
- A function $f : E \to \mathbb{R}$ is called a *real-valued function* on E.
- A function $f : E \to \overline{\mathbb{R}} := [-\infty, +\infty]$ is called a *numerical function* on E.
- An $\mathcal{E}/\overline{\mathcal{B}}$-measurable numerical function f is said to be $\mathcal{E}$-*measurable*; we write $f \in \mathcal{E}$. Also, $\mathcal{E}_+$ is the class of *positive* $\mathcal{E}$-measurable functions.

Theorem 1.20 shows that to check if $f \in \mathcal{E}$, all we have to verify is that

$$\{f \le r\} = f^{-1}[-\infty, r] \in \mathcal{E} \quad \text{for all} \quad r \in \mathbb{R}.$$

■ **Example 1.23 (Indicator Function)** Let $(E, \mathcal{E})$ be a measurable space. It will turn out that the building block for all $\mathcal{E}$-measurable functions is the *indicator function* (in real analysis also called *characteristic function* — not to be confused with the characteristic function defined in Section 2.6.2). The indicator function of a set $A \subseteq E$ is defined as

$$\mathbb{1}_A(x) := \begin{cases} 1 & \text{if } x \in A, \\ 0 & \text{if } x \notin A. \end{cases}$$

If $A \in \mathcal{E}$, then $\mathbb{1}_A$ is $\mathcal{E}$-measurable (in our notation: $\mathbb{1}_A \in \mathcal{E}$). However, if $A \notin \mathcal{E}$, then $\mathbb{1}_A$ is *not* $\mathcal{E}$-measurable. ■

The moral of the previous example is that not all numerical functions are measurable, but if we start from measurable indicator functions, we should be able to construct a large class of measurable functions. We will show this next.

The first extension is to take linear combinations of measurable indicator functions. A numerical function f on E is called *simple* if there exists an $n \in \{1, 2, \ldots\}$ and subsets $A_i \in \mathcal{E}$, $i = 1, \ldots, n$, such that f can be written as

$$f = \sum_{i=1}^{n} a_i \mathbb{1}_{A_i},$$

with each $a_i \in \mathbb{R}$. Any simple function can be written in a "canonical" form, where the sets $\{A_i\}$ form a *partition* of E; that is, they do not overlap and their union is E. It is easy to check that simple functions are measurable. Moreover, with f and g simple functions, any of the functions

$$f + g, \quad f - g, \quad fg, \quad f/g, \quad \underbrace{f \wedge g}_{\min\{f,g\}}, \quad \underbrace{f \vee g}_{\max\{f,g\}}$$

is again a simple function. For the function f/g we require that $g(x) \ne 0$ for all $x \in E$. We can further expand the collection of measurable functions that can be obtained from simple functions by taking limits of sequences of measurable functions. Let us first recall some definitions regarding sequences of real numbers.

Definition 1.24: Infimum and Supremum

Let (a_n) be a sequence of real numbers.

1. The *infimum*, $\inf a_n$, is the greatest element in $\overline{\mathbb{R}}$ that is less than or equal to all a_n. It can be $-\infty$.

2. The *supremum*, $\sup a_n$, is the smallest element in $\overline{\mathbb{R}}$ that is greater than or equal to all a_n. It can be ∞.

If for some $N \in \mathbb{N}$ it holds that $\inf a_n = a_N$, then we can write $\inf a_n$ also as $\min a_n$. In that case, the set of indexes for which the minimum is attained is denoted by $\operatorname{argmin} a_n$. In the same way, when the supremum is attained by some $N \in \operatorname{argmax} a_n$, we can write $\max a_n$ instead of $\sup a_n$. Similarly, the infimum and supremum can be defined for any set $A \subseteq \overline{\mathbb{R}}$. In particular, if $A := \{a_n, n \in \mathbb{N}\}$, then $\inf A = \inf a_n$.

Definition 1.25: Liminf and Limsup

Let (a_n) be a sequence of real numbers.

1. The *limit inferior*, $\liminf a_n$, is the eventual lower bound for the sequence (a_n); that is, $\liminf a_n := \lim_{m \to \infty} \inf_{n \geq m} a_n = \sup_m \inf_{n \geq m} a_n$.

2. The *limit superior*, $\limsup a_n$, is the eventual upper bound for the sequence (a_n); that is, $\limsup a_n := \lim_{m \to \infty} \sup_{n \geq m} a_n = \inf_m \sup_{n \geq m} a_n$.

For a sequence (f_n) of numerical functions on a set E, we can define functions

$$\inf f_n, \quad \sup f_n, \quad \liminf f_n, \quad \limsup f_n$$

pointwise, considering for each x the sequence (a_n) defined by $a_n := f_n(x)$.

Obviously, $\liminf f_n \leq \limsup f_n$ (pointwise). If the two are equal, we write $\lim f_n$ for the limit, and write $f_n \to f$. If (f_n) is a sequence of functions that increases pointwise, then the limit $f := \lim f_n$ always exists. In this case we write $f_n \uparrow f$. A similar notation $f_n \downarrow f$ holds for a decreasing sequence of functions.

Proposition 1.26: Limits of Measurable Functions

Let (f_n) be a sequence of $\mathcal{E}$-measurable numerical functions. Then, the numerical functions $\sup f_n$, $\inf f_n$, $\limsup f_n$, and $\liminf f_n$ are also $\mathcal{E}$-measurable. In particular, limits of measurable functions (when $\limsup f_n = \liminf f_n$) are measurable as well.

Proof.

1. Let $f := \sup f_n$. Since sets of the form $[-\infty, r]$ form a p-system that generates the Borel σ-algebra on $\overline{\mathbb{R}}$, it suffices to show, by Theorem 1.20, that $f^{-1}[-\infty, r] \in \mathcal{E}$ for all r. The latter follows from the observation that

    $$f^{-1}[-\infty, r] = \cap_n f_n^{-1}[-\infty, r],$$

    where $f_n^{-1}[-\infty, r] \in \mathcal{E}$ by the measurability of f_n, and the fact that $\mathcal{E}$ is closed under countable intersections.

2. To show $\inf f_n \in \mathcal{E}$, we simply observe that $\inf f_n = -\sup(-f_n)$.

3. Since $\limsup f_n = \inf_m \sup_{n \geq m} f_n$, we can use Steps 1 and 2 above to conclude that $\limsup f_n \in \mathcal{E}$.

4. Similarly, since $\liminf f_n = \sup_m \inf_{n \geq m} f_n$, we have $\liminf f_n \in \mathcal{E}$.

$\square$

The next theorem shows that, for any measurable space $(E, \mathcal{E})$, every positive numerical function on E can be obtained as an increasing limit of simple functions. Recall that the set of all positive measurable functions on E is denoted by $\mathcal{E}_+$.

> ## Theorem 1.27: Approximation from Below
>
> A *positive* numerical function on E is $\mathcal{E}$-measurable if and only if it is the (pointwise) limit of an increasing sequence of simple functions.

Proof. For each $n \in \{1, 2, \dots\}$, define

(1.28) $$d_n(r) := \begin{cases} \frac{k-1}{2^n} & \text{if} \quad \frac{k-1}{2^n} \leq r < \frac{k}{2^n} \text{ for some } k \in \{1, \dots, n2^n\}, \\ n & \text{if} \quad r \geq n. \end{cases}$$

Then, d_n is an increasing right-continuous simple function that takes values in $\mathbb{R}_+$, and $d_n(r)$ increases to r for each r as $n \to \infty$; see Figure 1.29 for the case $n = 3$.

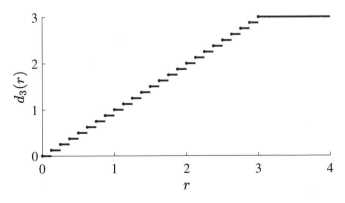

Figure 1.29: The graph of the function d_3.

Take a function $f \in \mathcal{E}_+$. We are going to use the function d_n to define the sequence (f_n) via $f_n := d_n \circ f$; that is, f_n is the composition of d_n and f, meaning $f_n(x) = d_n(f(x))$ for all x. Note that, by construction, $f_n \uparrow f$, because $d_n(r) \uparrow r$. Moreover, each f_n is positive and only takes a finite number of values, so f_n is a simple function. It is also measurable, since it is the composition of two measurable functions. This completes the necessity part of the proof. Sufficiency follows from Proposition 1.26.   □

The previous theorem characterizes the functions in $\mathcal{E}_+$, but what about the functions in $\mathcal{E}$? Fortunately, the solution is easy: each numerical function f on E can be written as the difference of its positive and negative part:

$$f = f^+ - f^-,$$

where $f^+ = f \vee 0$ is the positive part and $f^- = -(f \wedge 0) = (-f) \vee 0$ is the negative part of f; see Figure 1.30 for an illustration.

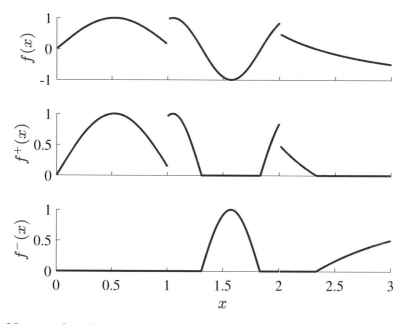

Figure 1.30: Any function can be decomposed as the difference of its positive and negative part.

We now have the following characterization:

Theorem 1.31: Positive and Negative Part of a Function

A function f is $\mathcal{E}$-measurable if and only if both f^+ and f^- are.

Proof. To prove sufficiency, suppose f^+ and f^- are measurable. Sketch graphs of f^+ and f^- to verify the following: For $r \geq 0$, we have $\{f \leq r\} = \{f^+ \leq r\}$. For $r < 0$, we have $\{f \leq r\} = \{f^- \geq -r\}$. In both cases the inverse image of f is in $\mathcal{E}$. Hence, f is $\mathcal{E}$-measurable.

Necessity is proved in a similar way; e.g., $\{f^+ \leq r\} = \{f \leq r\} \in \mathcal{E}$ for $r \geq 0$ and is equal to the empty set (also in $\mathcal{E}$) for $r < 0$. □

The important point of the preceding results is that every measurable function can be viewed as the pointwise limit of simple functions, which themselves are constructed as linear combinations of indicator functions. As a consequence, arithmetic operations on $\mathcal{E}$-measurable functions f and g, such as $f + g$, $f - g$, fg, f/g, $f \wedge g$, and $f \vee g$ are $\mathcal{E}$-measurable as well, as long as these functions are well-defined (for example, $\infty - \infty$ and ∞/∞ are undefined).

As an example, if $f, g \in \mathcal{E}$, then f^+, f^-, g^+ and g^- are in $\mathcal{E}_+$. Moreover, each of these functions is the limit of a sequence of simple functions f_n^+, f_n^-, g_n^+, and g_n^-, respectively. Thus, $f - g$ is the limit of the simple function $f_n^+ - f_n^- - g_n^+ + g_n^-$, which lies in $\mathcal{E}$. Hence, the limit is also in $\mathcal{E}$.

We conclude this section with the notion of a monotone class of functions.

Definition 1.32: Monotone Class of Functions

Let $(E, \mathcal{E})$ be a measurable space. A collection $\mathcal{M}$ of numerical functions on E is called a *monotone class* if it satisfies:
1. $\mathbb{1}_E \in \mathcal{M}$.
2. If $f, g \in \mathcal{M}$ are bounded and $a, b \in \mathbb{R}$, then $af + bg \in \mathcal{M}$.
3. If (f_n) is a sequence of positive functions in $\mathcal{M}$ that increases to some f, then $f \in \mathcal{M}$.

The following theorem is useful in proving that a certain property holds for all positive measurable functions:

Theorem 1.33: Monotone Class Theorem for Functions

Let $\mathcal{M}$ be a monotone class of functions on E and let C be a p-system that generates $\mathcal{E}$. If $\mathbb{1}_A \in \mathcal{M}$ for every $A \in C$, then $\mathcal{M}$ includes
- all positive $\mathcal{E}$-measurable functions,
- all bounded $\mathcal{E}$-measurable functions.

Proof. We first show that $\mathcal{M}$ contains all indicator functions in $\mathcal{E}$; it already contains all indicator functions in C. To this end, define the collection of sets

$$\mathcal{D} := \{A \in \mathcal{E} : \mathbb{1}_A \in \mathcal{M}\}.$$

This is a d-system. Namely:

1. $E \in \mathcal{D}$.

2. Take $A, B \in \mathcal{D}$ with $A \subseteq B$. Then, by Property 2 of $\mathcal{M}$, the function $\mathbb{1}_B - \mathbb{1}_A \in \mathcal{M}$. But this is the indicator function of $B \setminus A$. So $B \setminus A \in \mathcal{E}$.

3. Let (A_n) be an increasing sequence of sets in $\mathcal{D}$. Then, the sequence of indicators $(\mathbb{1}_{A_n})$ increases to $\mathbb{1}_A$, where $A = \cup_n A_n$. By Property 3 of $\mathcal{M}$, $\mathbb{1}_A \in \mathcal{M}$, and hence $A \in \mathcal{D}$.

Because $\mathcal{D}$ is a d-system that contains the p-system $\mathcal{C}$, it must contain $\mathcal{E}$, by the Monotone Class Theorem 1.12. Thus, $\mathbb{1}_A \in \mathcal{M}$ for all $A \in \mathcal{E}$.

Next, let $f \in \mathcal{E}_+$ (i.e., a positive $\mathcal{E}$-measurable function). By Theorem 1.27, there is a sequence of simple functions (f_n) in $\mathcal{E}_+$ that increases to f. Each f_n is a linear combination of indicator functions, which by the previous step all lie in $\mathcal{M}$. It follows by Property 2 of $\mathcal{M}$ that each f_n is a positive function in $\mathcal{M}$ as well, and so, by Property 3, $f \in \mathcal{M}$.

Finally, let f be bounded and $\mathcal{E}$-measurable. Then, its positive and negative parts are $\mathcal{E}$-measurable and thus in $\mathcal{M}$ via the preceding step, and are bounded obviously. By Property 2 of $\mathcal{M}$, we have $f = f^+ - f^- \in \mathcal{M}$.   □

1.3 Measures

Let $(E, \mathcal{E})$ be a measurable space. We wish to assign a measure to all sets in $\mathcal{E}$ that has the same properties as an area or length measure. In particular, any measure should have the property that the measure of the union of a disjoint (i.e., non-overlapping) collection of sets is equal to the sum of the measures for the individual sets, as illustrated in Figure 1.34.

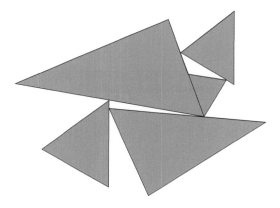

Figure 1.34: Any measure has the same properties as the "area" measure. For example, the total area of the non-overlapping triangles is the sum of the areas of the individual triangles.

Definition 1.35: Measure

Let $(E, \mathcal{E})$ be a measurable space. A *measure* μ on $\mathcal{E}$ is a set function

$$\mu : \mathcal{E} \rightarrow [0, \infty]$$

with $\mu(\emptyset) = 0$, such that for every sequence (A_n) of disjoint sets in $\mathcal{E}$,

(1.36) $$\mu(\cup_n A_n) = \sum_n \mu(A_n).$$

The main property (1.36) is called *countable additivity* or *σ-additivity*. The triple $(E, \mathcal{E}, \mu)$ is called a *measure space*. If $\mu(E) < \infty$, then μ is called *finite*. If $\mu(E) = 1$, it is called a *probability measure*. A measure μ is called *σ-finite* if there exists a sequence of sets (E_n) in $\mathcal{E}$ such that $\cup_n E_n = E$ and $\mu(E_n) < \infty$ for all n. Exercise 11 gives an equivalent way to verify countable additivity.

It is easy to check (Exercise 20) that if μ and ν are measures on $(E, \mathcal{E})$, then $a\mu + b\nu$ with $a, b \geq 0$ is a measure on $(E, \mathcal{E})$ as well. In fact, for any sequence (μ_n) of measures, $\sum_n \mu_n$ is also a measure. Such a measure is said to be *Σ-finite* if $\mu_n(E) < \infty$ for each n. Clearly, any σ-finite measure is Σ-finite.

■ **Example 1.37 (Dirac Measure)** Perhaps the simplest measure is one that assigns a measure of 1 to any set in $\mathcal{E}$ that contains a specific element $x \in E$ and 0 otherwise. This is called the *Dirac measure* at x, and is written as δ_x. In particular, for any $A \in \mathcal{E}$:

$$\delta_x(A) := \mathbb{1}_A(x) = \begin{cases} 1 & \text{if } x \in A, \\ 0 & \text{if } x \notin A. \end{cases}$$

■

■ **Example 1.38 (Counting Measure)** Let D be a countable subset in $\mathcal{E}$. The *counting measure* on D counts for each $A \in \mathcal{E}$ how many points of A fall in D; that is, it is the measure ν defined by

$$\nu(A) := \sum_{x \in D} \delta_x(A), \quad A \in \mathcal{E}.$$

■

We can further generalize counting measures to discrete measures as follows:

■ **Example 1.39 (Discrete Measure)** Let D be a countable subset in $\mathcal{E}$ and for each $x \in D$, let $m(x)$ be a positive number. Any measure μ of the form

$$\mu(A) := \sum_{x \in D} m(x)\, \delta_x(A), \quad A \in \mathcal{E}$$

is called a *discrete measure*. If E is countable and $\mathcal{E}$ is the set of all subsets of E (i.e., the power set of E), then *all* measures on $(E, \mathcal{E})$ are of this form. ∎

The preceding example indicates that for *discrete measurable spaces*, where E is countable and $\mathcal{E} := 2^E$, measures are easy. However, for uncountable sets such as $\mathbb{R}$ and $\mathbb{R}^n$ or subsets thereof, the existence of measures may be complicated, as we have seen in Example 1.3. Fortunately, the following theorem comes to the rescue. It guarantees that, under certain mild conditions, there *exists* exactly one measure μ on $\sigma(\mathcal{E}_0)$ that coincides with a *pre-measure* μ_0 on an algebra $\mathcal{E}_0$. The latter is a countably additive set function $\mu_0 : \mathcal{E}_0 \to [0, \infty]$, with $\mu_0(\emptyset) = 0$. We cannot call μ_0 a measure, as $\mathcal{E}_0$ is not a σ-algebra.

Theorem 1.40: Extension Theorem (Carathéodory)

Let E be a set with an algebra $\mathcal{E}_0$ thereon. Every pre-measure $\mu_0 : \mathcal{E}_0 \to [0, \infty]$ can be extended to a measure μ on $(E, \sigma(\mathcal{E}_0))$.

Proof. Appendix C provides a complete proof for the case where μ_0 is finite; see Theorem C.21. This is usually all that is required. □

∎ **Example 1.41 (Lebesgue Measure)** Let $\mathcal{E}_0$ be the algebra in Example 1.7; that is, every non-empty set in $\mathcal{E}_0$ is a finite union of intervals of the form $(a, b]$. Let $\mathcal{B}_{(0,1]}$ denote the σ-algebra that is generated by $\mathcal{E}_0$. The natural length of the set in $\mathcal{E}_0$ formed by the union $(a_1, b_1] \cup \cdots \cup (a_n, b_n]$ of disjoint intervals is of course

$$\mu_0((a_1, b_1] \cup \cdots \cup (a_n, b_n]) := \sum_{k=1}^{n} (b_k - a_k).$$

In Theorem C.6 we prove that μ_0 is countably additive (this is not trivial). Hence, by the extension theorem there *exists* a measure on $\mathcal{B}_{(0,1]}$ that coincides with μ_0 on $\mathcal{E}_0$. We will show shortly why there can be only *one* such measure. We have thus found the natural "length" measure on $\mathcal{B}_{(0,1]}$. It is called the *Lebesgue measure* (on $(0,1]$). In the same manner we can prove the existence of unique "volume" measures on $(\mathbb{R}^n, \mathcal{B}^n), n = 1, 2, \ldots$. These measures are also called Lebesgue measures. All these measures are σ-finite. ∎

As a direct consequence of the definition of a measure, we have the following properties for any measure μ:

Proposition 1.42: Properties of a Measure

Let μ be a measure on a measurable space $(E, \mathcal{E})$, and let A, B and $A_1, A_2, \ldots$ be measurable sets. Then, the following hold:

1. *(Monotonicity)*: $A \subseteq B \implies \mu(A) \leq \mu(B)$.

2. *(Continuity from below)*: If $A_1, A_2, \ldots$ is increasing (i.e., $A_1 \subseteq A_2 \subseteq \cdots$), then

   (1.43) $$\lim \mu(A_n) = \mu(\cup_n A_n).$$

3. *(Countable subadditivity)*: $\mu(\cup_n A_n) \leq \sum_n \mu(A_n)$.

Proof.

1. Suppose $A \subseteq B$. Then, we can write $B = A \cup (A^c \cap B)$. The countable additivity and positivity of μ imply that

$$\mu(B) = \mu(A) + \mu(A^c \cap B) \geq \mu(A),$$

which proves monotonicity.

2. Suppose (A_n) is an increasing sequence of measurable sets. From the definition of a σ-algebra it follows that $A := \cup_n A_n$ is again a measurable set. Now consider Figure 1.44.

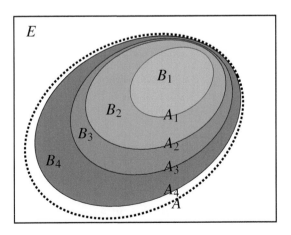

Figure 1.44: Sequential continuity from below.

Define sets $B_1, B_2, \ldots$ via $B_1 := A_1$ and $B_n := A_n \setminus A_{n-1}$, $n = 2, 3, \ldots$. The "rings" $B_1, B_2, \ldots$ are disjoint, with

$$\cup_{i=1}^{n} B_i = \cup_{i=1}^{n} A_i = A_n \quad \text{and} \quad \cup_{i=1}^{\infty} B_i = A.$$

Hence, from the countable additivity property (1.36) it follows that

$$\mu(A) = \mu\left(\cup_{i=1}^{\infty} B_i\right) = \sum_{i=1}^{\infty} \mu(B_i)$$
$$= \lim \sum_{i=1}^{n} \mu(B_i) = \lim \mu(\cup_{i=1}^{n} B_i) = \lim \mu(A_n),$$

which proves the sequential continuity from below.

3. Let (A_n) be a sequence of measurable sets, not necessarily disjoint. Define $B_1 := A_1$ and $B_n := A_n \setminus \cup_{i=1}^{n-1} B_i$ for $n = 2, 3, \ldots$. We have

$$\mu(\cup_n A_n) = \mu(\cup_n B_n) = \sum_n \mu(B_n) \le \sum_n \mu(A_n).$$

$\square$

Often we are dealing with a measurable space $(E, \mathcal{E})$ in which $\mathcal{E}$ can be generated by a p-system C; i.e., a collection that is closed under finite intersections (such as an algebra). In that case, any finite measure μ on $(E, \mathcal{E})$ is completely specified by its value on C. This is a consequence of the following theorem:

Theorem 1.45: Uniqueness of Finite Measures

Let C be a p-system that generates $\mathcal{E}$. If μ and ν are two measures on $(E, \mathcal{E})$ with $\mu(E) = \nu(E) < \infty$, then

$$\mu(A) = \nu(A) \text{ for all } A \in C \implies \mu(A) = \nu(A) \text{ for all } A \in \mathcal{E}.$$

Proof. Consider the collection

$$\mathcal{D} := \{A \in \mathcal{E} : \mu(A) = \nu(A)\}.$$

This is a d-system, because:

1. $E \in \mathcal{D}$ by the assumption of the theorem.

2. Take $A, B \in \mathcal{D}$ such that $A \supseteq B$. Then, $A \setminus B \in \mathcal{D}$, because

$$\mu(A \setminus B) = \mu(A) - \mu(B) = \nu(A) - \nu(B) = \nu(A \setminus B).$$

3. Take an increasing sequence of sets (A_n) in $\mathcal{D}$. Then, for all n, $\mu(A_n) = \nu(A_n)$ and, by continuity from below, $\mu(\cup A_n) = \lim_n \mu(A_n) = \lim_n \nu(A_n) = \nu(\cup A_n)$, so that $\cup_n A_n \in \mathcal{D}$.

Since $\mathcal{D}$ is a d-system that contains the p-system C, it must also contain the σ-algebra generated by C (see the Monotone Class Theorem 1.12), which is $\mathcal{E}$. $\square$

■ **Example 1.46 (Uniqueness of Lebesgue Measures)** Continuing Example 1.41, consider the Lebesgue measure on the measurable space $((0,1], \mathcal{B}_{(0,1]})$. Existence is guaranteed by Carathéodory's extension Theorem 1.40. Moreover, Theorem 1.45 ensures that there can only be *one* measure for which its value (length) of an interval $(a,b]$ is $b-a$ for all $0 \le a < b \le 1$.

To prove uniqueness for the Lebesgue measure λ on $(\mathbb{R}, \mathcal{B})$ we need an extra step, as λ is not finite. Define $E_n := (n, n+1]$, $n \in \mathbb{Z}$. The collection $\{E_n\}$ forms a partition of $\mathbb{R}$. Consider the algebra C of finite unions of intervals of the form $(a,b]$ in $\mathbb{R}$. This is a p-system that generates $\mathcal{B}$. Suppose that μ is a measure that coincides with λ on C. We want to show that $\mu = \lambda$ on $\mathcal{B}$. Take $A \in \mathcal{B}$. Then, $\{A \cap E_n\}$ forms a partition of A. For each measurable space $(E_n, \mathcal{B}_{E_n})$ we can invoke Theorem 1.45 to conclude that $\mu(A \cap E_n) = \lambda(A \cap E_n)$. Hence,

$$\mu(A) = \sum_n \mu(A \cap E_n) = \sum_n \lambda(A \cap E_n) = \lambda(A),$$

which had to be shown.   ■

Suppose we have two measure spaces $(E, \mathcal{E}, \mu)$ and $(F, \mathcal{F}, \nu)$. We can define a measure $\mu \otimes \nu$ on the product space (see Definition 1.14) $(E \times F, \mathcal{E} \otimes \mathcal{F})$ as follows:

Definition 1.47: Product Measure

Let $(E, \mathcal{E}, \mu)$ and $(F, \mathcal{F}, \nu)$ be two measure spaces. The *product measure* of μ and ν on the product space $(E \times F, \mathcal{E} \otimes \mathcal{F})$ is defined as the measure π with

(1.48) $\pi(A \times B) := \mu(A)\,\nu(B), \quad A \in \mathcal{E}, \ B \in \mathcal{F}.$

This measure π is often written as $\mu \otimes \nu$.

Note that the collection of rectangles $\{A \times B, A \in \mathcal{E}, B \in \mathcal{F}\}$ is a p-system that generates $\mathcal{E} \otimes \mathcal{F}$, so the product measure is unique if it is finite. This uniqueness property can be extended to the σ-finite case by considering a partition of $E \times F$ on which the measure is finite.

■ **Example 1.49 (Lebesgue Measure on $(\mathbb{R}^2, \mathcal{B}^2)$)** We can think of the Lebesgue measure λ on $(\mathbb{R}^2, \mathcal{B}^2)$ in two ways. First, it is the natural area measure on this measurable space. Second, it is the product measure of the Lebesgue measures on the coordinate spaces $(\mathbb{R}, \mathcal{B})$. That is, if μ is the Lebesgue (length) measure on $(\mathbb{R}, \mathcal{B})$, then $\lambda = \mu \otimes \mu$.   ■

1.4 Integrals

In this section, we combine measurable functions and measures to define integrals in a unified and elegant way. The procedure for building up proofs for measurable functions from indicator functions, simple functions, and limits thereof is often referred to as "the standard machine".

1.4.1 Definition of an Integral

In elementary mathematics, the integral of a function is thought of as the area underneath the graph of the function. We wish to extend this idea to a broad class of functions and measures, using the standard machine. In particular, let $(E, \mathcal{E}, \mu)$ be a measure space and let f be a (numerical) $\mathcal{E}$-measurable function. We define the (Lebesgue) *integral* of f with respect to μ in four stages (given next) and denote the integral by any of

$$\mu f, \quad \int_E f \, d\mu, \quad \text{or} \quad \int_E \mu(dx) \, f(x).$$

The index E under the integral sign is often omitted. The four defining stages are as follows:

1. If $f = \mathbb{1}_A$, then
$$\mu f := \mu(A).$$

2. If f is *simple* and positive, and its canonical form is $f = \sum_{i=1}^n a_i \mathbb{1}_{A_i}$, then (with $\infty \cdot 0 := 0 =: 0 \cdot \infty$)
$$\mu f := \sum_{i=1}^n a_i \, \mu(A_i).$$

3. If f is positive, put $f_n := d_n \circ f$, with d_n defined in (1.28). Then, (f_n) is a sequence of simple positive functions such that $f_n(x) \uparrow f(x)$ for all x. The sequence (μf_n) is increasing and so $\lim \mu f_n$ exists (possibly $+\infty$). We define
$$\mu f := \lim \mu f_n.$$

4. For general (not necessarily positive) $\mathcal{E}$-measurable numerical functions we can often also define the integral. Namely, each such function f can be written as
$$f = f^+ - f^-,$$
where $f^+ = f \vee 0$ and $f^- = (-f) \vee 0$ are both positive and $\mathcal{E}$-measurable (Theorem 1.31). We define in this case
$$\mu f := \mu f^+ - \mu f^-,$$

provided that at least one of the terms on the right-hand side is finite. Functions for which the above integral is *finite* are called (μ-) *integrable*. This is equivalent to $\mu f^+ < \infty$ and $\mu f^- < \infty$, and thus to $\mu|f| = \mu(f^+ + f^-) < \infty$.

■ **Example 1.50 (Integrals for Discrete Measures)** Let μ be a discrete measure (see Example 1.39) on $(E, \mathcal{E})$ of the form

(1.51) $$\mu := \sum_{x \in D} m(x)\, \delta_x$$

for some countable set D and positive masses $\{m(x), x \in D\}$. Then, for any $f \in \mathcal{E}_+$:

$$\mu f = \sum_{x \in D} m(x)\, f(x).$$

In particular, if E is a countable set and $\mathcal{E} := 2^E$ is the *discrete σ-algebra* (power set of E), then all measures on $(E, \mathcal{E})$ are of the form (1.51), with $D = E$ and $m(x) = \mu\{x\}$. Thus, for every $f \in \mathcal{E}_+$ we have $\mu f = \sum_{x \in E} \mu\{x\} f(x)$. The notation μf is thus similar to the one used in linear algebra regarding the multiplication of a row vector μ with a column vector f, to yield a number μf.   ■

■ **Example 1.52 (Lebesgue Integrals)** Let λ be the Lebesgue measure on $(\mathbb{R}, \mathcal{B})$ and let f be a $\mathcal{B}$-measurable function. The integral $\lambda f = \int f\, d\lambda$ (provided it exists, e.g., if $f \geq 0$) is called the *Lebesgue integral* of f. We can view this number as the area under the graph of f. The crucial difference between the Lebesgue integral and the Riemann integral is illustrated in Figure 1.53. In Riemann integration the domain of the function is discretized, whereas in Lebesgue integration the (real) range of the function is discretized. In order to keep the notation the same as in elementary (Riemann) integration, we write $\int f\, d\lambda$ as any of

$$\int_{\mathbb{R}} f(x)\, dx, \quad \int_{-\infty}^{\infty} f(x)\, dx, \quad \text{or} \quad \int dx\, f(x).$$

The integral of f over the interval (a, b) is written as

$$\int_{(a,b)} f(x)\, dx, \quad \int_a^b f(x)\, dx, \quad \text{or} \quad \int_a^b dx\, f(x).$$

For a general $A \in \mathcal{B}$, we write

$$\int_A f(x)\, dx \quad \text{for} \quad \int \mathbb{1}_A f\, d\lambda.$$

In the same way, we can define the Lebesgue integral $\int f\, d\lambda$ of a $\mathcal{B}^n$-measurable function on $\mathbb{R}^n$ with respect to the Lebesgue measure λ on $(\mathbb{R}^n, \mathcal{B}^n)$. We often use the notation

$$\int \cdots \int f(x_1, \ldots, x_n)\, dx_1 \ldots dx_n, \quad \int f(\boldsymbol{x})\, d\boldsymbol{x}, \quad \text{or} \quad \int d\boldsymbol{x}\, f(\boldsymbol{x})$$

instead of λf or $\int f \, \mathrm{d}\lambda$.

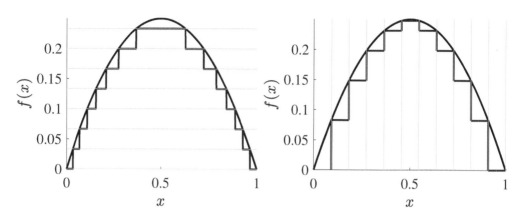

Figure 1.53: In Lebesgue integration the range (y-axis) is discretized and in Riemann integration the domain (x-axis) is discretized.

Continuing the notation in the last example, we define the integral of an $\mathcal{E}$-measurable function f over a set $C \in \mathcal{E}$ as the integral of $f \mathbb{1}_C$, and write it as any of the following:

$$\mu(f \mathbb{1}_C) = \int_C \mu(\mathrm{d}x) f(x) = \int_C f \, \mathrm{d}\mu.$$

The natural property holds that if $A = B \cup C$, with $B \in \mathcal{E}$ and $C \in \mathcal{E}$ disjoint, then

$$\mu(f \mathbb{1}_A) = \mu(f \mathbb{1}_B) + \mu(f \mathbb{1}_C).$$

Let $(E, \mathcal{E}, \mu)$ be a measure space. A set $A \subset E$ is said to be *negligible* if there exists a $B \in \mathcal{E}$ that contains A and for which $\mu(B) = 0$. It is possible to "complete" a measure space to contain all negligible sets. Two functions f and g are said to be equal μ-*almost everywhere* if the set of points for which they differ is negligible. The following shows that such functions have the same integral:

Proposition 1.54: Insensitivity of the Integral

Let $(E, \mathcal{E}, \mu)$ be a measure space and let f and g be positive $\mathcal{E}$-measurable functions that are equal μ-almost everywhere. Then, $\mu f = \mu g$.

Proof. The set $A := \{f \neq g\}$ has measure 0. We show that $\mu(f \mathbb{1}_A) = 0$. If $f \in \mathcal{E}_+$ is simple, then by definition of the integral (Steps 1 and 2), $\mu(f \mathbb{1}_A) = 0$. For general $f \in \mathcal{E}_+$ take the sequence (f_n) of simple functions increasing to f (in Step 3 of the definition). All the integrals $\mu(f_n \mathbb{1}_A)$ are 0 and so their limit, $\mu(f \mathbb{1}_A)$, is 0 as well. For the same reason $\mu(g \mathbb{1}_A) = 0$. Thus, $\mu f = \mu(f \mathbb{1}_{A^c}) = \mu(g \mathbb{1}_{A^c}) = \mu g$.   □

1.4.2 Properties of an Integral

Below, the numerical functions refer to a measurable space $(E, \mathcal{E})$. Let $\mathcal{E}_+$ be the set of *positive* $\mathcal{E}$-measurable numerical functions. The following theorem lists the main properties of an integral. Property 3 is called the *Monotone Convergence Theorem* and is the main reason why this integral definition is so powerful.

Theorem 1.55: Properties of an Integral

Below a, b are in $\mathbb{R}_+$ and f, g, f_n are in $\mathcal{E}_+$.

1. *(Positivity and Monotonicity)*: $\mu f \geq 0$, and $f = 0 \Rightarrow \mu f = 0$. Also, $f \leq g \Rightarrow \mu f \leq \mu g$.

2. *(Linearity)*: $\mu(af + bg) = a\mu f + b\mu g$.

3. *(Monotone convergence)*: If $f_n \uparrow f$, then $\mu f_n \uparrow \mu f$.

Proof.

1. That $\mu f \geq 0$ and that the integral of the zero function is 0 follows directly from the definition. Next, take $f \leq g$. With d_n as in (1.28), let $f_n := d_n \circ f$ and $g_n := d_n \circ g$ be simple measurable functions that increase to f and g, respectively. Then, for each n, $\mu f_n \leq \mu g_n$ (check this for simple functions). Now take the limit for $n \to \infty$ to obtain $\mu f \leq \mu g$.

2. Linearity of the integral for simple functions in $\mathcal{E}_+$ is easily checked. For general functions $f, g, \in \mathcal{E}_+$, we take, as above, simple measurable functions that increase to f and g, respectively; thus, $af_n + bg_n$ increases to $af + bg$. We have $\mu(af_n + bg_n) = a\mu f_n + b\mu g_n$. The limit of the left-hand side is $\mu(af + bg)$ and the limit of the right-hand side is $a\mu f + b\mu g$ by Step 3 of the definition of an integral.

3. The result is true, by definition, if f_n is of the form $f_n = d_n \circ f$ with $f \in \mathcal{E}_+$. But the point of the monotone convergence property is that it also holds for any increasing sequence (f_n) in $\mathcal{E}_+$ with limit $f = \lim f_n$. Since (f_n) is increasing, the limit f is well-defined, and $f \in \mathcal{E}_+$. So μf is well-defined. Since (f_n) is increasing, the integrals (μf_n) form an increasing sequence of numbers (by the monotonicity property of integrals) and so $\lim \mu f_n$ exists. We want to show that this limit is equal to μf. Since $f \geq f_n$, we have $\mu f \geq \mu f_n$ and hence $\mu f \geq \lim \mu f_n$. We want to show that also $\lim \mu f_n \geq \mu f$. We do this in three steps.

   (a) Let $b \geq 0$ and suppose that $f(x) > b$ for every $x \in B$, where $B \in \mathcal{E}$. Define $B_n := B \cap \{f_n > b\}$. Then, $B_n \uparrow B$ and $\lim \mu(B_n) = \mu(B)$ by the sequential continuity from below property of μ; see (1.43). Since

$$f_n \mathbb{1}_B \geq f_n \mathbb{1}_{B_n} \geq b \mathbb{1}_{B_n},$$

the monotonicity and linearity of the integral yield that $\mu(f_n \mathbb{1}_B) \geq b\mu(B_n)$ and, by taking limit for $n \to \infty$, that

$$\lim \mu(f_n \mathbb{1}_B) \geq b\,\mu(B).$$

This remains true if instead $f(x) \geq b$ on B. This is trivially true for $b = 0$. For $b > 0$, choose a sequence (b_m) strictly increasing to b. Then, $\lim_n \mu(f_n \mathbb{1}_B) \geq b_m \mu(B)$. Now take the limit for $m \to \infty$.

(b) Let $g = \sum_{i=1}^m b_i \mathbb{1}_{B_i}$ be a positive simple function, in canonical form, such that $f \geq g$. Thus, $f(x) \geq b_i$ for every $x \in B_i$, $i = 1, \ldots, m$, and part (a) above yields

$$\lim_n \mu(f_n \mathbb{1}_{B_i}) \geq b_i\,\mu(B_i), \quad i = 1, \ldots, m.$$

Hence,

$$\lim \mu f_n = \lim_n \sum_{i=1}^m \mu(f_n \mathbb{1}_{B_i}) = \sum_{i=1}^m \lim_n \mu(f_n \mathbb{1}_{B_i}) \geq \sum_{i=1}^m b_i\,\mu(B_i) = \mu g.$$

(c) Taking g in (b) equal to $d_m \circ f$ and letting $m \to \infty$, we find $\lim \mu f_n \geq \mu f$. $\square$

In our definition of the integral, we used a specific approximating sequence (f_n), given by $f_n := d_n \circ f$. An important consequence of the Monotone Convergence Theorem is that we could have taken *any* sequence (f_n) with $f_n \uparrow f$.

We leave it as an exercise to show that the linearity of the integral extends to integrable $f, g \in \mathcal{E}$ and arbitrary $a, b \in \mathbb{R}$.

■ **Remark 1.56 (Monotone Convergence and Insensitivity)** A consequence of the insensitivity property (Property 1.54) of the integral is that Theorem 1.55 also holds in an *almost everywhere* sense. For example, if $f_n(x) \uparrow f(x)$ for $x \in E_0$ with $\mu(E \setminus E_0) = 0$, then $\mu f_n \uparrow \mu f$. Namely, let $\widetilde{f_n} := f_n \mathbb{1}_{E_0} + f \mathbb{1}_{E \setminus E_0}$. Then, $\mu f_n = \mu \widetilde{f_n}$ by the insensitivity of the integral, and the Monotone Convergence Theorem yields $\mu \widetilde{f_n} \uparrow \mu \lim \widetilde{f_n} = \mu f$. ■

Any positive linear functional $L(f)$ on the set of positive measurable functions can be represented as an integral μf if it possesses the three main properties of the integral (positivity, linearity, and monotone convergence), as shown in the next theorem.

> **Theorem 1.57: Measures from Linear Functionals**
>
> Let $L : \mathcal{E}_+ \to \overline{\mathbb{R}}_+$. Then, there exists a unique measure μ on $(E, \mathcal{E})$ such that $L(f) = \mu f$ if and only if
> 1. $f = 0 \Rightarrow L(f) = 0$,
> 2. $f, g \in \mathcal{E}_+$ and $a, b \in \mathbb{R}_+ \Rightarrow L(af + bg) = aL(f) + bL(g)$,
> 3. $(f_n) \in \mathcal{E}_+$ and $f_n \uparrow f \Rightarrow L(f_n) \uparrow L(f)$.

Proof. Necessity is immediate from the properties of integration. To show sufficiency, suppose L has the properties stated above. The obvious candidate for μ is the set function $\mu : \mathcal{E} \to \overline{\mathbb{R}}_+$ via $\mu(A) := L(\mathbb{1}_A)$. We first show that μ satisfies the properties of a measure on $(E, \mathcal{E})$. First, $\mu(\emptyset) = 0$, because $\mu(\emptyset) = L(0) = 0$. Second, let (A_n) be a sequence of disjoint sets in $\mathcal{E}$ with union A. Define $B_n := \cup_{i=1}^n A_i$. Then, (B_n) increases to A and $\mathbb{1}_{B_n} = \sum_{i=1}^n \mathbb{1}_{A_i}$. The sequence $(\mathbb{1}_{B_n})$ increases to $\mathbb{1}_A$, and hence by the monotone convergence (Property 3) of L and the linearity (Property 2) of L, we have

$$L(\mathbb{1}_A) = \lim_n L\left(\sum_{i=1}^n \mathbb{1}_{A_i}\right) = \lim_n \sum_{i=1}^n L(\mathbb{1}_{A_i}) = \sum_{i=1}^\infty L(\mathbb{1}_{A_i}).$$

In other words, $\mu(A) = \sum_{i=1}^\infty \mu(A_i)$, and so μ is a measure.

To show that $L(f) = \mu f$ for all $f \in \mathcal{E}_+$, first observe that this is true for simple $f \in \mathcal{E}_+$, i.e., of the form $\sum_{i=1}^n b_i \mathbb{1}_{B_i}$, by using the linearity of L and the linearity property of an integral. For a general $f \in \mathcal{E}_+$, take a sequence (f_n) of simple functions in $\mathcal{E}_+$ that increases to f. For all f_n we have $L(f_n) = \mu f_n$. Taking the limit on the left-hand side gives $L(f)$ by the monotone convergence property of L, and taking the limit on the right-hand side gives μf by the monotone convergence property of integrals with respect to μ. Hence, $L(f) = \mu f$ for all $f \in \mathcal{E}_+$. $\square$

1.4.3 Indefinite Integrals, Image Measures, and Measures with Densities

Let μ be a measure on $(E, \mathcal{E})$ and p a positive $\mathcal{E}$-measurable function. We have defined the integral of p over a subset $A \in \mathcal{E}$ as

$$\mu(p\mathbb{1}_A) = \int_A p \, d\mu = \int \mathbb{1}_A p \, d\mu.$$

The mapping

$$A \mapsto \int_A p \, d\mu$$

defines a measure ν on $(E, \mathcal{E})$, which is called the *indefinite integral* of p with respect to μ. We leave the proof as an exercise; see Exercise 25. Conversely, we say that ν has *density* p with respect to μ, and write $\nu = p\mu$.

Theorem 1.58: Integrals with Densities

Let $\nu := p\mu$. Then, for every $f \in \mathcal{E}_+$, we have $\nu f = \mu(pf)$.

Proof. Let $L(f) := \mu(pf)$. L satisfies the properties in Theorem 1.57, so there is a unique measure $\widehat{\nu}$ on $(E, \mathcal{E})$ with $\widehat{\nu}f = \mu(pf)$. Taking $f := \mathbb{1}_A$ for $A \in \mathcal{E}$, we see that $\widehat{\nu}(A) = L(\mathbb{1}_A) = \mu(p\mathbb{1}_A) = \nu(A)$. So $\widehat{\nu} = \nu$. $\qquad\square$

Let ν and μ be two measures on $(E, \mathcal{E})$. Measure ν is said to be *absolutely continuous* with respect to μ if for all $A \in \mathcal{E}$,

$$\mu(A) = 0 \quad \text{implies} \quad \nu(A) = 0.$$

We write $\nu \ll \mu$. To answer the question whether a certain measure ν can be represented as the indefinite integral with respect to a given measure μ, we only need to verify $\nu \ll \mu$. That is the purport of the celebrated Radon–Nikodym theorem. A martingale-based proof will be given in Section 5.5.3.

Theorem 1.59: Radon–Nikodym

Let ν and μ be two measures on $(E, \mathcal{E})$ such that μ is σ-finite and $\nu \ll \mu$. Then, there *exists* a $p \in \mathcal{E}_+$ such that $\nu = p\mu$. Moreover, p is unique up to equivalence; that is, if there is another $\widetilde{p} \in \mathcal{E}_+$ such that $\nu = \widetilde{p}\mu$, then $p = \widetilde{p}$, μ-almost everywhere.

The function p is referred to as the *Radon–Nikodym derivative* of ν with respect to μ, and is frequently written as $p = d\nu/d\mu$. The functions p and $\widetilde{p}$ above are said to be *versions* of each other.

Another way of creating measures from integrals is via a change of variable. Specifically, let $(E, \mathcal{E})$ and $(F, \mathcal{F})$ be measurable spaces and let $h : E \to F$ be an $\mathcal{E}/\mathcal{F}$-measurable function. Suppose on $(E, \mathcal{E})$ we have a measure μ. Then, we can create a measure ν on $(F, \mathcal{F})$ by defining

$$(1.60) \qquad \nu(B) := \mu(h^{-1}(B)) = \mu(\{x \in E : h(x) \in B\}).$$

It is called the *image measure of μ under h*; the name *pushforward measure* is also used. Suggestive notation $\nu = \mu \circ h^{-1}$ (or also $h(\mu)$); see Figure 1.61. The figure also illustrates how integration with respect to image measures works, as detailed in Theorem 1.62.

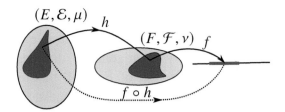

Figure 1.61: The image measure ν of the dark blue set in F is the same as the measure μ of the dark blue set in E. Also, the integral of f with respect to ν is the same as the integral of $f \circ h$ with respect to μ.

Theorem 1.62: Integration with Respect to Image Measures

Let $\nu = \mu \circ h^{-1}$ be the image measure of μ under h. Then, for every positive $\mathcal{F}$-measurable function f,

$$\nu f = (\mu \circ h^{-1}) f = \int_F f \, \mathrm{d}\nu = \int_E f \circ h \, \mathrm{d}\mu = \mu(f \circ h).$$

Proof. The idea is again to use the characterization from Theorem 1.57. Define $L(f) := \mu(f \circ h)$. Check yourself that L satisfies the conditions of Theorem 1.57. Thus, $L(f) = \widetilde{\nu} f$ for some measure $\widetilde{\nu}$. Taking $f := \mathbb{1}_B$ shows that $\widetilde{\nu}(B) = \mu(h^{-1}B) = (\mu \circ h^{-1})(B)$. Thus, $\widetilde{\nu} = \nu$. □

1.4.4 Kernels and Product Spaces

Recall that a measure μ and a measurable function f can be treated, via Theorem 1.57, as a row vector (linear functional) and a column vector, respectively. Their integral μf is then viewed as a "product" of the two. Continuing this analogy, we wish to consider vector-matrix products such as μK and $K f$. The corresponding matrix-like object in measure theory and probability is the *transition kernel*.

Definition 1.63: Transition Kernel

Let $(E, \mathcal{E})$ and $(F, \mathcal{F})$ be measurable spaces. A *transition kernel* from $(E, \mathcal{E})$ into $(F, \mathcal{F})$ is a mapping $K : E \times \mathcal{F} \to \overline{\mathbb{R}}_+$ such that:

1. $K(\cdot, B)$ is $\mathcal{E}$-measurable for every $B \in \mathcal{F}$.

2. $K(x, \cdot)$ is a measure on $(F, \mathcal{F})$ for every $x \in E$.

If $(F, \mathcal{F}) = (E, \mathcal{E})$ and $K(x, \cdot)$ is a probability measure for every $x \in E$, K is said to be a *probability transition kernel* on $(E, \mathcal{E})$.

As $K(x, \cdot)$ is a measure for every $x \in E$, we can consider its integral with $f \in \mathcal{F}_+$. This will define a function in E. Is it measurable? Similarly, as $K(\cdot, B)$ is $\mathcal{E}$-measurable for every $B \in \mathcal{F}$, we can consider its integral with respect to a measure μ on $(E, \mathcal{E})$, seen as a function of B. Will this result in a measure on $(F, \mathcal{F})$? The following theorem confirms both these questions:

Theorem 1.64: Measures and Functions from Kernels

Let K be a transition kernel from $(E, \mathcal{E})$ into $(F, \mathcal{F})$.

1. For all $f \in \mathcal{F}_+$ the function Kf defined by $(Kf)(x) := \int_F K(x, dy) f(y)$, $x \in E$, is in $\mathcal{E}_+$.

2. For every measure μ on $(E, \mathcal{E})$ the set function μK defined by $(\mu K)(B) := \int_E \mu(dx) K(x, B)$, $B \in \mathcal{F}$, is a measure on $(F, \mathcal{F})$.

3. For every measure μ on $(E, \mathcal{E})$ and $f \in \mathcal{F}_+$,

$$(\mu K) f = \mu(Kf) = \int_E \mu(dx) \int_F K(x, dy) f(y).$$

Proof.
1. Take $f \in \mathcal{F}_+$. Clearly, the function Kf is well-defined and positive. To show that $Kf \in \mathcal{E}_+$, first consider simple functions; i.e., $f = \sum_{i=1}^n b_i \mathbb{1}_{B_i}$. Then, $(Kf)(x) = \sum_{i=1}^n b_i K(x, B_i)$, and since each $K(\cdot, B_i)$ is $\mathcal{E}$-measurable, we have $Kf \in \mathcal{E}_+$. For general $f \in \mathcal{F}_+$, we know there exists a sequence of simple functions $f_n \in \mathcal{F}_+$ increasing to f. By the preceding, each $Kf_n \in \mathcal{E}_+$. Moreover, for each fixed x, $K(x, \cdot)$ is a measure on $(F, \mathcal{F})$ and so by the Monotone Convergence Theorem applied to $K(x, \cdot)$ and f_n, we have that the sequence of integrals $(K(x, \cdot) f_n)$ converges to $K(x, \cdot) f = (Kf)(x)$. In other words, the function Kf is the limit of $\mathcal{E}_+$-measurable functions and is therefore $\mathcal{E}_+$-measurable as well.

2. Take a measure μ on $(E, \mathcal{E})$. The set function μK is well-defined. We want to show that it is a measure by identifying its actions on functions $f \in \mathcal{F}_+$. Thereto, consider the functional $L : \mathcal{F}_+ \to \overline{\mathbb{R}}_+$ defined by

$$L(f) := \mu(Kf).$$

It satisfies the properties of Theorem 1.57:

(a) If $f = 0$, then $Kf = 0$ and hence $\mu(Kf) = 0$.

(b) If $f, g \in \mathcal{F}_+$ and $a, b \in \mathbb{R}_+$, then $L(af + bg) = \mu(K(af + bg))$. Note that $(K(af + bg))(x)$ is defined as the integral of $af + bg$ with respect to the measure $K(x, \cdot)$. Hence, by the linearity of the integral, $K(af + bg) =$

$aKf + bKg$. The integral of this function with respect to μ is $\mu(aKf + bKg)$ which, again by linearity of the integral, is equal to $a\mu(Kf) + b\mu(Kg)$. But the latter can be written as $aL(f) + bL(g)$. So L is linear.

(c) Take a sequence (f_n) in $\mathcal{F}_+$, with $f_n \uparrow f$. By the Monotone Convergence Theorem applied to each $K(x, \cdot)$, we have $(Kf_n)(x) \uparrow (Kf)(x)$. By the same theorem, applied to μ and Kf_n, we have $\mu(Kf_n) \uparrow \mu(Kf)$. In other words $L(f_n) \uparrow L(f)$.

It follows from Theorem 1.57 that there exists a measure ν on $(F, \mathcal{F})$ such that $L(f) = \nu f$, $f \in \mathcal{F}_+$. But this measure is μK, as can be found by taking $f = \mathbb{1}_B$, $B \in \mathcal{F}_+$. Namely: $\nu(B) = \mu(K\mathbb{1}_B) = \mu K(\cdot, B) = (\mu K)(B)$. So μK is indeed a measure.

3. A byproduct from the proof in Step 2 is that we have established that $(\mu K)f = \nu f = L(f) = \mu(Kf)$.

□

Let $(E, \mathcal{E})$ and $(F, \mathcal{F})$ be measurable spaces, and let K be a transition kernel from $(E, \mathcal{E})$ into $(F, \mathcal{F})$. The above shows that by starting with a measure μ on $(E, \mathcal{E})$, we can obtain a measure μK on $(F, \mathcal{F})$. Using μ and K, we can also construct a measure $\mu \otimes K$ on the product space $(E \times F, \mathcal{E} \otimes \mathcal{F})$ by defining

$$(1.65) \qquad (\mu \otimes K)(A \times B) := \int_A \mu(\mathrm{d}x) K(x, B), \quad A \in \mathcal{E}, \ B \in \mathcal{F}.$$

The notation

$$(\mu \otimes K)(\mathrm{d}x, \mathrm{d}y) = \mu(\mathrm{d}x) K(x, \mathrm{d}y)$$

is also used. This notation is compatible with the one that we used for the product measure $\mu \otimes \nu$ in the case where $K(x, \mathrm{d}y) = \nu(\mathrm{d}y)$ does not depend on x.

Instead of (1.65), one can also define the measure by specifying its integrals with respect to functions $f \in (\mathcal{E} \otimes \mathcal{F})_+$, via

$$(1.66) \qquad (\mu \otimes K)f := \int_E \mu(\mathrm{d}x) \int_F K(x, \mathrm{d}y) f(x, y), \quad f \in (\mathcal{E} \otimes \mathcal{F})_+.$$

More generally, the same construction can be employed to define measures on higher-dimensional product spaces. For example, using a kernel L from $(F, \mathcal{F})$ to a measurable space $(G, \mathcal{G})$, the integrals

$$(\mu \otimes K \otimes L)f := \int_E \mu(\mathrm{d}x) \int_F K(x, \mathrm{d}y) \int_G L(y, \mathrm{d}z) f(x, y, z), \quad f \in (\mathcal{E} \otimes \mathcal{F} \otimes \mathcal{G})_+,$$

define a measure on $(E \times F \times G, \mathcal{E} \otimes \mathcal{F} \otimes \mathcal{G})$.

For (1.65) and (1.66) to be well-defined, we need some form of finiteness restrictions on K and μ. Mild conditions are that μ should be σ-finite and K should be σ-bounded, meaning that there is a measurable partition (F_n) of F such that $K(\cdot, F_n)$ is bounded for each n. You can check with Exercises 26 and 27 that the right-hand side of (1.66) indeed defines a measure on $(E \times F, \mathcal{E} \otimes \mathcal{F})$, when K is a bounded kernel.

For a product measure $\mu \otimes \nu$, the order in which the integral (1.66) is evaluated should be irrelevant. This is the message of Fubini's theorem: under mild conditions, an integral with respect to a product measure, i.e., a multiple integral, can be computed by *repeated integration*, where the order of integration does not matter. Recall that a Σ-finite measure μ is of the form $\mu = \sum_n \mu_n$, where $\mu_n(E) < \infty$ for each n. Such a μ is therefore σ-finite as well.

Theorem 1.67: Fubini

Let μ and ν be Σ-finite measures on $(E, \mathcal{E})$ and $(F, \mathcal{F})$, respectively.

1. There exists a unique Σ-finite measure π on the product space $(E \times F, \mathcal{E} \otimes \mathcal{F})$ such that for every positive f in $\mathcal{E} \otimes \mathcal{F}$,

$$(1.68) \quad \pi f = \int_E \mu(dx) \int_F \nu(dy) f(x, y) = \int_F \nu(dy) \int_E \mu(dx) f(x, y).$$

2. If $f \in \mathcal{E} \otimes \mathcal{F}$ and is π-integrable, then $f(x, \cdot)$ is ν-integrable for μ-almost every x, and $f(\cdot, y)$ is μ-integrable for ν-almost every y and (1.68) holds again.

Proof. Assume first that μ and ν are finite measures. From Exercise 27, with $K(\cdot, B) = \nu(B)$, the first integral in (1.68) defines a measure $\pi := \mu \otimes \nu$ on the product space $(E \times F, \mathcal{E} \otimes \mathcal{F})$. With the same reasoning, the second integral defines a measure, $\widehat{\pi} := \nu \otimes \mu$ on $(F \times E, \mathcal{F} \otimes \mathcal{E})$. Defining $\widehat{f}(y, x) := f(x, y)$, we need to show that

$$\pi f = \widehat{\pi}\widehat{f}.$$

The "swap" mapping $s : (x, y) \mapsto (y, x)$ is obviously $(\mathcal{E} \otimes \mathcal{F})/(\mathcal{F} \otimes \mathcal{E})$-measurable, and for any $A \in \mathcal{E}$ and $B \in \mathcal{F}$, we have

$$\pi \circ s^{-1}(B \times A) = \pi(A \times B) = \mu(A)\mu(B) = \widehat{\pi}(B \times A).$$

Since the rectangles $A \times B$ form a p-system that generates $\mathcal{E} \otimes \mathcal{F}$, it follows by Theorem 1.45 that $\widehat{\pi}$ is simply the image measure of π under s; that is $\widehat{\pi} = \pi \circ s^{-1}$. It follows then from Theorem 1.62 that

$$\widehat{\pi}\widehat{f} = (\pi \circ s^{-1})\widehat{f} = \pi(\widehat{f} \circ s) = \pi f.$$

We can generalize this to Σ-finite measures $\mu := \sum \mu_i$ and $\nu := \sum \nu_j$, where the $\{\mu_i\}$ and $\{\nu_j\}$ are finite measures. Namely, in this case we have $\pi f = \sum_{i,j} (\mu_i \otimes \nu_j) f$

and $\widehat{\pi f} = \sum_{i,j} (v_j \otimes \mu_i) \widehat{f}$, so that again $\pi f = \widehat{\pi} \widehat{f}$, because $(\mu_i \otimes v_j) f = (v_j \otimes \mu_i) \widehat{f}$ for every pair (μ_i, v_j) of finite measures.

For the second part of the theorem, suppose f is π-integrable, then (1.68) holds for f, because it holds for both f^+ and f^-, and $\pi f = \pi f^+ - \pi f^-$, where both terms are finite. From Exercise 12, we know that the *section* $f(x, \cdot)$ is $\mathcal{F}$-measurable, and hence for each x the integral $\int_F v(dy) f(x, y)$ is well-defined. Moreover, from the integrability of f, this integral must be finite for μ-almost every x; that is, $f(x, \cdot)$ is v-integrable for μ-almost every x. Similarly, $f(\cdot, y)$ is μ-integrable for v-almost every y.   □

Exercises

1. Prove that a σ-algebra is also closed under countable intersections; that is, if $A_1, A_2, \ldots \in \mathcal{E}$, then also $\cap_n A_n \in \mathcal{E}$.

2. Let A, B, C be a partition of E. Describe the *smallest* σ-algebra containing the sets A, B, and C.

3. Let E be a sample set with n elements. If $\mathcal{F} = 2^E$ (i.e., the collection of all subsets of E), how many sets does $\mathcal{F}$ contain?

4.* Let C be a countable partition of E. Show that every set in σC is a countable union of sets in C.

5. Let $C := \{\{x\}, x \in \mathbb{R}\}$. Show that every set in σC is either countable or has a countable complement.

6. Show that $\mathcal{E}_0$ in Example 1.7 is not a σ-algebra.

7. Let $\{\mathcal{E}_i, i \in I\}$ be an arbitrary (countable or uncountable) collection of σ-algebras on E. Show that the intersection $\cap_{i \in I} \mathcal{E}_i$ is also a σ-algebra on E.

8. Let $(E, \mathcal{E})$ be a measurable space. Let $D \subseteq E$ (not necessarily in $\mathcal{E}$). Define $\mathcal{D}$ as the collection of sets of the form $A \cap D$, where $A \in \mathcal{E}$. Show that $\mathcal{D}$ is a σ-algebra on D. The measurable space $(D, \mathcal{D})$ is called the *trace* of $(E, \mathcal{E})$ on D.

9.* Let E be a set and $(F, \mathcal{F})$ a measurable space. For $f : E \to F$, define

$$f^{-1}\mathcal{F} = \{f^{-1}B : B \in \mathcal{F}\},$$

where $f^{-1}B$ is the inverse image of B. Show that $f^{-1}\mathcal{F}$ is a σ-algebra on E. It is the smallest σ-algebra on E such that f is measurable relative to it and $\mathcal{F}$. It is called the σ-*algebra generated by* f.

10. Prove that Definition 1.10 of a d-system is equivalent to Definition 1.6 of a σ-algebra, replacing (in the latter definition) $\mathcal{E}$ with $\mathcal{D}$ and Item 3. with:

   3'. If $A_1, A_2, \ldots \in \mathcal{D}$ are *disjoint*, then also $\cup_n A_n \in \mathcal{D}$.

11.* Let $(E, \mathcal{E})$ be a measurable space and μ a mapping from $\mathcal{E}$ to $[0, 1]$ such that

- $\mu(E) = 1$;
- for every *finite* sequence $A_1, \ldots, A_n$ of disjoint sets in $\mathcal{E}$ it holds that $\mu(\cup_{i=1}^n A_i) = \sum_{i=1}^n \mu(A_i)$;
- for every decreasing sequence $A_1, A_2, \ldots$, of sets in $\mathcal{E}$ that converges to the empty set, i.e., $A_n \downarrow \emptyset$, it holds that $\lim_n \mu(A_n) = 0$.

   Prove that μ is a probability measure on $(E, \mathcal{E})$.

12. Let $f : E \times F \to G$ be $(\mathcal{E} \otimes \mathcal{F})/\mathcal{G}$-measurable, where $\mathcal{E}, \mathcal{F}$, and $\mathcal{G}$ are σ-algebras on E, F, and G. Show that the function $y \mapsto f(x_0, y)$ is $\mathcal{F}/\mathcal{G}$-measurable for any fixed $x_0 \in E$. Such a function is called a *section* of f.

13. Prove the two assertions in Example 1.23 rigorously.

14.* Show that because $(-\infty, x]$, $x \in \mathbb{R}$, are elements of $\mathcal{B}$, the sets of the form $(a, b]$, (a, b), and $\{a\}$ are also in $\mathcal{B}$.

15.* What is the Lebesgue measure of $\mathbb{Q}$, the set of rational numbers?

16. Let $(E, \mathcal{E}, \mu)$ be a measure space and let $D \in \mathcal{E}$. Define $\nu(A) := \mu(A \cap D), A \in \mathcal{E}$. Show that ν is a measure on $(E, \mathcal{E})$. It is called the *trace* of μ on D.

17. Let $(E, \mathcal{E}, \mu)$ be a measure space and let $(D, \mathcal{D})$ be the trace of $(E, \mathcal{E})$ on $D \in \mathcal{E}$ (see Exercise 8). Define ν by

$$\nu(A) := \mu(A), \quad A \in \mathcal{D}.$$

Show that $(D, \mathcal{D}, \nu)$ is a measure space. The measure ν is called the *restriction* of μ to $(D, \mathcal{D})$.

18. Let $(E, \mathcal{E})$ be a measurable space and let $(D, \mathcal{D})$ be the trace of $(E, \mathcal{E})$ on $D \in \mathcal{E}$. Let ν be a measure on $(D, \mathcal{D})$. Define μ by

$$\mu(A) := \nu(A \cap D), \quad A \in \mathcal{E}.$$

Show that μ is a measure on $(E, \mathcal{E})$. It is called the *extension* of ν to $(E, \mathcal{E})$.

19.* Prove that the Cantor set in Example 1.1 has Lebesgue measure 0, and argue why it has as many points as the interval $[0, 1]$.

20. Prove that if μ and ν are measures, then $a\mu + b\nu$ with $a, b \geq 0$ is a measure as well. Hint: The main thing to verify is the countable additivity.

21. Let $(E, \mathcal{E}, \mu)$ be a measure space and f an $\mathcal{E}$-measurable (numerical) function. Prove that if f is integrable, then f must be real-valued μ-almost everywhere. Hint: show this first for $f \in \mathcal{E}_+$.

22. Let $\mu : \mathcal{B}_{(0,1]} \to [0, \infty]$ be defined by $\mu(A) := |A|$; that is, the number of elements in A. Show that μ is a measure, but is not Σ-finite.

23. Verify that ν in (1.60) is indeed a measure.

24. Let c be an increasing right-continuous function from $\mathbb{R}_+$ into $\overline{\mathbb{R}}_+$. Define

$$a(u) := \inf\{t \in \mathbb{R}_+ : c(t) > u\}, \quad u \in \mathbb{R}_+,$$

with the usual convention that $\inf \emptyset = \infty$.

  (a) Show that the function $a : \mathbb{R}_+ \to \overline{\mathbb{R}}_+$ is increasing and right-continuous, and that

$$c(t) = \inf\{u \in \mathbb{R}_+ : a(u) > t\}, \quad t \in \mathbb{R}_+.$$

Thus, a and c are right-continuous *functional inverses* of each other.

  (b) Suppose $c(t) < \infty$. Show that $a(c(t)) \geq t$, with equality if and only if $c(t + \varepsilon) > c(t)$ for every $\varepsilon > 0$.

25. Let μ be a measure on $(E, \mathcal{E})$ and p a positive $\mathcal{E}$-measurable function. Show that the mapping

$$\nu : A \mapsto \mu(\mathbb{1}_A\, p) = \int \mathbb{1}_A\, p\, \mathrm{d}\mu, \quad A \in \mathcal{E}$$

defines a measure ν on $(E, \mathcal{E})$.

26.* Let K be a *bounded* transition kernel from $(E, \mathcal{E})$ into $(F, \mathcal{F})$; i.e., $K(\cdot, F) < c$ for some $c \in \mathbb{R}_+$. Consider the transformation $T : (\mathcal{E} \otimes \mathcal{F})_+ \to \mathcal{E}_+$ defined by $(Tf)(x) := \int_F K(x, \mathrm{d}y) f(x, y)$.

  (a) Show that $T(af + bg) = aTf + bTg$ for all $f, g \in (\mathcal{E} \otimes \mathcal{F})_+$ and $a, b \in \mathbb{R}_+$.

  (b) Show that if (f_n) is a sequence in $(\mathcal{E} \otimes \mathcal{F})_+$ increasing to f, then $Tf_n \uparrow Tf$.

  (c) Prove that Tf is a positive $\mathcal{E}$-measurable numerical function for every $f \in (\mathcal{E} \otimes \mathcal{F})_+$. Hint: show that

$$\mathcal{M} := \{f \in (\mathcal{E} \otimes \mathcal{F}), f \text{ is positive or bounded} : \quad Tf \in \mathcal{E}\}$$

is a monotone class that includes $\mathbb{1}_{A \times B}$ for every $A \in \mathcal{E}$ and $B \in \mathcal{F}$.

27. Using Theorem 1.57 and Exercise 26, show that the right-hand side of (1.66) indeed defines a measure on $(E \times F, \mathcal{E} \otimes \mathcal{F})$, when K is a bounded kernel. Hint: define $L(f) := \mu(Tf)$.

28. The Σ-finiteness condition is necessary in Fubini's theorem. For example, take $E = F = (0, 1]$ and $\mathcal{E} = \mathcal{F} = \mathcal{B}_{(0,1]}$. Let μ be the Lebesgue measure on $(E, \mathcal{E})$ and ν the counting measure defined in Exercise 22. Show that, with $f(x, y) := \mathbb{1}_{\{x=y\}}$, $(x, y) \in E \times F$,

$$\int_E \mu(dx) \int_F \nu(dy) f(x, y) \neq \int_F \nu(dy) \int_E \mu(dx) f(x, y).$$

CHAPTER 2

PROBABILITY

The purpose of this chapter is to put various results from elementary probability theory in the more rigorous and mature framework of measure theory. In this framework we will treat concepts such as random variables, expectations, distributions, etc., in a *unified* manner, without necessarily having to introduce the dichotomy between the "discrete" and "continuous" case, as is customary in elementary probability.

2.1 Modeling Random Experiments

Probability theory is about modeling and analysing *random experiments*: experiments whose outcome cannot be determined in advance, but are nevertheless still subject to analysis. Mathematically, we can model a random experiment by defining a specific measure space $(\Omega, \mathcal{H}, \mathbb{P})$, which we call a *probability space*, where the three components are: a *sample space*, a collection of *events*, and a *probability measure*.

The *sample space* Ω of a random experiment is the set of all possible outcomes of the random experiment. Sample spaces can be countable, such as $\mathbb{N}$, or uncountable, such as $\mathbb{R}^n$ or $\{0, 1\}^\infty$. The sets in the σ-algebra $\mathcal{H}$ are called *events*. These are the subsets to which we wish to assign a probability. We can view $\mathcal{H}$ as the **h**oard of events. An event A *occurs* if the outcome of the experiment is one of the elements in A. The properties of $\mathcal{H}$ are natural in the context of random experiments and events:

- If A and B are events, the set $A \cup B$ is also an event, namely the event that A *or* B *or* both occur. For a sequence $A_1, A_2, \ldots$ of events, their union is the event that *at least one* of the events occurs.
- If A and B are events, the set $A \cap B$ is also an event, namely the event that A

and B both occur. For a sequence $A_1, A_2, \ldots$ of events, their intersection is the event that *all* of the events occur.

- If A is an event, its complement A^c is also an event, namely the event that A does *not* occur.
- The set Ω itself is an event, namely the *certain* event (it always occurs). Similarly $\emptyset$ is an event, namely the *impossible* event (it never occurs).

When the sample space Ω is $\mathbb{R}$, the natural σ-algebra $\mathcal{H}$ is the *Borel σ-algebra* $\mathcal{B}$. Recall that this is the smallest σ-algebra on $\mathbb{R}$ that contains all the intervals of the form $(-\infty, x]$ for $x \in \mathbb{R}$. This σ-algebra of sets is big enough to contain all important sets and small enough to allow us to assign a probability to all events.

The third ingredient in the model for a random experiment is the specification of the probability of the events. In fact, this is the crucial part of the model. Mathematically, we are looking for a *measure* $\mathbb{P}$ that assigns to each event A a number $\mathbb{P}(A)$ in $[0, 1]$ describing how likely or probable that event is.

Definition 2.1: Probability Measure

A *probability measure* is a measure on $(\Omega, \mathcal{H})$ with $\mathbb{P}(\Omega) = 1$. Thus, $\mathbb{P}$ is a mapping from $\mathcal{H}$ to $[0, 1]$ with the following properties:

1. $\mathbb{P}(\emptyset) = 0$ and $\mathbb{P}(\Omega) = 1$.

2. For any sequence $A_1, A_2, \ldots$ of disjoint events,

$$\mathbb{P}(\cup_n A_n) = \sum_n \mathbb{P}(A_n).$$

An event A is said to occur *almost surely* (a.s.) if $\mathbb{P}(A) = 1$. The following properties follow directly from the properties of a general measure in Proposition 1.42:

Theorem 2.2: Properties of a Probability Measure

Let $\mathbb{P}$ be a probability measure on a measurable space $(\Omega, \mathcal{H})$, and let A, B and $A_1, A_2, \ldots$ be events. Then, the following hold:

1. *(Monotonicity)*: $A \subseteq B \implies \mathbb{P}(A) \leq \mathbb{P}(B)$.

2. *(Continuity from below)*: If $A_1, A_2, \ldots$ is an increasing sequence of events, i.e., $A_1 \subseteq A_2 \subseteq \cdots$, then

(2.3) $$\lim \mathbb{P}(A_n) = \mathbb{P}(\cup_n A_n).$$

3. *(Countable subadditivity)*: $\mathbb{P}(\cup_n A_n) \leq \sum_n \mathbb{P}(A_n)$.

Additional properties arise from the fact that $\mathbb{P}$ is a finite measure.

Theorem 2.4: Additional Properties of a Probability Measure

Let $\mathbb{P}$ be a probability measure on a measurable space $(\Omega, \mathcal{H})$, and let A, B and $A_1, A_2, \ldots$ be events. Then, the following hold:

1. *(Complement):* $\mathbb{P}(A^c) = 1 - \mathbb{P}(A)$.

2. *(Union):* $\mathbb{P}(A \cup B) = \mathbb{P}(A) + \mathbb{P}(B) - \mathbb{P}(A \cap B)$.

3. *(Continuity from above):* If $A_1, A_2, \ldots$ is a decreasing sequence of events, i.e., $A_1 \supseteq A_2 \supseteq \cdots$, then

$$(2.5) \qquad\qquad \lim \mathbb{P}(A_n) = \mathbb{P}(\cap_n A_n).$$

Proof. Properties 1 and 2 follow directly from the (finite) additivity of $\mathbb{P}$ and the fact that $\mathbb{P}(\Omega) = 1$. It remains to prove Property 3. Let $A_1, A_2, \ldots$ be a sequence of events that decreases to $A := \cap_n A_n$. Then, $A_1^c, A_2^c, \ldots$ is a sequence of events that increases to $\cup_n A_n^c = (\cap_n A_n)^c = A^c$. By the continuity from below property, $\lim \mathbb{P}(A_n^c) = \mathbb{P}(A^c)$. Now apply Property 1 to conclude that $\lim \mathbb{P}(A_n) = \mathbb{P}(A)$. □

Thus, our model for a random experiment consists of specifying a probability space. We give two fundamental examples.

■ **Example 2.6 (Discrete Probability Space)** Let $\Omega := \{a_1, a_2, \ldots\}$ and let $\{p_1, p_2, \ldots\}$ be a set of positive numbers summing up to 1. Take $\mathcal{H}$ to be the power set of Ω. Then, any $\mathbb{P} : \mathcal{H} \to [0, 1]$ defined by $\mathbb{P}(A) := \sum_{i:a_i \in A} p_i$, $A \subseteq \Omega$, is a probability measure on $(\Omega, \mathcal{H})$ and, conversely, any probability measure on $(\Omega, \mathcal{H})$ is of this form. Thus, we can specify $\mathbb{P}$ by specifying only the probabilities of the *elementary events* $\{a_i\}$; see Figure 2.7 for an illustration.

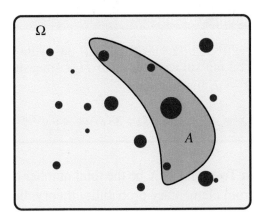

Figure 2.7: A discrete probability space.

■ **Example 2.8** (**Lebesgue Measure**) We select randomly a point in the interval $(0, 1]$ such that each point is equally likely to be drawn. Let $\Omega := (0, 1]$. How should we choose $\mathcal{H}$ and $\mathbb{P}$? We know that the probability that a randomly selected point falls in the interval $[a, b]$ should be proportional to the length of that interval; that is,

$$\mathbb{P}([a, b]) = b - a,$$

for $0 < a < b \leq 1$. Does such a $\mathbb{P}$ exist? Yes, take $\mathcal{H} := \mathcal{B}_{(0,1]}$ and $\mathbb{P} := \mathrm{Leb}_{(0,1]}$ as respectively the Borel σ-algebra and Lebesgue measure on $(0, 1]$. Both are well-defined; see Example 1.41. ■

Specifying an explicit probability space may not always be easy or necessary, and in practice we often leave the probability space in the background and choose to formulate and analyse the model via *random variables* instead.

2.2 Random Variables

Usually the most convenient way to describe quantities of interest connected with random experiments is by using random variables. Random variables allow us to use intuitive notations for certain events, such as $\{X \in A\}$, $\{\max(X, Y) \leq Z\}$, etc. Intuitively, we can think of a random variable X as a measurement on a random experiment that will become available *tomorrow*. However, all the think work can be done *today*. If for each possible outcome ω, the measurement is $X(\omega)$, we can specify today probabilities such as $\mathbb{P}(X \in A)$.

We can translate our intuitive notion of a random variable into rigorous mathematics by defining a random variable to be a measurable function from $(\Omega, \mathcal{H})$ to a measurable space $(E, \mathcal{E})$.

Definition 2.9: Random Variable

Let $(\Omega, \mathcal{H}, \mathbb{P})$ be a probability space and $(E, \mathcal{E})$ a measurable space. A *random variable* X taking values in E is an $\mathcal{H}/\mathcal{E}$-measurable function; that is, a mapping $X : \Omega \to E$ that satisfies

$$(2.10) \quad X^{-1}A = \{X \in A\} = \{\omega \in \Omega : X(\omega) \in A\} \in \mathcal{H} \quad \text{for all } A \in \mathcal{E}.$$

■ **Example 2.11** (**Coin Tosses**) Let X be the total number of heads in 20 tosses of a fair coin. We know from elementary probability theory that

$$(2.12) \qquad \mathbb{P}(X = k) = \binom{20}{k} \frac{1}{2^{20}}, \quad k = 0, 1, \ldots, 20.$$

We have left the probability space completely in the background. How can we reconcile this result with our mathematical model? First, for the sample space we could choose

$$\Omega := \{0, 1\}^{20},$$

i.e., the set of vectors $[x_1, \ldots, x_{20}]$ where each $x_i = 0$ or 1, indicating the result of the ith toss (1 = heads, 0 = tails). Since Ω has a finite number of elements, we can define $\mathcal{H}$ as the set of *all* subsets of Ω. The measure $\mathbb{P}$ can now be specified by assigning probability $1/2^{20}$ to each elementary event $\{[x_1, \ldots, x_{20}]\}$. This finalizes our probability space.

Let X be the function that assigns the total number of heads to each outcome ω. In particular, for $\omega = [x_1, \ldots, x_{20}]$, we have

$$X(\omega) := \sum_{i=1}^{20} x_i.$$

Now consider the set $\{\omega \in \Omega : X(\omega) = k\}$. The probability of this event is given by the right-hand side of (2.12). Thus, if we *abbreviate* $\{\omega \in \Omega : X(\omega) = k\}$ to $\{X = k\}$ and further abbreviate $\mathbb{P}(\{X = k\})$ to $\mathbb{P}(X = k)$, then we have justified (2.12)! ∎

We often are interested in *numerical* random variables; that is, random variables taking values in $E := \overline{\mathbb{R}}$ (the extended real line), with σ-algebra $\mathcal{E} := \overline{\mathcal{B}}$. However, note that Definition 2.9 allows for a general measurable space $(E, \mathcal{E})$. The meaning of (2.10) becomes clear in a probabilistic context: we want to be able to assign a probability to each set $\{X \in A\}$, and so these sets should be events.

■ **Example 2.13 (Indicator Functions and Simple Functions)** The simplest example of a numerical random variable is an *indicator function* of an event A:

$$\mathbb{1}_A(\omega) := \begin{cases} 1 & \text{if } \omega \in A, \\ 0 & \text{if } \omega \notin A. \end{cases}$$

Positive linear combinations of indicator random variables, i.e., $\sum_i a_i \mathbb{1}_{A_i}$ with $a_i \in \mathbb{R}_+$ and $A_i \in \mathcal{H}$ for all i, are again positive numerical random variables. In fact, any positive numerical random variable is the increasing limit of a sequence of such *simple* random variables. See also Theorems 1.27 and 1.31 for a characterization of numerical random variables. ∎

The fact that random variables are measurable functions has some nice consequences. For example, any measurable function of X is again a random variable, by Theorem 1.21. In particular, if X is a real-valued random variable and $g : \mathbb{R} \to \mathbb{R}$ is measurable with respect to the Borel σ-algebra $\mathcal{B}$ on $\mathbb{R}$, then $g(X) = g \circ X$ is again a numerical random variable.

■ **Example 2.14** (**Quadratic Function**) Let X be a real-valued random variable and let g be the function $x \mapsto x^2, x \in \mathbb{R}$. The function g is continuous, meaning that the inverse image $g^{-1}(O)$ of any open set O is again open. Since the open sets in $\mathbb{R}$ generate $\mathcal{B}$, we have $g^{-1}(O) \in \mathcal{B}$ and thus, by Theorem 1.20, g is $\mathcal{B}/\mathcal{B}$-measurable. It follows then from Theorem 1.21 that $g \circ X$ is $\mathcal{H}/\mathcal{B}$-measurable; that it, it is again a real-valued random variable. This random variable maps $\omega \in \Omega$ to the number $g(X(\omega)) = (X(\omega))^2$. Therefore, we write $g \circ X$ also as X^2. ■

Often a random experiment is described via more than one random variable. Let $(\Omega, \mathcal{H}, \mathbb{P})$ be a probability space and let $\mathbb{T}$ be an arbitrary index set, which may be countable or uncountable. We can think of $\mathbb{T}$ as a set of times. For each t, let $(E_t, \mathcal{E}_t)$ be a measurable space. Generalizing Definition 1.14, the *product space* of $\{(E_t, \mathcal{E}_t), t \in \mathbb{T}\}$ is the measurable space $(\times_{t \in \mathbb{T}} E_t, \otimes_{t \in \mathbb{T}} \mathcal{E}_t)$. That is, each element $\boldsymbol{x} \in \times_{t \in \mathbb{T}} E_t$ is of the form $\boldsymbol{x} = (x_t, t \in \mathbb{T})$. Often $(E_t, \mathcal{E}_t)$ is one and the same measurable space $(E, \mathcal{E})$, and in that case $\boldsymbol{x}$ can be viewed as a function from $\mathbb{T}$ to E. The σ-algebra on $\times_{t \in \mathbb{T}} E_t$ is the σ-algebra generated by the rectangles

$$(2.15) \qquad \underset{t \in \mathbb{T}}{\times} A_t = \left\{ \boldsymbol{x} \in \underset{t \in \mathbb{T}}{\times} E_t : x_t \in A_t \text{ for each } t \text{ in } \mathbb{T} \right\},$$

where $A_t = E_t$ except for a finite number of t.

Definition 2.16: Stochastic Process

A collection of random variables $\boldsymbol{X} := \{X_t, t \in \mathbb{T}\}$, where $\mathbb{T}$ is any index set, is called a *stochastic process*. When $\mathbb{T}$ is finite, $\boldsymbol{X}$ is called a *random vector*.

The point of the following theorem is that we can think of a stochastic process $\boldsymbol{X} := \{X_t, t \in \mathbb{T}\}$ in two equivalent ways.

Theorem 2.17: Stochastic Process

$\boldsymbol{X} := \{X_t, t \in \mathbb{T}\}$ is a stochastic process taking values in $\times_{t \in \mathbb{T}} E_t$ with σ-algebra $\otimes_{t \in \mathbb{T}} \mathcal{E}_t$ if and only if each X_t is a random variable taking values in E_t with σ-algebra $\mathcal{E}_t$.

Proof. Let $\boldsymbol{X} := \{X_t, t \in \mathbb{T}\}$ be a stochastic process taking values in $\times_{t \in \mathbb{T}} E_t$ with σ-algebra $\otimes_{t \in \mathbb{T}} \mathcal{E}_t$. We wish to show that each X_t is a random variable taking values in E_t with σ-algebra $\mathcal{E}_t$. In $\otimes_{t \in \mathbb{T}} \mathcal{E}_t$ consider rectangles of the form $\times_{u \in \mathbb{T}} A_u$, where $A_u = E_u$ for all $u \neq t$ and $A_t = B_t \in \mathcal{E}_t$. The inverse image of this set is an event, since $\boldsymbol{X}$ is a stochastic process. But this inverse image is exactly $\{X_t \in B_t\}$. As this is true for all $B_t \in \mathcal{E}_t$, X_t is a random variable. To prove the converse, suppose all the X_t are random variables. Now take a rectangle of the form (2.15). Its inverse image under X is the intersection of $\{X_t \in A_t\}$ for a finite number of t and is thus an event. Since these rectangles generate $\otimes_{t \in \mathbb{T}} \mathcal{E}_t$, X is a stochastic process.   □

From Section 1.2, we see that measurable functions of stochastic processes again yield stochastic processes. In particular, when dealing with numerical random variables X and Y, the functions $X + Y$, XY, X/Y are random variables as well, as long as they are well-defined. Also, if (X_n) is a sequence of numerical random variables then, by Proposition 1.26, $\sup X_n$, $\inf X_n$, $\limsup X_n$, and $\liminf X_n$ are all random variables. In particular, if (X_n) converges (that is, pointwise) to X, then X is again a random variable. Two random variables are said to be equal *almost surely*, if they are the same $\mathbb{P}$-almost everywhere; that is, if the set of points for which they differ has probability 0.

2.3   Probability Distributions

Let $(\Omega, \mathcal{H}, \mathbb{P})$ be a probability space and let X be a random variable taking values in a set E with σ-algebra $\mathcal{E}$. To simplify our usage, we also say that "X takes values in $(E, \mathcal{E})$". If we wish to describe our experiment via X, then we need to specify the probabilities of events such as $\{X \in B\}$, $B \in \mathcal{E}$. In elementary probability theory we usually made a distinction between "discrete" and "continuous" random variables. However, it is sometimes better to analyse random variables in a more unified manner; many definitions and properties of random variables do not depend on whether they are "discrete" or "continuous". Moreover, some random variables are neither discrete nor continuous.

Recall that by definition *all* the sets $\{X \in B\}$ with $B \in \mathcal{E}$ are events. So we may assign a probability to each of these events. This leads to the notion of the *distribution* of a random variable.

Definition 2.18: Probability Distribution

Let $(\Omega, \mathcal{H}, \mathbb{P})$ be a probability space and let X be a random variable taking values in $(E, \mathcal{E})$. The *(probability) distribution* of X is the image measure μ of $\mathbb{P}$ under X. That is,

$$(2.19) \qquad\qquad \mu(B) := \mathbb{P}(X \in B), \quad B \in \mathcal{E}.$$

The probability distribution of X therefore carries all the "information" about X that is known. Note that above we should have written $\mathbb{P}(\{X \in B\})$ instead of $\mathbb{P}(X \in B)$. Since this abbreviation is harmless, we will use it from now on.

A distribution μ defined by (2.19) is thus a probability measure on $(E, \mathcal{E})$ and is completely specified by its values on a p-system that generates $\mathcal{E}$, as shown in

Theorem 1.45. In particular, for a numerical random variable, it suffices to specify $\mu([-\infty, x]) = \mathbb{P}(X \leq x)$ for all $x \in \mathbb{R}$. This leads to the following definition:

Definition 2.20: Cumulative Distribution Function

The *(cumulative) distribution function* of a numerical random variable X is the function $F : \mathbb{R} \rightarrow [0, 1]$ defined by

$$F(x) := \mathbb{P}(X \leq x), \quad x \in \mathbb{R}.$$

We usually abbreviate *cumulative distribution function* to *cdf*. In Figure 2.21 the graph of an arbitrary cdf is depicted.

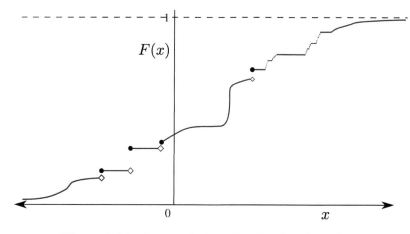

Figure 2.21: A cumulative distribution function.

The following properties for a cdf F are a direct consequence of the properties of the probability measure $\mathbb{P}$. We leave the proof as an exercise; see Exercise 3.

Proposition 2.22: Properties of a Cumulative Distribution Function

1. *(Bounded)*: $0 \leq F(x) \leq 1$.

2. *(Increasing)*: $x \leq y \Rightarrow F(x) \leq F(y)$.

3. *(Right-continuous)*: $\lim_{h \downarrow 0} F(x + h) = F(x)$.

The uniqueness result in Theorem 1.45 shows that to each distribution μ of a numerical random variable X there corresponds exactly one cdf F and vice versa. The distribution is said to be *proper* if $\lim_{x \to -\infty} F(x) = 0$ and $\lim_{x \to \infty} F(x) = 1$.

In many cases we specify distributions of random variables as measures with a density with respect to some other measure, usually a counting measure or Lebesgue measure; see Section 1.4.3 for the measure-theoretic background.

Definition 2.23: Discrete Distribution

We say that X has a *discrete* distribution μ on $(E, \mathcal{E})$ if μ is a discrete measure; that is, it is of the form

$$\mu = \sum_{x \in D} f(x)\, \delta_x$$

for some countable set D and positive masses $\{f(x), x \in D\}$.

We see that f is the density of μ with respect to the counting measure on D, and that $f(x) = \mu(\{x\}) = \mathbb{P}(X = x)$. The number $f(x)$ represents the amount of probability "mass" at x. In elementary probability f is often called the *probability mass function* (pmf) of the *discrete* random variable X.

Definition 2.24: Distribution with a Density

We say that X has an *absolutely continuous* distribution μ on $(E, \mathcal{E})$ with respect to a measure λ on $(E, \mathcal{E})$ if there exists a positive $\mathcal{E}$-measurable function f such that $\mu = f\,\lambda$; that is,

$$\mu(B) = \mathbb{P}(X \in B) = \int_B \lambda(\mathrm{d}x)\, f(x), \quad B \in \mathcal{E}.$$

The function f is thus the *(probability) density function* (pdf) of X with respect to the measure λ — often taken to be the Lebesgue measure. Note that μ is absolutely continuous with respect to λ in the sense that for all $B \in \mathcal{E}$: $\lambda(B) = 0 \Rightarrow \mu(B) = 0$.

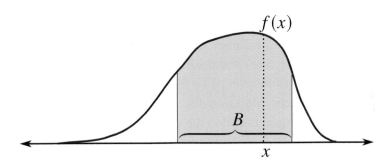

Figure 2.25: A probability density function of a numerical random variable.

Describing an experiment via a random variable and its probability distribution or density is often more convenient than specifying the probability space. In fact, the probability space usually stays in the background. The question remains, however, whether there *exists* a probability space $(\Omega, \mathcal{H}, \mathbb{P})$ and a numerical random variable X for a given probability distribution. Fortunately, existence can be established for all practical probability models. We will have a closer look at this in Section 4.4.

Here is a simple example: there exists a probability space and a numerical random variable X such that X has a given cdf F. Namely, take $\Omega := \overline{\mathbb{R}}$, $\mathcal{H} := \overline{\mathcal{B}}$, and let $\mathbb{P}$ be the measure on $\overline{\mathcal{B}}$ that is defined by

$$\mathbb{P}([-\infty, x]) := F(x).$$

Now, let X be the *identity* function $x \mapsto x$ on $\overline{\mathbb{R}}$. Then, X is a random variable on $(\Omega, \mathcal{H}, \mathbb{P})$ with cdf F, because

$$\mathbb{P}(X \le x) = \mathbb{P}([-\infty, x]) = F(x).$$

Some probability distributions $(\mathbb{R}, \mathcal{B})$ are neither discrete nor absolutely continuous with respect to the Lebesgue measure; see the example below. In elementary probability the pdf f of a "continuous" random variable is often taken to be the derivative of its cdf F. This is not true in general. In Exercise 6 a *continuous* cdf F is constructed for a distribution that does *not* have a pdf with respect to the Lebesgue measure, even though the derivative of F exists almost everywhere.

■ **Example 2.26 (Mixture Distribution)** Let F_d and F_c be distribution functions of a discrete and absolutely continuous (with respect to the Lebesgue measure) distribution on $(\mathbb{R}, \mathcal{B})$, respectively. For any $0 < \alpha < 1$, the function F defined by

$$F(x) := \alpha F_d(x) + (1 - \alpha) F_c(x), \quad x \in \mathbb{R},$$

is again a distribution function, but the corresponding distribution is neither discrete nor absolute continuous with respect to the Lebesgue measure. This is an example of a *mixture* of distributions. ■

Tables 2.1 and 2.2 list a number of important absolutely continuous (with respect to the Lebesgue measure) and discrete distributions on $(\mathbb{R}, \mathcal{B})$. Note that in Table 2.1, Γ is the gamma function: $\Gamma(\alpha) := \int_0^\infty dx\, e^{-x} x^{\alpha-1}$, $\alpha > 0$. The normal distribution is often called the *Gaussian* distribution — we will use both names. The Gamma$(n/2, 1/2)$ distribution is called the *chi-squared distribution* with n degrees of freedom, denoted χ_n^2. The t_1 distribution is also called the *Cauchy* distribution.

We write

$$X \sim \text{Dist} \quad \text{or} \quad X \sim f$$

to indicate that the random variable X has distribution Dist or density f.

Table 2.1: Commonly used continuous distributions.

Name	Notation	$f(x)$	$x \in$	Parameters
Uniform	$\mathcal{U}[\alpha, \beta]$	$\dfrac{1}{\beta - \alpha}$	$[\alpha, \beta]$	$\alpha < \beta$
Normal	$\mathcal{N}(\mu, \sigma^2)$	$\dfrac{1}{\sigma\sqrt{2\pi}} e^{-\frac{1}{2}\left(\frac{x-\mu}{\sigma}\right)^2}$	$\mathbb{R}$	$\sigma > 0, \mu \in \mathbb{R}$
Gamma	$\mathrm{Gamma}(\alpha, \lambda)$	$\dfrac{\lambda^\alpha x^{\alpha-1} e^{-\lambda x}}{\Gamma(\alpha)}$	$\mathbb{R}_+$	$\alpha, \lambda > 0$
Inverse Gamma	$\mathrm{InvGamma}(\alpha, \lambda)$	$\dfrac{\lambda^\alpha x^{-\alpha-1} e^{-\lambda x^{-1}}}{\Gamma(\alpha)}$	$\mathbb{R}_+$	$\alpha, \lambda > 0$
Exponential	$\mathrm{Exp}(\lambda)$	$\lambda e^{-\lambda x}$	$\mathbb{R}_+$	$\lambda > 0$
Beta	$\mathrm{Beta}(\alpha, \beta)$	$\dfrac{\Gamma(\alpha+\beta)}{\Gamma(\alpha)\Gamma(\beta)} x^{\alpha-1}(1-x)^{\beta-1}$	$[0, 1]$	$\alpha, \beta > 0$
Weibull	$\mathrm{Weib}(\alpha, \lambda)$	$\alpha\lambda\,(\lambda x)^{\alpha-1} e^{-(\lambda x)^\alpha}$	$\mathbb{R}_+$	$\alpha, \lambda > 0$
Pareto	$\mathrm{Pareto}(\alpha, \lambda)$	$\alpha\lambda\,(1+\lambda x)^{-(\alpha+1)}$	$\mathbb{R}_+$	$\alpha, \lambda > 0$
Student	t_ν	$\dfrac{\Gamma(\frac{\nu+1}{2})}{\sqrt{\nu\pi}\,\Gamma(\frac{\nu}{2})} \left(1 + \dfrac{x^2}{\nu}\right)^{-(\nu+1)/2}$	$\mathbb{R}$	$\nu > 0$
F	$F(m, n)$	$\dfrac{\Gamma(\frac{m+n}{2})\,(m/n)^{m/2} x^{(m-2)/2}}{\Gamma(\frac{m}{2})\,\Gamma(\frac{n}{2})\,[1+(m/n)x]^{(m+n)/2}}$	$\mathbb{R}_+$	$m, n \in \mathbb{N}_+$

Table 2.2: Commonly used discrete distributions.

Name	Notation	$f(x)$	$x \in$	Parameters
Bernoulli	$\mathrm{Ber}(p)$	$p^x(1-p)^{1-x}$	$\{0, 1\}$	$0 \leq p \leq 1$
Binomial	$\mathrm{Bin}(n, p)$	$\binom{n}{x} p^x(1-p)^{n-x}$	$\{0, 1, \ldots, n\}$	$0 \leq p \leq 1, n \in \mathbb{N}$
Discrete uniform	$\mathcal{U}\{1, \ldots, n\}$	$\dfrac{1}{n}$	$\{1, \ldots, n\}$	$n \in \{1, 2, \ldots\}$
Geometric	$\mathrm{Geom}(p)$	$p(1-p)^{x-1}$	$\{1, 2, \ldots\}$	$0 \leq p \leq 1$
Poisson	$\mathrm{Poi}(\lambda)$	$e^{-\lambda}\dfrac{\lambda^x}{x!}$	$\mathbb{N}$	$\lambda > 0$
Negative binomial	$\mathrm{NegBin}(n, p)$	$\binom{n+x-1}{n-1} p^n(1-p)^x$	$\mathbb{N}$	$0 \leq p \leq 1, n \in \mathbb{N}$

Definition 2.18 is very general and applies, in particular, to any numerical random vector $X := [X_1, \ldots, X_n]^\top$. In this case, the distribution μ of X is defined by

$$\mu(B) := \mathbb{P}(X \in B), \quad B \in \overline{\mathscr{B}}^n.$$

Where possible, we characterize the distribution of a numerical random vector via densities with respect to counting or Lebesgue measures. For example, if the distribution of X has a density f with respect to the Lebesgue measure, then

$$\mathbb{P}(X \in B) = \int_B \mathrm{d}x\, f(x), \quad B \in \overline{\mathscr{B}}^n.$$

If instead X is a *discrete* random vector, i.e., one that can only take values in some countable set D, then the distribution of X is most easily specified via its probability mass function $f(x) := \mathbb{P}(X = x)$, $x \in D$, which is in this case a function of n variables.

The concept of independence is of great importance in the study of random experiments and the construction of probability distributions. Loosely speaking, independence is about the lack of shared information between random objects.

Definition 2.27: Independent Random Variables

Let X and Y be random variables taking values in $(E, \mathcal{E})$ and $(F, \mathcal{F})$, respectively. They are said to be *independent* if for any $A \in \mathcal{E}$ and $B \in \mathcal{F}$ it holds that

(2.28) $$\mathbb{P}(X \in A, Y \in B) = \mathbb{P}(X \in A)\,\mathbb{P}(Y \in B).$$

Intuitively, this means that information regarding X does not affect our knowledge of Y and vice versa. Mathematically, (2.28) simply states that the distribution, μ say, of (X, Y) is given by the *product measure* of the distributions μ_X of X and μ_Y of Y. We can extend the independence concept to a finite or infinite collection of random variables as follows:

Definition 2.29: Independency

We say that a collection of random variables $\{X_t, t \in \mathbb{T}\}$, with each X_t taking values in $(E_t, \mathcal{E}_t)$, is an *independency* if for any finite choice of indexes $t_1, \ldots, t_n \in \mathbb{T}$ and sets $A_{t_1} \in \mathcal{E}_{t_1}, \ldots, A_{t_n} \in \mathcal{E}_{t_n}$ it holds that

$$\mathbb{P}(X_{t_1} \in A_{t_1}, \ldots, X_{t_n} \in A_{t_n}) = \mathbb{P}(X_{t_1} \in A_{t_1}) \cdots \mathbb{P}(X_{t_n} \in A_{t_n}).$$

■ **Remark 2.30 (Independence as a Model Assumption)** More often than not, the independence of random variables is a model *assumption*, rather than a consequence.

Instead of describing a random experiment via an explicit description of Ω and $\mathbb{P}$, we will usually model the experiment through one or more (independent) random variables.   ∎

Suppose $X_1, \ldots, X_n$ are independent random variables taking values in $(E, \mathcal{E})$. Denote their individual distributions — the so-called *marginal* distributions — by $\mu_1, \ldots, \mu_n$, and suppose that these have densities $f_1, \ldots, f_n$ with respect to some measure λ on $\mathcal{E}$ (think Lebesgue measure or counting measure). Then, the product distribution $\mu := \mu_1 \otimes \cdots \otimes \mu_n$ has a density f with respect to the product measure $\lambda \otimes \cdots \otimes \lambda$, and

$$f(x_1, \ldots, x_n) = f_1(x_1) \cdots f_n(x_n).$$

If the $\{X_i\}$ are independent and have the *same* distribution, they are said to be *independent and identically distributed*, abbreviated as iid. We write

$$X_1, \ldots, X_n \overset{\text{iid}}{\sim} f \quad \text{or} \quad X_1, \ldots, X_n \overset{\text{iid}}{\sim} \text{Dist},$$

to indicate that the random vectors are iid with a density f or distribution Dist.

∎ **Example 2.31 (Bernoulli Trials)** Consider the experiment where we throw a coin repeatedly. The easiest way to model this is by using random variables, via:

$$X_1, X_2, \ldots \overset{\text{iid}}{\sim} \text{Ber}(p).$$

We interpret $\{X_i = 1\}$ as the event that the ith toss yields a success, e.g., heads. Here, we (naturally) assume that the $X_1, X_2, \ldots$ are *independent*. Note that the model depends on a single parameter p, which may be known or remain unspecified; e.g., $p = 1/2$ when the coin is fair. The collection of random variables $\{X_i\}$ is called a *Bernoulli process* with *success* parameter p. It is the most important stochastic process in the study of probability and serves as the basis for many more elaborate stochastic processes.

For example, let S_n denote the number of successes (i.e., the number of 1s) in the first n trials; that is,

$$S_n := \sum_{i=1}^{n} X_i, \quad n = 1, 2, \ldots.$$

Put $S_0 := 0$. The stochastic process $\{S_n, n \in \mathbb{N}\}$ is an example of a *random walk*. Verify yourself that $S_n \sim \text{Bin}(n, p)$. For $p = 1/2$, let $Y_n := 2S_n - n$. The process $\{Y_n, n \in \mathbb{N}\}$ is called the *symmetric random walk on the integers*. A typical path is given in Figure 2.32.

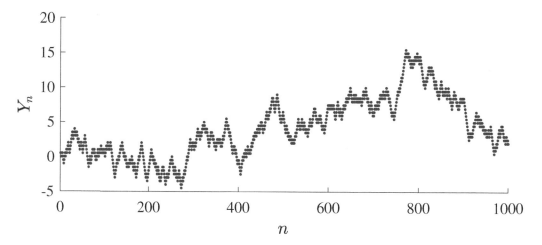

Figure 2.32: Symmetric random walk on the integers.

Another process derived from the Bernoulli process is the process $\{T_k, k \in \mathbb{N}\}$, where $T_0 := 0$ and T_k is the time of kth success, $k = 1, 2, \ldots$. Verify yourself that the intersuccess times $T_k - T_{k-1}$, $k = 1, 2, \ldots$ are independent and have a $\mathsf{Geom}(p)$ distribution.

Finally, note that in our model for the coin toss experiment, we have completely ignored the probability space $(\Omega, \mathcal{H}, \mathbb{P})$. The question arises whether there *exists* a probability space on which we can define independent $\mathsf{Ber}(p)$ random variables $X_1, X_2, \ldots$. Exercise 7 gives a concrete probability space for the case $p = 1/2$. ∎

We will come back to the concept of independence in Section 2.7 and define it for various other probabilistic objects.

2.4 Expectation

The expectation of a (numerical) random variable is a weighted average of all the values that the random variable can take. In elementary probability theory the expectation of a *discrete* random variable X is defined as

$$\mathbb{E}X := \sum_x x\, \mathbb{P}(X = x)$$

and for a *continuous* random variable X with pdf f it is defined as

$$\mathbb{E}X := \int_{-\infty}^{\infty} \mathrm{d}x\, x\, f(x).$$

Having to define the expectation in two ways is not very satisfactory. Also, we have seen that there exists random variables whose distribution is neither discrete nor continuous. How should we define the expectation for these random variables? Again, measure theory comes to the rescue. Here is the formal definition:

Definition 2.33: Expectation

Let $(\Omega, \mathcal{H}, \mathbb{P})$ be a probability space and X a numerical random variable. The *expectation* of X is defined as the integral

$$\mathbb{E}X := \mathbb{P}X = \int X \, d\mathbb{P}.$$

Below, we could have copied the four defining steps for a general integral, given in Section 1.4.1, but we slightly simplified these using the linearity and monotone convergence properties of general integrals, which were given and proved in Theorem 1.55.

1. If X is an *indicator function*, $X = \mathbb{1}_A$ of an event A, then we define

$$\mathbb{E}X := \mathbb{P}(A).$$

2. If X is a *simple function*, $X = \sum_{i=1}^{n} a_i \mathbb{1}_{A_i}$ for events $\{A_i\}$, then we define

$$\mathbb{E}X := \sum_{i=1}^{n} a_i \mathbb{P}(A_i).$$

3. If X is a *positive* random variable, then, see Theorem 1.27, X is the (pointwise) limit of an increasing sequence of positive *simple* random variables $X_1, X_2, \ldots$, and by the Monotone Convergence Theorem (see Theorem 1.55), we define

$$\mathbb{E}X := \lim \mathbb{E}X_n.$$

4. For general X, write $X = X^+ - X^-$, where $X^+ := \max\{X, 0\}$ and $X^- := \max\{-X, 0\}$. Then, X^+ and X^- are both positive random variables, for which the integral is defined. We now define

$$\mathbb{E}X := \mathbb{E}X^+ - \mathbb{E}X^-,$$

provided that the right-hand side is well-defined.

Thus, for positive random variables the expectation always exists (can be $+\infty$). For general random variables, it only fails to exist if both the positive and negative parts of a random variable have infinite integrals. Next, we list various properties of the expectation. The first list of properties is simply a restatement of Theorem 1.55.

Theorem 2.34: Properties of an Expectation

Let $a, b \in \mathbb{R}$ and let X, Y, and (X_n) be numerical random variables for which the expectations below are well-defined. Then, the following hold:

1. *(Monotonicity)*: If $X \leq Y$, then $\mathbb{E}X \leq \mathbb{E}Y$.

2. *(Linearity)*: $\mathbb{E}(aX + bY) = a\,\mathbb{E}X + b\,\mathbb{E}Y$.

3. *(Monotone convergence)*: If $X_n \uparrow X$, then $\mathbb{E}X_n \uparrow \mathbb{E}X$.

The next property, Fatou's lemma, is helpful in various technical proofs. Simply replace $\mathbb{E}$ with μ and X_n with f_n, to obtain the result for a general measure μ and positive numerical functions (f_n).

Lemma 2.35: Fatou

For any sequence (X_n) of positive numerical random variables,

$$\mathbb{E} \liminf X_n \leq \liminf \mathbb{E}X_n.$$

Proof. Define $Y_m := \inf_{n \geq m} X_n$. Then, $Y_m \uparrow \liminf X_n$. Hence, by Theorem 2.34 (monotone convergence):

$$\lim \mathbb{E}Y_m = \mathbb{E} \liminf X_n.$$

But we also have $Y_m \leq X_n$ for all $n \geq m$, so that $\mathbb{E}Y_m \leq \mathbb{E}X_n$ for $n \geq m$ by the monotonicity property of the expectation. Thus,

$$\mathbb{E}Y_m \leq \inf_{n \geq m} \mathbb{E}X_n.$$

Combining the two displayed results above gives $\mathbb{E} \liminf X_n \leq \liminf \mathbb{E}X_n$.   □

We can use Fatou's lemma to prove the following instrumental properties:

Theorem 2.36: Dominated and Bounded Convergence

Let (X_n) be a sequence of numerical random variables for which $\lim X_n = X$ exists. Then, the following hold:

1. *(Dominated convergence)*: If, for each n, $|X_n| \leq Y$ for some random variable Y with $\mathbb{E}Y < \infty$, then $\mathbb{E}X_n \to \mathbb{E}X < \infty$.

2. *(Bounded convergence)*: If $|X_n| \leq c$ for all n for some $c \in \mathbb{R}$, then $\mathbb{E}X_n \to \mathbb{E}X < \infty$.

Proof. If $\mathbb{E}Y < \infty$, then Y is real-valued almost surely and $X_n + Y \geq 0$ almost surely. Hence, by Fatou's lemma,

$$\mathbb{E}\liminf(X_n + Y) \leq \liminf \mathbb{E}(X_n + Y),$$

which shows that

$$\mathbb{E}\liminf X_n \leq \liminf \mathbb{E}X_n.$$

Similarly, $Y - X_n \geq 0$ almost surely, so that

$$\mathbb{E}\liminf(Y - X_n) \leq \liminf \mathbb{E}(Y - X_n),$$

leading to

$$\limsup \mathbb{E}X_n \leq \mathbb{E}\limsup X_n.$$

Combining, we have

$$\mathbb{E}\liminf X_n \leq \liminf \mathbb{E}X_n \leq \limsup \mathbb{E}X_n \leq \mathbb{E}\limsup X_n.$$

Since $\lim X_n =: X$ exists, $\liminf X_n = \limsup X_n$, and so we have equality of the four terms above. In particular, by the monotonicity property of $\mathbb{E}$, this limit must have a finite expectation, $\mathbb{E}X$, which lies between $-\mathbb{E}Y$ and $\mathbb{E}Y$. In the case where (X_n) is bounded by a constant c, we can take $Y := c$ (i.e., $Y(\omega) := c$ for all $\omega \in \Omega$) to obtain the bounded convergence result. $\qquad\square$

In Theorem 2.36, replace $\mathbb{E}$ with μ and X_n, X, Y with f_n, f, g to obtain the corresponding results for integration with respect to a general measure μ. For bounded convergence, this measure needs to be finite.

The following theorem — which is just Theorem 1.62 — is the workhorse of the theory. It enables us to actually calculate the expectation of a function of a random variable.

Theorem 2.37: Expectation and Image Measure

Let X be a random variable taking values in $(E, \mathcal{E})$ with distribution μ and let h be an $\mathcal{E}$-measurable function. Then, provided the integral exists:

$$\mathbb{E}h(X) = \int h(X)\,\mathrm{d}\mathbb{P} = \int_{\mathbb{R}} \mu(\mathrm{d}x)\,h(x) = \mu h.$$

Combining Theorem 2.37 with Theorem 1.58 for measures with a density, gives the familiar results for the expectation of a function of a random variable. In particular, when X has a discrete distribution, we have

$$\mathbb{E}h(X) = \sum_x h(x)\,\mathbb{P}(X = x).$$

Similarly, when X has a density f with respect to the Lebesgue measure, we have

$$\mathbb{E}\, h(X) = \int_{-\infty}^{\infty} \mathrm{d}x\, h(x) f(x).$$

The generalization to random vectors is immediate. Namely, if $X := [X_1, \ldots, X_n]^\top$ is a random vector with (n-dimensional) pdf f, and h a measurable numerical function on $\mathbb{R}^n$, then

$$\mathbb{E}\, h(X) = \int_{\mathbb{R}^n} \mathrm{d}x\, h(x) f(x).$$

Once we have specified a probabilistic model in terms of random variables and their distributions (possibly including independence assumptions), we may wish to explore the properties of certain functions of the random variables in the model. For example, the following theorem is useful for computing probability densities of functions of random variables. See Exercises 8–14 for more examples.

Theorem 2.38: Transformation Rule

Let X be a random vector with density f_X with respect to the Lebesgue measure on $(\mathbb{R}^n, \mathcal{B}^n)$. Let $Z := g(X)$, where $g : \mathbb{R}^n \to \mathbb{R}^n$ is invertible with inverse g^{-1} and *Jacobian*[1] $\left| \frac{\partial z}{\partial x} \right|$. Then, at $z = g(x)$ the random vector Z has the following density with respect to the Lebesgue measure on $(\mathbb{R}^n, \mathcal{B}^n)$:

$$(2.39) \qquad f_Z(z) := \frac{f_X(x)}{\left| \frac{\partial z}{\partial x} \right|} = f_X(g^{-1}(z)) \left| \frac{\partial x}{\partial z} \right|, \qquad z \in \mathbb{R}^n.$$

Proof. By Theorem 2.37, we have for any function $h \in \mathcal{B}^n$:

$$\mathbb{E} h(Z) = \int \mathrm{d}z\, h(z) f_Z(z) = \int \mathrm{d}x\, h(g(x)) f_X(x).$$

Evaluating the last integral via a multidimensional change of variables, with $z = g^{-1}(x)$, gives

$$\int \mathrm{d}x\, h(g(x)) f_X(x) = \int \mathrm{d}z\, h(z) \frac{f_X(g^{-1}(z))}{\left| \frac{\partial z}{\partial x} \right|}.$$

As h is arbitrary, the first equation in (2.39) follows. The second equation is because the matrix $\frac{\partial z}{\partial x}$ of partial derivatives of g is the inverse of the matrix $\frac{\partial x}{\partial z}$ of partial derivatives of g^{-1}, and so their determinants (and Jacobians) are reciprocal. □

---

[1] The Jacobian is the absolute value of the determinant of the matrix of partial derivatives of g.

■ **Example 2.40 (Linear Transformation)** Let $X := [X_1, \ldots, X_n]^\top$ be a numerical random column vector with density f_X with respect to the Lebesgue measure on $(\mathbb{R}^n, \mathcal{B}^n)$. Consider the linear transformation:

$$Z = AX,$$

where A is an *invertible* matrix. Then, by Theorem 2.38, Z has density

$$f_Z(z) := \frac{f_X(A^{-1}z)}{|A|}, \quad z \in \mathbb{R}^n.$$

■

We conclude this section with a review of some common notions from elementary probability and statistics that involve expectations.

The *variance* of a random variable X is defined by

$$\operatorname{Var} X := \mathbb{E}(X - \mathbb{E}X)^2.$$

The variance measures the "spread" in the values of X — the average squared distance from the mean. The square root of the variance is called the *standard deviation*. The standard deviation measures the spread in the same units as the random variable — unlike the variance, which uses squared units. The expectation $\mathbb{E}X^n$ is called the *nth moment* of X. The *covariance* of two random variables X and Y with expectations μ_X and μ_Y, respectively, is defined as

$$\operatorname{Cov}(X, Y) := \mathbb{E}[(X - \mu_X)(Y - \mu_Y)].$$

This is a measure of the amount of linear dependency between the variables. For $\sigma_X^2 := \operatorname{Var} X$ and $\sigma_Y^2 := \operatorname{Var} Y$, a scaled version of the covariance is given by the *correlation coefficient*,

$$\varrho(X, Y) := \frac{\operatorname{Cov}(X, Y)}{\sigma_X \sigma_Y}.$$

Table 2.3 shows properties of the variance and covariance, which follow directly from their definitions. We leave the proof as an exercise; see Exercise 15.

As a consequence of Properties 2, 8, and 9 in Table 2.3, we have that for any sequence of *independent* random variables $X_1, \ldots, X_n$ with variances $\sigma_1^2, \ldots, \sigma_n^2$,

$$\operatorname{Var}(a_1 X_1 + a_2 X_2 + \cdots + a_n X_n) = a_1^2 \sigma_1^2 + a_2^2 \sigma_2^2 + \cdots + a_n^2 \sigma_n^2$$

for any choice of constants $a_1, \ldots, a_n$.

For random vectors, it is convenient to write the expectations and covariances in vector and matrix form. For a random column vector $X := [X_1, \ldots, X_n]^\top$, we define its *expectation vector* as the vector of expectations:

$$\mu := [\mu_1, \ldots, \mu_n]^\top := [\mathbb{E}X_1, \ldots, \mathbb{E}X_n]^\top.$$

Table 2.3: Properties of the variance and covariance.

1. $\mathbb{V}\mathrm{ar}\, X = \mathbb{E}X^2 - \mu_X^2$.
2. $\mathbb{V}\mathrm{ar}(aX + b) = a^2\, \sigma_X^2$.
3. $\mathbb{C}\mathrm{ov}(X, Y) = \mathbb{E}[XY] - \mu_X\, \mu_Y$.
4. $\mathbb{C}\mathrm{ov}(X, Y) = \mathbb{C}\mathrm{ov}(Y, X)$.
5. $-\sigma_X\sigma_Y \le \mathbb{C}\mathrm{ov}(X, Y) \le \sigma_X\sigma_Y$.
6. $\mathbb{C}\mathrm{ov}(aX + bY, Z) = a\, \mathbb{C}\mathrm{ov}(X, Z) + b\, \mathbb{C}\mathrm{ov}(Y, Z)$.
7. $\mathbb{C}\mathrm{ov}(X, X) = \sigma_X^2$.
8. $\mathbb{V}\mathrm{ar}(X + Y) = \sigma_X^2 + \sigma_Y^2 + 2\, \mathbb{C}\mathrm{ov}(X, Y)$.
9. If X and Y are independent, then $\mathbb{C}\mathrm{ov}(X, Y) = 0$.

Similarly, letting the expectation of a matrix be the matrix of expectations, we define for two random vectors $X \in \mathbb{R}^n$ and $Y \in \mathbb{R}^m$ their $n \times m$ *covariance matrix* by:

$$\mathbb{C}\mathrm{ov}(X, Y) := \mathbb{E}[(X - \mathbb{E}X)(Y - \mathbb{E}Y)^\top],$$

with (i, j)th element $\mathbb{C}\mathrm{ov}(X_i, Y_j) = \mathbb{E}[(X_i - \mathbb{E}X_i)(Y_j - \mathbb{E}Y_j)]$. A consequence of this definition is that

$$\mathbb{C}\mathrm{ov}(\mathbf{A}X, \mathbf{B}Y) = \mathbf{A}\, \mathbb{C}\mathrm{ov}(X, Y)\, \mathbf{B}^\top,$$

where $\mathbf{A}$ and $\mathbf{B}$ are two matrices with n and m columns, respectively.

The covariance matrix of the vector X is defined as the $n \times n$ matrix $\mathbb{C}\mathrm{ov}(X, X)$. Any such covariance matrix $\mathbf{\Sigma}$ is *positive semidefinite*, meaning that $x^\top \mathbf{\Sigma} x \ge 0$ for all $x \in \mathbb{R}^n$. To see this, write

$$x^\top \mathbf{\Sigma} x = \mathbb{E}\Big[\underbrace{x^\top(X - \mathbb{E}X)}_{Y}\, \underbrace{(X - \mathbb{E}X)^\top x}_{Y} \Big] = \mathbb{E}Y^2 \ge 0.$$

Any positive semidefinite matrix $\mathbf{\Sigma}$ can be written as

$$\mathbf{\Sigma} = \mathbf{B}\mathbf{B}^\top$$

for some real matrix $\mathbf{B}$, which can be obtained, for example, by using the *Cholesky square-root* method; see, for example, Kroese et al. (2019, Algorithm A.6.2). Conversely, for any real matrix $\mathbf{B}$, the matrix $\mathbf{B}\mathbf{B}^\top$ is positive semidefinite. The covariance matrix is also denoted as $\mathbb{V}\mathrm{ar}X := \mathbb{C}\mathrm{ov}(X, X)$, by analogy to the scalar identity $\mathbb{V}\mathrm{ar}X = \mathbb{C}\mathrm{ov}(X, X)$.

A useful application of the cyclic property and linearity of the trace of a matrix is the following:

> **Proposition 2.41: Expectation of a Quadratic Form**
>
> Let $\mathbf{A}$ be an $n \times n$ matrix and X an n-dimensional random vector with expectation vector μ and covariance matrix Σ. The random variable $Y := X^\top \mathbf{A} X$ has expectation $\operatorname{tr}(\mathbf{A}\Sigma) + \mu^\top \mathbf{A}\mu$.

Proof. Since Y is a scalar, it is equal to its trace. Now, using the cyclic property:
$\mathbb{E}Y = \mathbb{E}\operatorname{tr}(Y) = \mathbb{E}\operatorname{tr}(X^\top \mathbf{A} X) = \mathbb{E}\operatorname{tr}(\mathbf{A} X X^\top) = \operatorname{tr}(\mathbf{A}\,\mathbb{E}[X X^\top]) = \operatorname{tr}(\mathbf{A}(\Sigma + \mu\mu^\top)) = \operatorname{tr}(\mathbf{A}\Sigma) + \operatorname{tr}(\mathbf{A}\mu\mu^\top) = \operatorname{tr}(\mathbf{A}\Sigma) + \mu^\top \mathbf{A}\mu.$ □

■ **Example 2.42 (Multivariate Normal (Gaussian) Distribution)** Let $Z_1, \ldots, Z_n$ be independent and standard normal (i.e., $\mathcal{N}(0,1)$-distributed) random variables. The joint pdf of $\mathbf{Z} := [Z_1, \ldots, Z_n]^\top$ is given by

$$f_{\mathbf{Z}}(z) := \prod_{i=1}^n \frac{1}{\sqrt{2\pi}} e^{-\frac{1}{2} z_i^2} = (2\pi)^{-\frac{n}{2}} e^{-\frac{1}{2} z^\top z}, \quad z \in \mathbb{R}^n.$$

We write $\mathbf{Z} \sim \mathcal{N}(\mathbf{0}, \mathbf{I}_n)$, where $\mathbf{I}_n$ is the n-dimensional identity matrix. Consider the affine transformation

$$X = \mu + \mathbf{B}\,\mathbf{Z}$$

for some $m \times n$ matrix $\mathbf{B}$ and m-dimensional vector μ. Note that X has expectation vector μ and covariance matrix $\Sigma := \mathbf{B}\mathbf{B}^\top$. We say that X has a *multivariate normal* or *multivariate Gaussian* distribution with mean vector μ and covariance matrix Σ. We write $X \sim \mathcal{N}(\mu, \Sigma)$. The multivariate normal distribution has many interesting properties; see, for example, Kroese et al. (2019, Section C.7). In particular:

1. Any *affine combination* of independent multivariate normal random vectors is again multivariate normal.

2. The (marginal) distribution of any *subvector* of a multivariate normal random vector is again multivariate normal.

3. The *conditional* distribution of a multivariate normal random vector, given any of its subvectors, is again multivariate normal.

4. Two jointly multivariate normal random vectors are *independent* if and only if their covariance matrix is the zero matrix.

■

2.5 L^p Spaces

Let X be a numerical random variable on $(\Omega, \mathcal{H}, \mathbb{P})$. For $p \in [1, \infty)$ define

$$\|X\|_p := (\mathbb{E}|X|^p)^{\frac{1}{p}}$$

and let

$$\|X\|_\infty := \inf\{x : \mathbb{P}(|X| \le x) = 1\}.$$

Definition 2.43: L^p Space

For each $p \in [1,\infty]$ the space L^p is comprised of all numerical random variables X for which $\|X\|_p < \infty$.

In particular, L^1 consists of all integrable random variables and L^2 is the space of all square-integrable random variables. If we identify random variables that are almost surely equal as one and the same, then L^p is a *Banach space*: a complete normed vector space. Completeness means that every Cauchy sequence converges, with respect to the norm $\|\cdot\|_p$. That is, if $\lim_{m,n\to\infty}\|X_m - X_n\|_p = 0$, then there exists an $X \in L^p$ such that $\lim_{n\to\infty}\|X_n - X\|_p = 0$. Of particular importance is L^2, which is in fact a *Hilbert space*, with inner product $\langle X, Y \rangle := \mathbb{E}[XY]$; see Appendix B for more details.

Theorem 2.47 summarizes important properties of L^p spaces through the properties of $\|\cdot\|_p$; the first three show that L^p is a vector space with norm $\|\cdot\|_p$. To prove these properties, we first need the following result on expectations of convex functions, which is of independent interest and is very useful in many applications. Let $X \subseteq \mathbb{R}^n$. A function $h : X \to \mathbb{R}$ is said to be *convex* on X if for each x in the interior of X there exists a vector v — called a *subgradient* of h — such that

$$(2.44) \qquad h(y) \ge h(x) + v^\top(y - x), \quad y \in X.$$

Lemma 2.45: Jensen's Inequality

Let $h : X \to \mathbb{R}$ be a convex function and let X be a random variable taking values in X, with expectation vector $\mathbb{E}X$. Then,

$$\mathbb{E}\,h(X) \ge h(\mathbb{E}X).$$

Proof. In (2.44) replace y with X and x with $\mathbb{E}X$, and then take expectations on both sides, to obtain:

$$\mathbb{E}\,h(X) \ge \mathbb{E}\,h(\mathbb{E}X) + v^\top \mathbb{E}(X - \mathbb{E}X) = h(\mathbb{E}X).$$

$\square$

■ **Example 2.46 (Convex Function)** A well-known property of a convex function h is that

$$h(\alpha u + (1 - \alpha)v) \le \alpha h(u) + (1 - \alpha)h(v), \quad \alpha \in [0, 1].$$

This is a simple consequence of Jensen's inequality, by taking X the random variable that takes the value $\boldsymbol{u}$ with probability α and the value $\boldsymbol{v}$ with probability $1 - \alpha$.

For convex functions on $\mathbb{R}$ we can further exploit the above inequality by taking $\boldsymbol{u} := U$ and $\boldsymbol{v} := V$ for any pair of real-valued random variables (U, V). Using the monotonicity of the expectation, we conclude that

$$\mathbb{E}h(\alpha U + (1 - \alpha)V) \le \alpha \, \mathbb{E}h(U) + (1 - \alpha) \, \mathbb{E}h(V), \quad \alpha \in [0, 1].$$

We use this device to prove various properties of the L^p norm.   ∎

Theorem 2.47: Properties of the L^p Norm

Let X and Y be numerical random variables on $(\Omega, \mathcal{H}, \mathbb{P})$. Then, the following hold:

1. *(Positivity):* $\|X\|_p \ge 0$, and $\|X\|_p = 0 \Leftrightarrow X = 0$ almost surely.

2. *(Multiplication by a constant):* $\|c\,X\|_p = |c|\,\|X\|_p$.

3. *(Minkowski's (triangle) inequality):*

$$\big| \|X\|_p - \|Y\|_p \big| \le \|X + Y\|_p \le \|X\|_p + \|Y\|_p.$$

4. *(Hölder's inequality):* For $p, q, r \in [1, \infty]$ with $\frac{1}{p} + \frac{1}{q} = \frac{1}{r}$,

   (2.48) $\|XY\|_r \le \|X\|_p \, \|Y\|_q.$

5. *(Monotonicity):* If $1 \le p < q \le \infty$, then $\|X\|_p \le \|X\|_q$.

Proof.

1. Obviously, $\mathbb{E}|X|^p \ge 0$ implies $\|X\|_p \ge 0$, and $X = 0$ (a.s.) $\Rightarrow \mathbb{E}|X|^p = 0 \Rightarrow \|X\|_p = 0$. Conversely, $\|X\|_p = 0 \Rightarrow \mathbb{E}|X|^p = 0$, which can only be true if $\mathbb{P}(X = 0) = 1$.

2. This follows from the linearity of the expectation for $p < \infty$. The identity is also trivial for $p = \infty$ and $c = 0$. In the case of $p = \infty$ and $|c| > 0$, we have

$$\begin{aligned}
\|c\,X\|_\infty &= \inf\{x : \mathbb{P}(|c\,X| \le x) = 1\} \\
&= \inf\{x : \mathbb{P}(|X| \le x/|c|) = 1\} \\
&= |c| \inf\{x/|c| : \mathbb{P}(|X| \le x/|c|) = 1\} \\
&= |c|\,\|X\|_\infty.
\end{aligned}$$

3. Assume $X, Y \in L^p$; otherwise, there is nothing to prove. For $p = \infty$, let $x := \|X\|_\infty$ and $y := \|Y\|_\infty$. Then, almost surely $|X+Y| \leq x+y$, and so $\|X+Y\|_\infty \leq x+y$. Next, consider the case $p \in [1, \infty)$. Since $x \mapsto |x|^p$ is convex, we have

$$|\alpha U + (1 - \alpha)V|^p \leq \alpha |U|^p + (1 - \alpha)|V|^p, \quad \alpha \in [0, 1]$$

for any $\alpha \in [0, 1]$ and random variables U and V. In particular, it holds for $U := X/\|X\|_p$, $V := Y/\|Y\|_p$, and $\alpha := \|X\|_p/(\|X\|_p + \|Y\|_p)$. Substituting and taking expectations on both sides gives the inequality:

$$\mathbb{E}\left|\frac{X + Y}{\|X\|_p + \|Y\|_p}\right|^p \leq \alpha \frac{\mathbb{E}|X|^p}{\|X\|_p^p} + (1 - \alpha)\frac{\mathbb{E}|Y|^p}{\|Y\|_p^p} = 1,$$

which after rearrangement, yields the second inequality in Minkowski's Property 3. The first inequality is a consequence of this, because $\|X\|_p = \|X + Y - Y\|_p \leq \|X + Y\|_p + \|Y\|_p$ and $\|Y\|_p = \|X + Y - X\|_p \leq \|X + Y\|_p + \|X\|_p$.

4. Let $X, Y \in L^p$. For $p = \infty$, we have $r = q$, and almost surely $|XY| \leq \|X\|_\infty |Y|$, so

$$\|XY\|_r = (\mathbb{E}|XY|^r)^{1/r} \leq (\|X\|_\infty^r \, \mathbb{E}|Y|^r)^{1/r} = \|X\|_\infty \, \|Y\|_r.$$

For $q = \infty$, we have $r = p$ and the same holds. Next, consider the case where both p and q are finite. Since the logarithmic function is concave (that is, $-g$ is convex), we have that for any $\alpha \in [0, 1]$ and positive u and v:

$$\alpha \ln u + (1 - \alpha) \ln v \leq \ln(\alpha u + (1 - \alpha)v).$$

We have thus proved the geometric and arithmetic mean inequality:

$$u^\alpha v^{1-\alpha} \leq \alpha u + (1 - \alpha)v.$$

This inequality remains valid, almost surely, if we replace u and v with positive random variables U and V. In particular, letting

$$\alpha := r/p, \quad U := \frac{|X|^{r/\alpha}}{\|X\|_p^{r/\alpha}}, \quad V := \frac{|Y|^{r/(1-\alpha)}}{\|Y\|_q^{r/(1-\alpha)}}$$

and taking expectations on both sides yields:

$$\mathbb{E}\frac{|XY|^r}{\|X\|_p^r \|Y\|_q^r} \leq \alpha \frac{\mathbb{E}|X|^p}{\|X\|_p^p} + (1 - \alpha)\frac{\mathbb{E}|Y|^q}{\|Y\|_q^q} = 1,$$

which, after rearrangement, yields Hölder's inequality.

5. Apply (2.48) with $Y := 1$, $r := p$, $p := q$, and $q := pq/(q - p)$.

□

For random variables in L^2 the concepts of variance and covariance have a geometric interpretation. Namely, if X and Y are zero-mean random variables (their expectation is 0), then

$$\mathbb{V}\text{ar}\, X = \|X\|_2^2 \quad \text{and} \quad \mathbb{C}\text{ov}(X,Y) = \langle X, Y \rangle.$$

In particular, from the Hölder's inequality with $p = q = 2$ and $r = 1$, we obtain the famous *Cauchy–Schwarz* inequality

(2.49) $$|\mathbb{C}\text{ov}(X,Y)| \leq \sqrt{\mathbb{V}\text{ar}\, X \, \mathbb{V}\text{ar}\, Y}.$$

2.6 Integral Transforms

Expectations are used in many probabilistic analyses. In this section, we highlight their use in integral transforms. Many calculations and manipulations involving probability distributions are facilitated by the use of such transforms. We describe two major types of transforms.

2.6.1 Moment Generating Functions

Definition 2.50: Moment Generating Function

The *moment generating function* (MGF) of a numerical random variable X with distribution μ is the function $M : \mathbb{R} \to [0, \infty]$, given by

$$M(s) := \mathbb{E}\, e^{sX} = \int_{-\infty}^{\infty} \mu(dx)\, e^{sx}, \quad s \in \mathbb{R}.$$

We sometimes write M_X to stress the role of X. Table 2.4 gives a list of MGFs for various important distributions that are absolutely continuous with respect to the Lebesgue measure.

The MGF of a *positive* random variable is called the *Laplace transform*. In most definitions the sign of s is then flipped. That is, the Laplace transform of X is the function $s \mapsto \mathbb{E}\, e^{-sX} = M(-s)$, $s \in \mathbb{R}$.

The set $I := \{s \in \mathbb{R} : M(s) < \infty\}$ always includes 0. For some distributions this may be the *only* point. However, the important case is where I contains a neighborhood of 0; i.e., there is an s_0 such that $M(s) < \infty$ for all $s \in (-s_0, s_0)$. In this case M completely determines μ. That is, two distributions are the same if and only if their MGFs are the same.

Theorem 2.51: Uniqueness of the MGF

A measure μ is determined by its MGF M if the latter is finite in a neighborhood of 0.

The proof relies on a similar uniqueness result for the characteristic function, which we will prove in Section 2.6.2; see also Billingsley, page 390.

Table 2.4: MGFs for common distributions.

Distr.	$f(x)$	$x \in$	$M(s)$	$s \in$
$\mathcal{U}[a,b]$	$\dfrac{1}{b-a}$	$[a,b]$	$\dfrac{e^{bs}-e^{as}}{s(b-a)}$	$\mathbb{R}$
$\mathrm{Exp}(\lambda)$	$\lambda\,e^{-\lambda x}$	$\mathbb{R}_+$	$\left(\dfrac{\lambda}{\lambda-s}\right)$	$(-\infty,\lambda)$
$\mathrm{Gamma}(\alpha,\lambda)$	$\dfrac{\lambda^\alpha x^{\alpha-1}e^{-\lambda x}}{\Gamma(\alpha)}$	$\mathbb{R}_+$	$\left(\dfrac{\lambda}{\lambda-s}\right)^\alpha$	$(-\infty,\lambda)$
$\mathcal{N}(\mu,\sigma^2)$	$\dfrac{1}{\sigma\sqrt{2\pi}}\,e^{-\frac{1}{2}\left(\frac{x-\mu}{\sigma}\right)^2}$	$\mathbb{R}$	$e^{\mu s+\frac{1}{2}\sigma^2 s^2}$	$\mathbb{R}$

For a discrete random variable X it is often convenient to express the MGF $M(s)$ in terms of $z = e^s$; that is, $\mathbb{E}\,e^{sX} = \mathbb{E}\,z^X$. For common discrete distributions this is done in Table 2.5. The function $z \mapsto \mathbb{E}\,z^X$ is called the *probability generating function* of X.

Table 2.5: MGFs $M(s)$, expressed in terms of $z = e^s$.

Distr.	$f(x)$	$x \in$	$M(s)$
$\mathrm{Ber}(p)$	$p^x(1-p)^{1-x}$	$\{0,1\}$	$1-p+pz$
$\mathrm{Bin}(n,p)$	$\binom{n}{x}p^x(1-p)^{n-x}$	$\{0,1,\ldots,n\}$	$(1-p+pz)^n$
$\mathrm{Poi}(\lambda)$	$e^{-\lambda}\dfrac{\lambda^k}{k!}$	$\{0,1,\ldots\}$	$e^{-\lambda(1-z)}$
$\mathrm{Geom}(p)$	$p(1-p)^{x-1}$	$\{1,2,\ldots\}$	$\dfrac{pz}{1-(1-p)z}$
$\mathrm{NegBin}(n,p)$	$\binom{n+x-1}{n-1}p^n(1-p)^x$	$\{0,1,\ldots\}$	$\left(\dfrac{p}{1-(1-p)z}\right)^n$

> **Theorem 2.52: Taylor's Theorem for MGFs**
>
> Suppose $M(s) := \mathbb{E}e^{sX}$ is finite in the neighborhood $(-s_0, s_0)$ of 0. Then, M has the Taylor expansion
>
> $$M(s) = \sum_{k=0}^{\infty} \frac{s^k}{k!} \mathbb{E}X^k, \quad |s| < s_0.$$
>
> In particular, all moments $\mathbb{E}X^k$, $k \in \mathbb{N}$ are finite.

Proof. Suppose $M(s) < \infty$ for $s \in (-s_0, s_0)$ for some $s_0 > 0$. The Taylor expansion for e^{sx} is $\sum_{k=0}^{\infty}(sx)^k/k!$, so we can write

$$M(s) = \mathbb{E}e^{sX} = \mathbb{E}\lim_{n}\sum_{k=0}^{n}\frac{(sX)^k}{k!}.$$

We want to use the Dominated Convergence Theorem (see Theorem 2.36) to swap the limit and expectation. This would give

$$M(s) = \mathbb{E}\lim_{n}\sum_{k=0}^{n}\frac{(sX)^k}{k!} = \lim_{n}\mathbb{E}\sum_{k=0}^{n}\frac{(sX)^k}{k!} = \lim_{n}\sum_{k=0}^{n}\frac{s^k\,\mathbb{E}X^k}{k!} = \sum_{k=0}^{\infty}\frac{s^k\,\mathbb{E}X^k}{k!},$$

which would prove the result. It remains to show that $\left|\sum_{k=0}^{n}\frac{(sX)^k}{k!}\right| \leq Y$ for all n, for some positive random variable Y with finite expectation. We can take $Y := e^{|sX|}$. For all $s \in (-s_0, s_0)$ the expectation of Y is finite, since $Y \leq e^{-sX} + e^{sX}$ and both of the right-hand side terms have finite expectations. □

Theorems 2.51 and 2.52 show that the positive integer moments of a random variable completely determine its distribution, *provided* that the MGF is finite in a neighborhood of 0.

The Taylor expansion in Theorem 2.52 is a power series, and we know from real and complex analysis that such infinite sums behave like finite sums as long as s is less than the radius of convergence — in particular, less than s_0. It follows that

$$\mathbb{E}X^k = M^{(k)}(0), \quad k \geq 1;$$

that is, the kth derivative of M, evaluated at 0. This allows us to derive moments if we have an expression for $M(s)$ in terms of a power series. Another very useful property is the following *convolution* theorem:

> **Theorem 2.53: Product of MGFs**
>
> If X and Y are independent, then $M_{X+Y}(s) = M_X(s)\,M_Y(s)$.

Proof. The distribution of (X, Y) is a product measure $\mu \otimes \nu$. Hence, by Fubini's Theorem 1.68:

$$\mathbb{E}e^{s(X+Y)} = \int (\mu \otimes \nu)(dx, dy)\, e^{s(x+y)} = \int \mu(dx) \int \nu(dy)\, e^{sx}\, e^{sy}$$

$$= \int \mu(dx)\, e^{sx} \int \nu(dy)\, e^{sy} = \int \mu(dx)\, e^{sx}\, \mathbb{E}e^{sY} = \mathbb{E}e^{sX}\, \mathbb{E}e^{sY}.$$

$\square$

■ **Example 2.54 (Sum of Poisson Random Variables)** Let $X \sim \mathsf{Poi}(\lambda)$ and $Y \sim \mathsf{Poi}(\mu)$ be independent. Then, the MGF of $X + Y$ is given by the product

$$e^{-\lambda(1-e^s)}\, e^{-\mu(1-e^s)} = e^{-(\lambda+\mu)(1-e^s)}, \quad s \in \mathbb{R}.$$

By the uniqueness of the MGF, it follows that $X + Y \sim \mathsf{Poi}(\lambda + \mu)$. ■

■ **Example 2.55 (Binomial Distribution)** There is an interesting method for deriving the density of $X \sim \mathsf{Bin}(n, p)$ using MGFs. First, the expression for the MGF of a $\mathsf{Ber}(p)$ random variable is $pe^s + q$, with $q := 1 - p$ for $s \in \mathbb{R}$. Since X can be viewed as the sum of n independent $\mathsf{Ber}(p)$ random variables, the expression for *its* MGF is $(pe^s + q)^n$. But using the well-known Newton's formula (binomial theorem), we have

$$(pe^s + q)^n = \sum_{k=0}^{n} \binom{n}{k}(pe^s)^k\, q^{n-k} = \sum_{k=0}^{n} \binom{n}{k} p^k\, q^{n-k}\, e^{sk},$$

which shows that X has density $k \mapsto \binom{n}{k} p^k\, q^{n-k}$ with respect to the counting measure on $\{0, 1, \ldots, n\}$. ■

■ **Example 2.56 (Linear Combinations of Normal Random Variables)** An important property of the normal distribution is that any affine combination of independent normals is again normal; that is, if $X_i \sim \mathcal{N}(\mu_i, \sigma_i^2)$, independently, for $i = 1, 2, \ldots, n$, then

$$Y := a + \sum_{i=1}^{n} b_i X_i \sim \mathcal{N}\left(a + \sum_{i=1}^{n} b_i \mu_i, \sum_{i=1}^{n} b_i^2 \sigma_i^2\right).$$

Note that the parameters follow easily from the rules for the expectation and variance.

We can prove this via the MGF. Namely,

$$\mathbb{E}e^{sY} = \exp(sa) \prod_{i=1}^{n} \mathbb{E}\exp(sb_i X_i) = \exp(sa) \prod_{i=1}^{n} \exp\left\{\mu_i s b_i + \frac{1}{2}\sigma_i^2 s^2 b_i^2\right\},$$

so that

$$\mathbb{E}e^{sY} = \exp\left\{s\left(a + \sum_{i=1}^{n} b_i \mu_i\right) + \frac{1}{2}s^2 \sum_{i=1}^{n} b_i^2 \sigma_i^2\right\},$$

which shows the desired result, by the uniqueness of the MGF. ■

2.6.2 Characteristic Functions

The ultimate integral transform is the characteristic function. Every random variable has a characteristic function that is finite, no matter how strange that random variable might be. It is closely related to the classical Fourier transform of a function and has analytical properties superior to those of the MGF.

Definition 2.57: Characteristic Function

The *characteristic function* of a numerical random variable X with distribution μ is the function $\psi : \mathbb{R} \to \mathbb{C}$, defined by

$$\psi(r) := \mathbb{E} e^{irX} = \mathbb{E}\cos(rX) + i\,\mathbb{E}\sin(rX) = \int_{-\infty}^{\infty} \mu(dx)\, e^{irx}, \quad r \in \mathbb{R}.$$

For a random vector $X \in \mathbb{R}^d$, the characteristic function is defined similarly as

$$\psi(r) := \mathbb{E} e^{i r^\top X}, \quad r \in \mathbb{R}^d.$$

The characteristic function has many interesting properties, some of which are similar to those of the generating functions already mentioned. The following establishes the uniqueness property:

Theorem 2.58: Inversion and Uniqueness of the Characteristic Function

If the probability measure μ has characteristic function ψ, and if $\mu\{a\} = \mu\{b\} = 0$, then

$$(2.59) \qquad \mu(a,b] = \lim_{t\to\infty} \frac{1}{2\pi} \int_{\mathbb{R}} dr\, \mathbb{1}_{[-t,t]}(r)\, \frac{e^{-ira} - e^{-irb}}{ir}\, \psi(r) =: \lim_{t\to\infty} I_t.$$

In particular, the characteristic function determines μ uniquely. If, in addition, $\int_{\mathbb{R}} dr\, |\psi(r)| < \infty$, then μ has a density f with respect to the Lebesgue measure, given by

$$(2.60) \qquad f(x) := \frac{1}{2\pi} \int_{-\infty}^{\infty} dr\, e^{-irx}\, \psi(r).$$

Proof. Let I_t be the quantity for which we take the limit, as defined in the theorem. Then, by expanding the definition of $\psi(r)$ and using Fubini's theorem, we may write

$$I_t = \frac{1}{2\pi} \int_{-\infty}^{\infty} \mu(dx) \int_{-t}^{t} dr\, \frac{e^{ir(x-a)} - e^{ir(x-b)}}{ir}.$$

Using $e^{iz} = \cos z + i \sin z$, we can compute the inner integral in terms of functions sin, cos, and $Si(z) := \int_0^z dx \, (\sin(x)/x)$ to obtain

$$I_t = \int_{-\infty}^{\infty} \mu(dx) \left(\frac{\text{sgn}(x-a)}{\pi} Si(t \, |x-a|) - \frac{\text{sgn}(x-b)}{\pi} Si(t \, |x-b|) \right).$$

Since $\lim_{z \to \infty} Si(z) = \pi/2$ and $\int_z^{\infty} dx \, (\sin(x)/x) = \cos(z)/z - \int_z^{\infty} dx \, (\cos(x)/x^2)$ for $z > 0$, we have

$$\left| Si(|z|) - \frac{\pi}{2} \right| \leq \int_{|z|}^{\infty} dx \, \frac{\sin x}{x} \leq \frac{2}{|z|}, \quad |z| > 0.$$

Hence, the integrand in the integral I_t above is bounded and converges as $t \to \infty$ to $g_{a,b}(x)$, given by

$$g_{a,b}(x) := \begin{cases} 0 & \text{if } x < a \text{ or } x > b, \\ 1/2 & \text{if } x = a \text{ or } x = b, \\ 1 & \text{if } a < x < b. \end{cases}$$

So, $g_{a,b}$ is almost-everywhere the indicator function of the interval $[a, b]$. Hence, $I_t \to \mu \, g_{a,b}$, which implies (2.59), if $\mu\{a\} = \mu\{b\} = 0$. Uniqueness follows from the fact that intervals of the form $(a, b]$ form a p-system that generate $\mathcal{B}$ and so determine a measure uniquely.

When $\int_{\mathbb{R}} dr \, |\psi(r)| < \infty$, then the integral in (2.59) can be extended to $\mathbb{R}$; that is, from $\left| (e^{-ira} - e^{-irb})/(ir) \right| = \left| \int_a^b du \, e^{-iru} \right| \leq |b - a|$ and Theorem 2.36, we have

$$\lim_{t \to \infty} \frac{1}{2\pi} \int_{\mathbb{R}} dr \, \mathbb{1}_{[-t,t]}(r) \frac{e^{-ira} - e^{-irb}}{ir} \psi(r) = \frac{1}{2\pi} \int_{\mathbb{R}} dr \, \frac{e^{-ira} - e^{-irb}}{ir} \psi(r),$$

so in terms of the cdf of μ:

$$\frac{F(x+h) - F(x)}{h} = \frac{1}{2\pi} \int_{\mathbb{R}} dr \, \frac{e^{-irx} - e^{-ir(x+h)}}{irh} \psi(r).$$

Taking the limit for $h \to 0$ gives (2.60).   □

Proposition 2.61 lists some more properties of the characteristic function.

Proposition 2.61: Properties of the Characteristic Function

1. *(Symmetry)*: The random variables X and $-X$ are identically distributed if and only if the characteristic function of X is real-valued.

2. *(Convolution)*: If X and Y are independent, then $\psi_{X+Y} = \psi_X \psi_Y$.

3. *(Affine transformation)*: If a and b are real numbers, then

$$\psi_{aX+b}(r) = e^{irb}\psi_X(ar), \quad r \in \mathbb{R}.$$

4. *(Taylor's theorem)*: If $\mathbb{E}|X^n| < \infty$, then

$$\psi(r) = \sum_{k=0}^{n} \frac{\mathbb{E}X^k}{k!}(ir)^k + o(r^n), \quad r \in \mathbb{R}.$$

Proof. The only non-trivial result is Taylor's theorem. We prove it by using the following useful inequality:

$$(2.62) \qquad \left| e^{ix} - \sum_{k=0}^{n} \frac{(ix)^k}{k!} \right| \leq \min\left\{ \frac{|x|^{n+1}}{(n+1)!}, \frac{2|x|^n}{n!} \right\}.$$

If X has a moment of order n, it follows that

$$\left| \psi(r) - \sum_{k=0}^{n} \frac{(ir)^k \mathbb{E}X^k}{k!} \right| \leq |r|^n \mathbb{E}\min\left\{ \frac{|r||X|^{n+1}}{(n+1)!}, \frac{2|X|^n}{n!} \right\} =: |r|^n \mathbb{E}Y_r.$$

The random variable Y_r defined above goes to 0 when $r \to 0$. Moreover, it is dominated by $2|X^n|/n!$, which has a finite expectation. Hence, by the Dominated Convergence Theorem 2.36, $\mathbb{E}Y_r \to 0$ as $r \to 0$. In other words, the error term is of order $o(r^n)$ for $r \to 0$. $\qquad\square$

Table 2.6 gives a list of the characteristic functions for various common distributions. In the table, we assume the geometric distribution with probability masses $(1-p)^{k-1}p, k = 1, 2, \ldots$ (so starting from 1, not 0). Note that for $X \sim \mathcal{U}[-\alpha, \alpha]$, where $\alpha > 0$, we have

$$\psi(r) = \begin{cases} \frac{\sin \alpha r}{\alpha r}, & r \neq 0, \\ 1, & r = 0. \end{cases}$$

It is also worth mentioning that the Cauchy distribution does not have an MGF that is finite in a neighborhood of 0, and so does not appear in Table 2.4 for the MGFs, whereas its characteristic function is perfectly well-defined.

Table 2.6: Characteristic functions for common distributions.

Distr.	$\psi(r)$	Distr.	$\psi(r)$
constant c	e^{irc}	$\mathcal{U}[a,b]$	$\dfrac{e^{ibr} - e^{iar}}{ir(b-a)}$
Ber(p)	$1 - p + pe^{ir}$	Exp(λ)	$\left(\dfrac{\lambda}{\lambda - ir}\right)$
Bin(n,p)	$(1 - p + pe^{ir})^n$	Gamma(α, λ)	$\left(\dfrac{\lambda}{\lambda - ir}\right)^{\alpha}$
Poi(λ)	$\exp(-\lambda(1 - e^{ir}))$	$\mathcal{N}(\mu, \sigma^2)$	$\exp\left(i\mu r - \dfrac{1}{2}\sigma^2 r^2\right)$
Geom(p)	$\dfrac{e^{ir}p}{1 - (1-p)e^{ir}}$	Cauchy	$e^{-|r|}$

2.7 Information and Independence

In this section, we have another look at the role of σ-algebras and independence in random experiments. As always, $(\Omega, \mathcal{H}, \mathbb{P})$ is our probability space in the background. Let X be a random variable taking values in some measurable space $(E, \mathcal{E})$. We are usually only interested in events of the form $\{X \in A\} = X^{-1}(A)$ for $A \in \mathcal{E}$. These events form a σ-algebra by themselves; see Exercise 1.9.

Definition 2.63: σ-Algebra Generated by a Random Variable

Let X be a random variable taking values in $(E, \mathcal{E})$. The sets $\{X \in A\}, A \in \mathcal{E}$ form a σ-algebra, called the *σ-algebra generated by* X, and we write σX.

Definition 2.63 also applies to a stochastic process $X := \{X_t, t \in \mathbb{T}\}$. Recall from Theorem 2.17 that X may be viewed equivalently as (1) a random variable taking values in $(E, \mathcal{E}) := (\times_{t \in \mathbb{T}} E_t, \otimes_{t \in \mathbb{T}} \mathcal{E}_t)$ and (2) a collection of random variables where X_t takes values in $(E_t, \mathcal{E}_t)$.

Theorem 2.64: σ-Algebra Generated by a Stochastic Process

For the stochastic process $X := \{X_t, t \in \mathbb{T}\}$, the σ-algebra generated by X is the smallest σ-algebra on Ω that is generated by the union of the σ-algebras $\sigma X_t, t \in \mathbb{T}$. We write $\bigvee_{t \in \mathbb{T}} \sigma X_t$ or $\sigma\{X_t, t \in \mathbb{T}\}$.

Proof. Let $\mathcal{H} = \sigma X$. Then, X is a random variable. By Theorem 2.17, it follows that each X_t is a random variable. This in turn implies that each σX_t is contained in

σX and, therefore, σX contains $\sigma\{X_t, t \in \mathbb{T}\}$. Conversely, let $\mathcal{H} = \sigma\{X_t, t \in \mathbb{T}\}$. Then, each X_t is a random variable. By Theorem 2.17, it follows that X is a random variable and, therefore, σX is contained in $\sigma\{X_t, t \in \mathbb{T}\}$. Hence, the two must be the same.   □

We can think of σX as a precise way of describing the information on a stochastic experiment modelled by a random variable or process X. What do σX-measurable numerical functions look like? The following theorem shows that they must be deterministic measurable functions of X. Recall that if h is $\mathcal{E}/\overline{\mathcal{B}}$-measurable, we write $h \in \mathcal{E}$.

Theorem 2.65: σX-Numerical Random Variables

Let X be a random variable taking values in a measurable space $(E, \mathcal{E})$. A mapping $V : \Omega \to \overline{\mathbb{R}}$ belongs to σX if and only if $V = h \circ X$ for some $h \in \mathcal{E}$.

Proof. Sufficiency follows from Theorem 1.21: If h and X are measurable with respect to $\mathcal{E}/\overline{\mathcal{B}}$ and $\sigma X/\mathcal{E}$, then the composition $V := h \circ X$ is measurable with respect to $\sigma X/\overline{\mathcal{B}}$, so $V \in \sigma X$.

For necessity, we invoke the Monotone Class Theorem 1.33. Define

$$\mathcal{M} := \{V = h \circ X : h \in \mathcal{E}\}.$$

This is a monotone class of random variables. Check this yourself. Now, consider an event $H \in \sigma X$. Since X is a random variable, there is a set $A \in \mathcal{E}$ such that $H = X^{-1}A$. Hence, $\mathbb{1}_H = \mathbb{1}_A \circ X$, which lies in $\mathcal{M}$. So $\mathcal{M}$ contains all indicator variables in σX. By the Monotone Class Theorem, it must therefore contain all positive random variables in σX.

Finally, for an arbitrary $V \in \sigma X$, we write $V = V^+ - V^-$, where both the positive and negative part can be written as measurable functions of X.   □

A corollary is that for a stochastic process $X := \{X_1, X_2, \ldots\}$ a random variable V belongs to σX if and only if it can be written as some measurable function of $X_1, X_2, \ldots$. In fact, we have the following generalization:

Theorem 2.66: Stochastic Process with Arbitrary Index Set

For each t in an arbitrary set $\mathbb{T}$, let X_t be a random variable taking values in a measurable set $(E_t, \mathcal{E}_t)$. Then, the mapping $V : \Omega \to \overline{\mathbb{R}}$ belongs to $\sigma\{X_t \in \mathbb{T}\}$ if and only if there exists a sequence (t_n) in $\mathbb{T}$ and a function h in $\otimes_n \mathcal{E}_{t_n}$ such that

$$V = h(X_{t_1}, X_{t_2}, \ldots).$$

We can view σX as the information that we have about the stochastic experiment, based on the measurement X. If we have more measurements on the same experiment, say Y and Z, we may gain more information about the experiment. So $\sigma\{X, Y, Z\}$ contains at least as many events of interest as σX — usually many more. For stochastic processes indexed by time, it makes sense to consider increasing collections of information as time increases. This leads to the following definition:

Definition 2.67: Filtration

Let $\mathbb{T}$ be a subset of $\mathbb{R}$. A *filtration* is an increasing collection $\{\mathcal{F}_t : t \in \mathbb{T}\}$ of sub-σ-algebras of $\mathcal{H}$; that is, $\mathcal{F}_s \subseteq \mathcal{F}_t$ whenever $s < t$.

Filtrations are often used when dealing with stochastic processes $\{X_t : t \in \mathbb{T}\}$, where $\mathbb{T} \subseteq \mathbb{R}$. The *natural filtration* is then the filtration defined by $\mathcal{F}_t := \sigma\{X_s : s \leq t, s \in \mathbb{T}\}$.

We can also use families of σ-algebras to model the notion of *independence* in a random experiment. We already defined independence for random variables. We now extend this to σ-algebras.

Definition 2.68: Independence for σ-Algebras

Let $\{\mathcal{F}_i, i = 1, \ldots, n\}$ be a collection of sub-σ-algebras of $\mathcal{H}$. The $\{\mathcal{F}_i\}$ are said to be (mutually) *independent* or form an *independency* if

(2.69)
$$\mathbb{E}V_1 \cdots V_n = \mathbb{E}V_1 \cdots \mathbb{E}V_n$$

for all positive random variables $V_i \in \mathcal{F}_i, i = 1, \ldots, n$.

We can extend the independence definition to an arbitrary collection of σ-algebras by requiring that every finite choice of σ-algebras forms an independency. The following extends the criterion for independence of random variables given in Definition 2.29:

Theorem 2.70: Independence of σ-Algebras

Sub-σ-algebras $\mathcal{F}_1, \ldots, \mathcal{F}_n$ of $\mathcal{H}$ generated by p-systems $\mathcal{C}_1, \ldots, \mathcal{C}_n$ are independent if and only if for all $H_i \in \mathcal{C}_i \cup \{\Omega\}, i = 1, \ldots, n$, it holds that:

(2.71)
$$\mathbb{P}(H_1 \cap \cdots \cap H_n) = \mathbb{P}(H_1) \cdots \mathbb{P}(H_n).$$

Proof. If (2.69) holds, then it holds in particular for $V_i = \mathbb{1}_{H_i}, i = 1, \ldots, n$, and so

(2.71) follows. This proves necessity.

Next, suppose that (2.71) holds for all $H_i \in C_i \cup \{\Omega\}$, $i = 1, \ldots, n$. Fix such $H_2, \ldots, H_n$ and define

$$\mathcal{D} := \{H_1 \in \mathcal{F}_1 : \mathbb{P}(H_1 \cap \cdots \cap H_n) = \mathbb{P}(H_1) \cdots \mathbb{P}(H_n)\}.$$

This is a d-system (check yourself). It contains C_1 and so by Theorem 1.12 it must contain $\mathcal{F}_1$. We can do the same process for $H_2, H_3, \ldots, H_n$ to see that (2.71) holds for $H_1 \in \mathcal{F}_1, \ldots, H_n \in \mathcal{F}_n$. Hence, (2.69) holds for all indicator functions $V_i := \mathbb{1}_{H_i}$. This extends to positive simple functions and from these to positive measurable functions via the Monotone Convergence Theorem 2.34. □

We now have an equivalent characterization of independence between random variables.

Theorem 2.72: Independence of Random Variables

Random variables X and Y taking values in $(E, \mathcal{E})$ and $(F, \mathcal{F})$ are independent if and only if

$$(2.73) \qquad \mathbb{E}g(X)\, h(Y) = \mathbb{E}g(X)\, \mathbb{E}h(Y), \quad g \in \mathcal{E}_+,\ h \in \mathcal{F}_+.$$

Proof. Suppose X and Y are independent in the sense of (2.69). Take any $g \in \mathcal{E}_+$ and $h \in \mathcal{F}_+$. By Theorem 2.65, $V := g(X)$ and $W := h(Y)$ are positive random variables in σX and σY, respectively. Thus, (2.69) implies (2.73). Conversely, suppose that (2.73) holds and take positive random variables $V \in \sigma X$ and $W \in \sigma Y$. Again, by Theorem 2.65, there must be a $g \in \mathcal{E}_+$ and $h \in \mathcal{F}_+$ such that $V = g \circ X$ and $W = h \circ Y$, and so (2.73) implies (2.69); i.e., $\mathbb{E}VW = \mathbb{E}V\, \mathbb{E}W$. □

We can rewrite equation (2.73) in terms of the joint distribution π of (X, Y) and marginal distributions μ and ν of X and Y as follows:

$$\int_{E \times F} \pi(\mathrm{d}x, \mathrm{d}y)\, g(x)\, h(y) = \int_E \mu(\mathrm{d}x)\, g(x) \int_F \nu(\mathrm{d}y)\, h(y).$$

In other words; π is equal to the product measure $\mu \otimes \nu$.

2.8 Important Stochastic Processes

We conclude this chapter by showcasing some important classes of stochastic processes, namely Gaussian processes, Poisson random measures, and Lévy processes. Together with Markov processes, they account for most of the study of stochastic processes. We will discuss Markov processes at the end of Chapter 4, after we have had more experience with the concept of conditioning.

2.8.1 Gaussian Processes

Gaussian processes can be thought of as generalizations of Gaussian random variables and vectors. Their distributional properties derive directly from the properties of the multivariate Gaussian distribution; see also Example 2.42.

Definition 2.74: Gaussian Process

A real-valued stochastic process $\{X_t, t \in \mathbb{T}\}$ is said to be *Gaussian* if all its finite-dimensional distributions are Gaussian (normal); that is, if the vector $[X_{t_1}, \ldots, X_{t_n}]^\top$ is multivariate normal for any choice of n and $t_1, \ldots, t_n \in \mathbb{T}$.

An equivalent condition for a process to be Gaussian is that every linear combination $\sum_{i=1}^n b_i X_{t_i}$ has a Gaussian distribution. The probability distribution of a Gaussian process is determined completely by its *expectation function*

$$\mu_t := \mathbb{E} X_t, \quad t \in \mathbb{T}$$

and *covariance function*

$$\gamma_{s,t} := \mathbb{C}\text{ov}(X_s, X_t), \quad s, t \in \mathbb{T}.$$

The latter is a *positive semidefinite* function; meaning that for every $n \geq 1$ and every choice of $\alpha_1, \ldots, \alpha_n \in \mathbb{R}$ and $t_1, \ldots, t_n \in \mathbb{T}$, it holds that

$$(2.75) \qquad \sum_{i=1}^n \sum_{j=1}^n \alpha_i \, \gamma_{t_i, t_j} \, \alpha_j \geq 0.$$

A *zero-mean* Gaussian process is one for which $\mu_t = 0$ for all t.

To simulate a realization of a Gaussian process with expectation function (μ_t) and covariance function $(\gamma_{s,t})$ at times $t_1, \ldots, t_n$, we can simply generate a multivariate normal random vector $\mathbf{Y} := [Y_1, \ldots, Y_n]^\top := [X_{t_1}, \ldots, X_{t_n}]^\top$ with mean vector $\boldsymbol{\mu} := [\mu_{t_1}, \ldots, \mu_{t_n}]^\top$ and covariance matrix $\boldsymbol{\Sigma}$, with $\Sigma_{i,j} := \gamma_{t_i, t_j}$. As such, the basic generation method is as follows:

■ **Algorithm 2.76 (Gaussian Process Generator)**

1. Construct the mean vector $\boldsymbol{\mu}$ and covariance matrix $\boldsymbol{\Sigma}$ as specified above.

2. Derive the Cholesky decomposition $\boldsymbol{\Sigma} = \mathbf{A}\mathbf{A}^\top$.

3. Simulate $Z_1, \ldots, Z_n \overset{\text{iid}}{\sim} \mathcal{N}(0, 1)$. Let $\mathbf{Z} := [Z_1, \ldots, Z_n]^\top$.

4. Output $\mathbf{Y} := \boldsymbol{\mu} + \mathbf{A}\mathbf{Z}$.

■ **Example 2.77 (Wiener Process and Brownian Motion)** The prototypical Gaussian process on $\mathbb{R}_+$ is the *Wiener process* $(W_t, t \geq 0)$, which will be discussed in detail in Chapter 6. For now, it suffices to define the Wiener process as a zero-mean Gaussian process with continuous sample paths and covariance function $\gamma_{s,t} = s \wedge t$ for $s, t \geq 0$. A typical path of the process is given in Figure 2.78, suggesting that the Wiener process can be viewed as a continuous version of the symmetric random walk in Figure 2.32.

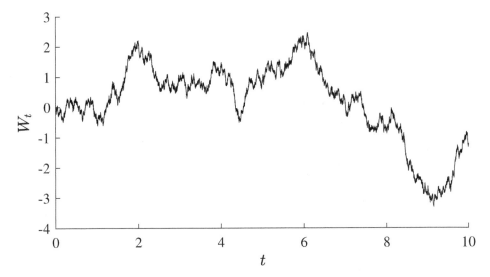

Figure 2.78: A typical realization of the Wiener process on the interval [0,10].

By applying the affine transformation

$$B_t = B_0 + at + cW_t, \quad t \geq 0,$$

where B_0 is independent of (W_t), we obtain a *Brownian motion* process $(B_t, t \geq 0)$, with drift[2] a and diffusion coefficient c. It is a Gaussian process only if B_0 is Gaussian. More generally, a d-dimensional Wiener process $\boldsymbol{W} := (\boldsymbol{W}_t, t \geq 0)$ is a stochastic process whose component processes are independent Wiener processes. Similarly, a d'-dimensional Brownian motion $\boldsymbol{B} := (\boldsymbol{B}_t, t \geq 0)$ is obtained from $\boldsymbol{W}$ via the affine transformation

$$\boldsymbol{B}_t = \boldsymbol{B}_0 + \boldsymbol{a}\,t + \mathbf{C}\boldsymbol{W}_t, \quad t \geq 0,$$

where $\mathbf{C}$ is a $d' \times d$ matrix and $\boldsymbol{a}$ a d'-dimensional vector, and $\boldsymbol{B}_0$ is independent of $\boldsymbol{W}$. ■

---

[2]Often μ and σ are used for the drift and diffusion coefficients, but in this chapter we would rather reserve the Greek letters for measures.

2.8.2 Poisson Random Measures and Poisson Processes

Poisson random measures are used to model random configurations of points in space and time. Specifically, let $E \subseteq \mathbb{R}^d$ and let $\mathcal{E}$ be the collection of Borel sets on E. To any collection of random points $\{X_i, i \in I\}$ in E corresponds a *random counting measure N* defined by

$$N(A) := \sum_{i \in I} \mathbb{1}_{\{X_i \in A\}}, \quad A \in \mathcal{E},$$

counting the random number of points in A. The quintessential random counting measure is the Poisson random measure.

Definition 2.79: Poisson Random Measure

A random measure N on $(E, \mathcal{E})$ is said to be a *Poisson random measure* with *mean measure μ* if the following properties hold:

1. $N(A) \sim \mathrm{Poi}(\mu(A))$ for any set $A \in \mathcal{E}$.

2. For any selection of disjoint sets $A_1, \ldots, A_n \in \mathcal{E}$, the random variables $N(A_1), \ldots, N(A_n)$ are independent.

In most practical cases, the mean measure has a density r with respect to the Lebesgue measure, called the *intensity* or *rate* function, so that

$$\mu(A) = \int_A \mathrm{d}\boldsymbol{x}\, r(\boldsymbol{x}).$$

In that case, μ is a *diffuse* measure; that is, $\mu(\{\boldsymbol{x}\}) = 0$ for all $\boldsymbol{x} \in E$. The Poisson random measure is said to be *homogeneous* if the rate function is constant.

The distribution of any random counting measure N is completely determined by its *Laplace functional*:

$$L(f) := \mathbb{E}\mathrm{e}^{-Nf}, \quad f \in \mathcal{E}_+,$$

where Nf is the random variable $\int N(\mathrm{d}\boldsymbol{x}) f(\boldsymbol{x})$. This is an example of a *stochastic integral*. For every outcome $\omega \in \Omega$ the integral is an ordinary Lebesgue integral in the sense of Section 1.4.1. For the Poisson random measure with mean measure μ, the Laplace functional is

$$\text{(2.80)} \qquad \mathbb{E}\mathrm{e}^{-Nf} = \mathrm{e}^{-\mu(1-\mathrm{e}^{-f})}, \quad f \in \mathcal{E}_+;$$

see Exercise 22 for a proof. Compare this with the Laplace transform of a Poisson random variable N with rate μ:

$$\mathbb{E}\mathrm{e}^{-Ns} = \mathrm{e}^{-\mu(1-\mathrm{e}^{-s})}, \quad s \geq 0.$$

We can think of N both in terms of a random counting measure, where $N(A)$ denotes the random number of points in $A \in \mathcal{E}$, or in terms of the (countable) collection $\{X_i, i \in I\}$ of random points (often called *atoms*) in E, where $I \subseteq \mathbb{N}$ may be finite and random, or countably infinite. In particular, we have

$$(2.81) \qquad Nf = \int_E N(\mathrm{d}x)f(x) = \sum_{i \in I} f(X_i), \quad f \in \mathcal{E}_+.$$

The following proposition describes one way of constructing a Poisson random measure on a product space. In its proof, the principle of *repeated conditioning* is used; see Theorem 4.5. An elementary understanding will suffice at this point.

Proposition 2.82: Poisson Random Measure on a Product Space

Let $X := \{X_i, i \in I\}$ be the atoms of a Poisson random measure N on $(E, \mathcal{E})$ with mean measure μ, and let $\{Y_i, i \in \mathbb{N}\}$ be iid random variables on $(F, \mathcal{F})$ with distribution π, independent of X. Then, $\{(X_i, Y_i), i \in I\}$ form the atoms of a Poisson random measure M on $(E \times F, \mathcal{E} \otimes \mathcal{F})$ with mean measure $\mu \otimes \pi$.

Proof. For any $f \in (\mathcal{E} \otimes \mathcal{F})_+$, we have $Mf = \sum_i f(X_i, Y_i)$, and

$$\mathbb{E}e^{-Mf} = \mathbb{E}\,\mathbb{E}_X \prod_i e^{-f(X_i,Y_i)} = \mathbb{E} \prod_i \mathbb{E}_X e^{-f(X_i,Y_i)} = \mathbb{E} \prod_i \int_F \pi(\mathrm{d}y)\, e^{-f(X_i,y)},$$

where repeated conditioning is used in the first equation, and the independence of Y_i of X is used in the second equation. Defining

$$e^{-g(x)} := \int_F \pi(\mathrm{d}y)\, e^{-f(x,y)},$$

we thus have

$$\mathbb{E}e^{-Mf} = \mathbb{E} \prod_i e^{-g(X_i)} = \mathbb{E}e^{-Ng} = e^{-\mu(1-e^{-g})},$$

where we have used (2.80) for the Laplace functional of N. Since

$$\mu(1 - e^{-g}) = \int_E \mu(\mathrm{d}x) \int \pi(\mathrm{d}y)(1 - e^{-f(x,y)}) = (\mu \otimes \pi)(1 - e^{-f}),$$

the Laplace functional of M is of the form (2.80), and hence M is a Poisson random measure with mean measure $\mu \otimes \pi$.   □

Let N be a Poisson random measure on $(E, \mathcal{E})$ with mean measure μ, with $\mu(E) =: c < \infty$. The total number of atoms of N has a Poisson distribution with mean c. The following theorem shows how the points are distributed conditional on the total number of points. Conditioning will be discussed in detail in Chapter 4.

> **Theorem 2.83: Conditioning on the Total Number of Points**
>
> Let N be a Poisson random measure on $(E, \mathcal{E})$ with mean measure μ satisfying $\mu(E) =: c < \infty$. Then, conditional on $N(E) = k$, the k atoms of N are iid with probability distribution μ/c.

Proof. Let $K := N(E) \sim \mathsf{Poi}(c)$, and on $\{K = k\}$ denote the atoms by $X_1, \dots, X_k$. Suppose that, conditionally on $\{K = k\}$, the atoms are iid with probability distribution μ/c. We want to show that the Laplace functional of such a random measure is precisely that of N. For $f \in \mathcal{E}_+$ we have

$$\mathbb{E} \exp\left(-\sum_{k=1}^{K} f(X_k)\right) = \mathbb{E}\,\mathbb{E}_K \exp\left(-\sum_{k=1}^{K} f(X_k)\right)$$

$$= \mathbb{E}\left(\int (\mu(\mathrm{d}x)/c)\, \mathrm{e}^{-f(x)}\right)^K = \mathbb{E}\left((\mu/c)\, \mathrm{e}^{-f}\right)^K =: \mathbb{E} z^K,$$

where $\mathbb{E}_K$ denotes the expectation conditional on K. In the first equation we are applying the repeated conditioning property; see Theorem 4.5. The second equation follows from the conditional independence assumption for the atoms. The third equation is just a simplification of notation, as in $vg = \int v(\mathrm{d}x)g(x)$. Since K has a $\mathsf{Poi}(c)$ distribution, it holds that

$$\mathbb{E} z^K = \mathbb{E}\mathrm{e}^{-c(1-z)} = \mathbb{E}\mathrm{e}^{-c(1-(\mu/c)\,\mathrm{e}^{-f})} = \mathbb{E}\mathrm{e}^{-\mu(1-\mathrm{e}^{-f})} = \mathbb{E}\mathrm{e}^{-Nf},$$

as had to be shown.   □

The preceding theorem leads directly to the following generic algorithm for simulating a Poisson random measure on E, assuming that $\mu(E) = \int_E \mathrm{d}x\, r(x) < \infty$:

■ **Algorithm 2.84 (Simulating a General Poisson Random Measure)**

1. Generate a Poisson random variable $K \sim \mathsf{Poi}(\mu(E))$.

2. Given $K = k$, draw $X_1, \dots, X_k \overset{\text{iid}}{\sim} g$, where $g := r/\mu(E)$ is the mean density, and return these as the atoms of the Poisson random measure.

■ **Example 2.85 (Convex Hull of a Poisson Random Measure)** Figure 2.86 shows six realizations of the point sets and their *convex hulls* of a homogeneous Poisson random measure on the unit square with rate 20. The MATLAB code is given below. A particular object of interest could be the random volume of the convex hull formed in this way.

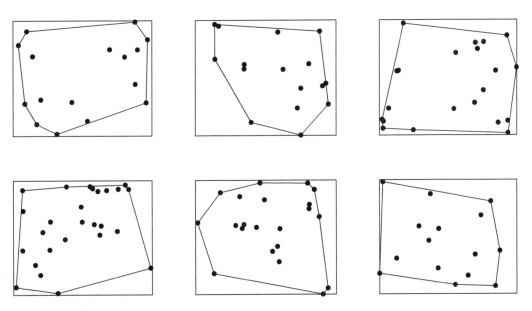

Figure 2.86: Realizations of a homogeneous Poisson random measure with rate 20. For each case the convex hull is also plotted.

```
for i=1:6
    K = poissrnd(20); %using the Statistics Toolbox
    x = rand(K,2);
    j = convhull(x(:,1),x(:,2));
    % [J,v] = convhulln(x); %v is the area
    subplot(2,3,i);
    plot(x(j,1),x(j,2),'r-',x(:,1),x(:,2),'b.')
end
```

For a homogeneous Poisson random measure on $\mathbb{R}_+$ with rate r a more direct simulation algorithm can be formulated. Denote the points by $0 < T_1 < T_2 < \cdots$, which are interpreted as *arrival* points of some sort, and let $A_i := T_i - T_{i-1}$ be the ith *interarrival* time, $i = 1, 2, \ldots$, setting $T_0 := 0$. It turns out that the interarrival times $\{A_i\}$ are iid and $\mathsf{Exp}(r)$ distributed. We will see an elegant proof of this in Example 5.38, once we have been introduced to martingales.

Let $N_t := N([0, t])$ be the number of arrivals in $[0, t]$ of the above homogeneous Poisson random measure N on $\mathbb{R}_+$. The process $(N_t, t \geq 0)$ is called a *Poisson counting process*, or simply *Poisson process*, with *rate r*. We can thus generate the points of this Poisson process on some interval $[0, t]$ as follows:

■ **Algorithm 2.87** (**Simulating a Poisson Process**)

1. Set $T_0 := 0$ and $n := 1$.

2. Generate $U \sim \mathcal{U}(0, 1)$.

3. Set $T_n := T_{n-1} - \frac{1}{r} \ln U$.

4. If $T_n > t$, stop; otherwise, set $n := n + 1$ and go to Step 2.

For an intuitive "proof" of the above, it is useful to view the properties of a Poisson process or, more generally, a Poisson random measure, through the glasses of the following *Bernoulli approximation*. Divide the time-axis into small intervals $[0, h), [h, 2h), \dots$. The numbers of arrivals in each of these intervals are independent and have a $\text{Poi}(rh)$ distribution. So, for small h, with a large probability there will be either no arrivals or 1 arrival in each time interval, with probability $1 - rh + O(h^2)$ and $rh + O(h^2)$, respectively. Next, consider a Bernoulli process $\{X_n\}$ with success parameter $p := rh$. Put $Y_0 := 0$ and let $Y_n := X_1 + \cdots + X_n$ be the total number of successes in n trials. Properties 1 and 2 of Definition 2.79 indicate that for small h the process $\{Y_i, i = 0, 1, \dots, n\}$ should have similar properties to the Poisson counting process $(N_s, 0 \le s \le t)$, if t and n are related via $n = t/h$. In particular:

1. For small h, N_t should have approximately the same distribution as Y_n. Hence,

$$\mathbb{P}(N_t = k) = \lim_{h \downarrow 0} \mathbb{P}(Y_n = k) = \lim_{h \downarrow 0} \binom{n}{k} (r\,h)^k (1 - (r\,h))^{n-k}$$

$$= \lim_{n \to \infty} \binom{n}{k} \left(\frac{rt}{n}\right)^k \left(1 - \frac{rt}{n}\right)^{n-k} = \frac{(rt)^k}{k!} \lim_{n \to \infty} \frac{n!}{n^k (n-k)!} \left(1 - \frac{rt}{n}\right)^{n-k}$$

$$= \frac{(rt)^k}{k!} \lim_{n \to \infty} \left(1 - \frac{rt}{n}\right)^n = \frac{(rt)^k}{k!} e^{-rt}.$$

This shows heuristically that N_t *must* have a Poisson distribution and that this is entirely due to the independence assumption in Property 2.

2. Let $U_1, U_2, \dots$ denote the times of success for the Bernoulli process X. We know that the intersuccess times $U_1, U_2 - U_1, \dots$ are independent and have a geometric distribution with parameter $p = rh$. The interarrival times $A_1, A_2, \dots$ of N should therefore also be iid. Moreover, for small h we have, again with $n = t/h$,

$$\mathbb{P}(A_1 > t) \approx \mathbb{P}(U_1 > n) = (1 - rh)^n \approx \left(1 - \frac{rt}{n}\right)^n \approx e^{-rt},$$

which is in accordance with the fact that $A_1 \sim \mathsf{Exp}(r)$. This can be made precise by observing that

$$\mathbb{P}(A_1 > t) = \mathbb{P}(N_t = 0) = e^{-rt}, \quad t \geq 0.$$

3. In higher dimensions we can make similar Bernoulli approximations. For example, we can partition a region $A \subset \mathbb{R}^2$ into a fine grid and associate with each grid cell a Bernoulli random variable — a success indicating a point in that cell; see Figure 2.88. In the limit, as the grid size goes to 0, the Bernoulli process tends to a Poisson random measure on A. The independence property follows directly from the independence of the Bernoulli random variables, and this independence leads also to the number of points in any region having a Poisson distribution. The Poisson random measure is homogeneous if all the Bernoulli success probabilities are the same.

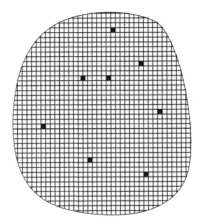

Figure 2.88: Bernoulli approximation of a two-dimensional Poisson random measure.

Let $(E, \mathcal{E}, \nu)$ be a measure space with a *finite* measure ν. Combining Proposition 2.82 with Algorithm 2.87 gives a convenient algorithm for simulating the points $\{(T_i, Y_i), i = 1, 2, \ldots\}$ of a Poisson random measure (PRM) on $\mathbb{R}_+ \times E$ with mean measure $\mathrm{Leb} \otimes \nu$. This algorithm is relevant for the construction of compound Poisson and Lévy processes, to be discussed next.

■ **Algorithm 2.89 (Simulating a PRM on $\mathbb{R}_+ \times E$ with Mean Measure $\mathrm{Leb} \otimes \nu$)**

1. Set $T_0 := 0$, $i := 1$, and $c := \nu(E)$.

2. Simulate $U \sim \mathcal{U}(0, 1)$ and set $T_i := T_{i-1} - c^{-1} \ln U$.

3. Simulate $Y_i \sim \nu/c$.

4. Set $i := i + 1$ and repeat from Step 2.

2.8.3   Compound Poisson Processes

Let N be a Poisson random measure on $\mathbb{R}_+ \times \mathbb{R}^d$ with mean measure $\text{Leb} \otimes \nu$ — we also write the latter as $\mathrm{d}t\, \nu(\mathrm{d}x)$. We assume that $c := \nu(\mathbb{R}^d) < \infty$. The process (K_t) with $K_t := N([0, t] \times \mathbb{R}^d)$ is a Poisson process with rate c. We now construct the following stochastic process as a stochastic integral with respect to N.

Definition 2.90: Compound Poisson Process

The process $(X_t, t \geq 0)$ defined by

$$X_t := \int_{[0,t] \times \mathbb{R}^d} N(\mathrm{d}s, \mathrm{d}x)\, x, \quad t \geq 0,$$

is the *compound Poisson process* corresponding to the measure ν.

A compound Poisson process can be thought of as a "batch" Poisson process, where arrivals occur according to a Poisson process with rate c, and each arrival adds a batch of size $Y \sim \nu/c$ to the total, so that we may also write

$$X_t = \sum_{k=1}^{K_t} Y_k,$$

where $Y_1, Y_2, \ldots \overset{\text{iid}}{\sim} \nu/c$ are independent of $K_t \sim \text{Poi}(ct)$. The compound Poisson process is an important example of a *Lévy process* — a stochastic process with independent and stationary increments; see Section 2.8.4. In this context the measure ν is called the *Lévy measure*.

The characteristic function of X_t can be found by conditioning on K_t:

$$\mathbb{E}\mathrm{e}^{\mathrm{i}r^\top X_t} = \mathbb{E}\,\mathbb{E}_{K_t} \mathrm{e}^{\mathrm{i}r^\top \sum_{k=1}^{K_t} Y_k} = \mathbb{E}\left(\mathbb{E}\mathrm{e}^{\mathrm{i}r^\top Y}\right)^{K_t} = \exp(-ct\,(1 - \mathbb{E}\mathrm{e}^{\mathrm{i}r^\top Y}))$$

$$= \exp\left(t \int \nu(\mathrm{d}y)(\mathrm{e}^{\mathrm{i}r^\top y} - 1)\right).$$

Denoting the jump times of the compound Poisson process by $\{T_k\}$ and the jump sizes by $\{Y_k\}$, we have the following simulation algorithm.

■ **Algorithm 2.91 (Simulating a Compound Poisson Process)**

1. Initialize $T_0 := 0$, $X_0 := \mathbf{0}$, and set $k := 1$.

2. Generate $A_k \sim \text{Exp}(c)$.

3. Generate $Y_k \sim \nu/c$.

4. Set $T_k := T_{k-1} + A_k$ and $X_{T_k} := X_{T_{k-1}} + Y_k$.

5. Set $k := k + 1$ and repeat from Step 2.

■ **Example 2.92 (Compound Poisson Process)** The bottom panel of Figure 2.93 shows a realization of the compound Poisson process $X := (X_t, t \geq 0)$ with Lévy measure

$$\nu(dx) := 5 e^{-|x|} dx, \quad x \in \mathbb{R}.$$

The top panel shows the atoms of the corresponding Poisson random measure N. For each atom (t, x) of N, the process jumps an amount x at time t.

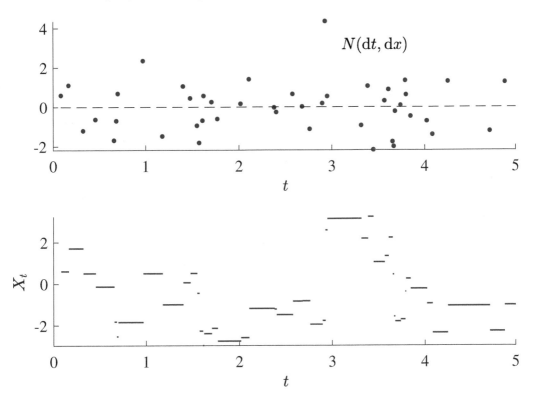

Figure 2.93: Realization of a compound Poisson process taking values in $\mathbb{R}$.

■

2.8.4 Lévy Processes

Lévy processes generalize the stationarity and independence of increments properties of the Wiener and compound Poisson processes. They can be completely characterized in terms of a Poisson random measure and a Brownian motion.

Let $(\Omega, \mathcal{H}, \mathbb{P})$ be a probability space and $X := (X_t, t \geq 0)$ a stochastic process with state space $\mathbb{R}^d$.

Definition 2.94: Lévy Process

The process X is said to be a *Lévy process* if

1. the paths of X are almost surely right-continuous and left-limited, with $X_0 = 0$, and

2. for every $t_1 < t_2 < t_3 < t_4$, the increments $X_{t_2} - X_{t_1}$ and $X_{t_4} - X_{t_3}$ are independent, and for every $t, u \in \mathbb{R}_+$, the increment $X_{t+u} - X_t$ has the same distribution as X_u.

Property 2 is summarized by stating that a Lévy process has *independent and stationary increments*. A Lévy process can be viewed as a continuous-time generalization of a random walk process. Indeed the process observed at times $0 =: t_0 < t_1 < t_2 < \cdots$ forms a random walk,

$$(2.95) \qquad\qquad X_{t_n} = \sum_{i=1}^{n} (X_{t_i} - X_{t_{i-1}}),$$

whose increments $\{X_{t_i} - X_{t_{i-1}}\}$ are independent. Moreover, if the times are chosen at an equal distance from each other, $t_i - t_{i-1} = h$, then the increments are iid, and so the distribution of the Lévy process is completely specified by its increment distribution in any time interval of length $h > 0$, for example by the distribution of X_1. Note also that linear combinations of independent Lévy processes are again Lévy. It is not difficult to show (see Exercise 24) that the characteristic function of X_t must be of the form

$$\mathbb{E}e^{i r^\top X_t} = e^{t\,\phi(r)}, \quad r \in \mathbb{R}^d$$

for some complex-valued function ϕ. This is called the *characteristic exponent* of the Lévy process.

■ **Example 2.96 (Brownian Motion and the Compound Poisson Process)** A d-dimensional Brownian motion process starting at $\mathbf{0}$ is the archetypal Lévy process with *continuous* sample paths; in fact, we will see it is the *only* Lévy process with this property.

To check independence and stationarity of the increments for the Brownian motion process, it suffices to verify these properties for a one-dimensional Wiener process. Namely, a linear drift process $t \mapsto \boldsymbol{a}t$ is a trivial Lévy process, and a Brownian motion minus its drift process and initial position is a linear transformation of a d-dimensional Wiener process, whose components, in turn, are independent one-dimensional Wiener processes. So, let $W := (W_t, t \geq 0)$ be a Wiener process. Since W is a zero-mean Gaussian process with covariance function $s \wedge t$, the increment $W_{t+u} - W_t$ has a Gaussian distribution with expectation 0 and variance

$$\mathbb{C}\text{ov}(W_{t+u} - W_t, W_{t+u} - W_t) = \mathbb{C}\text{ov}(W_{t+u}, W_{t+u}) + \mathbb{C}\text{ov}(W_t, W_t) - 2\,\mathbb{C}\text{ov}(W_{t+u}, W_t)$$
$$= t + u + t - 2t = u,$$

which does not depend on t. Next, take $t_1 < t_2 < t_3 < t_4$ and let $U := W_{t_2} - W_{t_1}$ and $V := W_{t_4} - W_{t_3}$. We have

$$\mathbb{C}\text{ov}(U, V) = \mathbb{C}\text{ov}(W_{t_2}, W_{t_4}) - \mathbb{C}\text{ov}(W_{t_2}, W_{t_3}) - \mathbb{C}\text{ov}(W_{t_1}, W_{t_4}) + \mathbb{C}\text{ov}(W_{t_1}, W_{t_3})$$
$$= t_2 - t_2 - t_1 + t_1 = 0.$$

This shows that U and V are independent, since $[U, V]^\top$ is a Gaussian random vector.

On the other side of the coin is the compound Poisson process

$$(2.97) \qquad\qquad X_t := \int_{[0,t] \times \mathbb{R}^d} N(\mathrm{d}s, \mathrm{d}x)\, x, \quad t \geq 0,$$

where N is a Poisson random measure on $\mathbb{R}_+ \times \mathbb{R}^d$ with mean measure $\mathrm{Leb} \otimes v$, satisfying $v(\mathbb{R}^d) < \infty$. This is the typical *pure jump* Lévy process. Independence and stationary of increments follow directly from the properties of the Poisson random measure. ∎

For a Lévy process, each increment can be written as the sum of n iid random variables, for every n. We say that the increments have an *infinitely divisible* distribution. Many of the common distributions are infinitely divisible; see Exercise 25.

For the compound Poisson process in (2.97), the requirement is that $v(\mathbb{R}^d) < \infty$. If this restriction on v is relaxed to

$$(2.98) \qquad\qquad \int_E v(\mathrm{d}x)\,(\|x\| \wedge 1) < \infty,$$

the process (X_t) in (2.97) still defines a pure-jump Lévy process, but not necessarily a compound Poisson process. In fact, any pure-jump Lévy process must be of this form for some Poisson random measure N, with v satisfying (2.98). In this context, N measures the times and magnitudes of the jumps, and v the expected number of jumps of a certain magnitude.

More precisely, for a Lévy process X, the *jump measure* N is such that $N([0, t] \times A)$ counts the number of jumps of X during the interval $[0, t]$ whose size lies in $A \in \mathcal{B}^d$, excluding $\mathbf{0}$. The mean measure of N is then $\mathrm{Leb} \otimes v$, where v is the *Lévy measure*. The expected number of jumps of X during $[0, 1]$ whose size lies in A is then $v(A) = \mathbb{E}N([0, 1] \times A)$. The main theorem on Lévy processes is the following.

Theorem 2.99: Lévy–Itô Decomposition

The process $X := (X_t, t \geq 0)$ is a Lévy process if and only if there exists a d-dimensional Brownian motion process $B := (B_t, t \geq 0)$ starting at $\mathbf{0}$ and a Poisson random measure N on $\mathbb{R}_+ \times \mathbb{R}^d$ that is independent of B and has mean measure Leb $\otimes\, \nu$, with ν satisfying

$$(2.100) \qquad \int_{\mathbb{R}^d} \nu(\mathrm{d}x)\,(\|x\|^2 \wedge 1) < \infty,$$

such that $X = B + U + Z$, where the summands are independent, $U := (U_t, t \geq 0)$ is the compound Poisson process

$$U_t := \int_0^t \int_{\|x\| > 1} N(\mathrm{d}s, \mathrm{d}x)\, x, \quad t \geq 0,$$

and $Z := (Z_t, t \geq 0)$ is the limit of compensated compound Poisson processes:

$$Z := \lim_{\delta \downarrow 0} Z^\delta \quad \text{with} \quad Z_t^\delta := \int_0^t \int_{\delta \leq \|x\| \leq 1} [N(\mathrm{d}s, \mathrm{d}x) - \mathrm{d}s\, \nu(\mathrm{d}x)]\, x, \; t \geq 0.$$

Proof. (Sketch). To show necessity, one needs to show (1) that adding independent Lévy processes gives another Lévy process, and (2) that each of the processes B, U, and Z is a Lévy process. The first statement follows easily from the definition of a Lévy process, and we already proved (2) for the Brownian motion B and the compound Poisson process U. Obviously, Z^δ is a Lévy process as well, for every $\delta > 0$. Thus, the main thing that remains to be proved is that the limit Z is well-defined under condition (2.100) and that it is Lévy. We omit the proof.

To show sufficiency, start with a Lévy process X and let N be its jump measure. This yields the Lévy processes U and Z via the integrals defined in the theorem. Let $\mathbb{B}$ be the unit ball on $\mathbb{R}^d$. As U is determined by the trace of N on $\mathbb{R}_+ \times \mathbb{B}^c$, and Z by the trace of N on $\mathbb{R}_+ \times \mathbb{B}$, the two processes are independent by the Poisson nature of N. By removing the jumps from X, the process $X - U - Z$ has continuous sample paths. Moreover, it is a Lévy process. It can be shown that the only Lévy processes with continuous sample paths are Brownian motions. Consequently, we have the decomposition $X = B + U + Z$. There remains to show that B is independent of the jump measure N (and hence independent of U and Z). This is the tricky part of the proof. A complete proof can be found in Çinlar (2011, Chapter 7). □

■ **Remark 2.101 (Truncation Level)** The choice of 1 for the truncation level in the above theorem is arbitrary. It may be changed to any positive number. However, this will change the drift of B. ■

The Lévy–Itô decomposition gives a complete and precise characterization of Lévy processes. The continuous part is described by a Brownian motion process, and the jump part by a Poisson random measure with Lévy measure ν satisfying (2.100). Note that this condition on ν is less restrictive than (2.98). We now have the following picture for the jumps of a Lévy process:

1. If $\nu(E) < \infty$, there are finitely many jumps in any time interval.

2. If $\nu(E) = \infty$, but (2.98) holds, then the process $\mathbf{Z}$ becomes

$$\mathbf{Z} = \int_0^t \int_{0 \leq \|x\| \leq 1} N(\mathrm{d}s, \mathrm{d}x)\, x - t \int_{0 \leq \|x\| \leq 1} \nu(\mathrm{d}x)\, x, \quad t \geq 0,$$

where the first integral can be combined with $\mathbf{U}$ to give a pure jump process of the form (2.97), and the second integral is a linear drift term. In each time interval there are infinitely many small jumps, tending to 0.

3. When $\nu(E) = \infty$, (2.98) does not hold, but (2.100) does hold, then the integral

$$\int_{0 \leq \|x\| \leq 1} \nu(\mathrm{d}x)\, x$$

is not finite (or does not exist). Nevertheless, as $\delta \downarrow 0$, the limit of $\mathbf{Z}_t^\delta$ still exists, leading to a process $\mathbf{Z}$ that has both an infinite number of jumps in any time interval as well as a drift.

A corollary of the Lévy–Itô decomposition is that the characteristic exponent ϕ of a Lévy process is of the form

$$\underbrace{-\frac{1}{2} r^\top \Sigma r + i\, r^\top a}_{\text{from } \mathbf{B}_t} + \underbrace{\int_{\|x\|>1} \nu(\mathrm{d}x) \left(e^{i\, r^\top x} - 1 \right)}_{\text{from } \mathbf{U}_t} + \underbrace{\int_{\|x\| \leq 1} \nu(\mathrm{d}x) \left(e^{i\, r^\top x} - 1 - i\, r^\top x \right)}_{\text{from } \mathbf{Z}_t}$$

for some vector a and covariance matrix $\Sigma = \mathbf{C}\mathbf{C}^\top$. It follows that each Lévy process is characterized by a *characteristic triplet* (a, Σ, ν).

■ **Example 2.102 (Gamma Process)** Let N be a Poisson random measure on $\mathbb{R}_+ \times \mathbb{R}_+$ with mean measure $\mathrm{d}t\, \nu(\mathrm{d}x) := \mathrm{d}t\, g(x)\, \mathrm{d}x$, where

$$g(x) := \frac{\alpha\, e^{-\lambda x}}{x}, \quad x > 0.$$

Define

$$X_t := \int_0^t \int_0^\infty N(\mathrm{d}s, \mathrm{d}x)\, x, \quad t \geq 0.$$

Then, $(X_t, t \geq 0)$ is, by construction, an increasing Lévy process. The characteristic exponent is

$$\phi(r) = \int dx \, (e^{i\,rx} - 1) \, g(x) = \alpha(\ln \lambda - \ln(\lambda - i\,r)),$$

which shows that $X_1 \sim \text{Gamma}(\alpha, \lambda)$, hence the name *gamma process*. The characteristic triplet is thus $(a, 0, \nu)$, with $a = \int_0^1 dx \, x g(x) = \frac{\alpha}{\lambda}(1 - e^{-\lambda})$. Note that an increment $X_{t+s} - X_t$ has a $\text{Gamma}(\alpha s, \lambda)$ distribution. A typical realization on $[0, 1]$ with $\alpha = 10$ and $\lambda = 1$ is given in Figure 2.103.

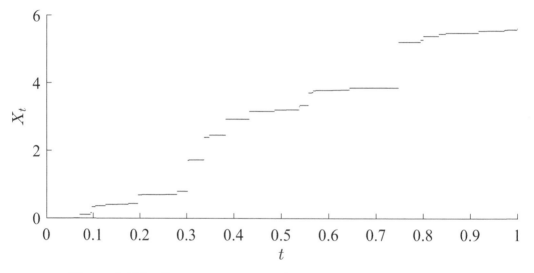

Figure 2.103: Gamma process realization for $\alpha = 10$ and $\lambda = 1$.

The simplest method to simulate certain Lévy processes is based on the random walk property (2.95). For this approach to work, the distribution of X_t needs to be known for all t.

■ **Algorithm 2.104 (Known Marginal Distributions)** Suppose X_t has a known distribution $\text{Dist}(t)$, $t \geq 0$. Generate a realization of the Lévy process at times $0 =: t_0 < t_1 < \cdots < t_n$ as follows:

1. Set $X_0 := 0$ and $k := 1$.

2. Draw $A \sim \text{Dist}(t_k - t_{k-1})$.

3. Set $X_{t_k} := X_{t_{k-1}} + A$.

4. If $k = n$ then stop; otherwise, set $k := k + 1$ and return to Step 2.

■ **Example 2.105 (Cauchy Process)** Let (X_t) be a Lévy process such that $X_1 \sim$ Cauchy. We use Algorithm 2.104 and the ratio-of-normals method for Cauchy random variables to simulate this process at times $t_k := k\Delta$, for $k \in \mathbb{N}$, and $\Delta := 10^{-5}$. Sample MATLAB code is given below, and a typical realization on $[0, 1]$ is given in Figure 2.106. Note that the process is a pure jump process with occasional very large increments.

```
Delta=10^(-5); N=10^5; times=(0:1:N).*Delta;
Z=randn(1,N+1)./randn(1,N+1);
Z=Delta.*Z; Z(1)=0;
X=cumsum(Z);
plot(times,X)
```

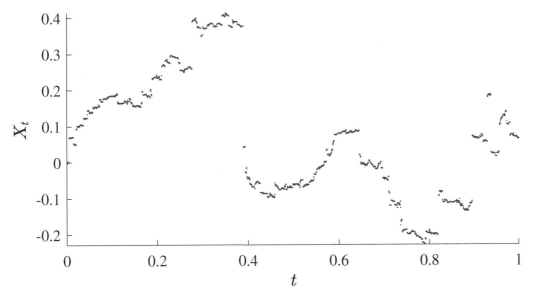

Figure 2.106: Cauchy process realization.

■

Exercises

1. We draw uniformly at random a point in the unit square $E := [0, 1]^2$. Give a probability space $(\Omega, \mathcal{H}, \mathbb{P})$ for this experiment such that the set $A := \{(x, y) \in E : x^2 + y^2 \leq 1\}$ is an event. Calculate $\mathbb{P}(A)$.

2. Let X be a random variable with $\mathbb{P}(X = k) := (1/2)^k, k = 1, 2, \ldots$. Show that such an object really exists. That is, construct an example of a probability space and a function X with the above property.

3. Prove Proposition 2.22.

4. Let F be a cumulative distribution function. Is $\lim_{x \to \infty} F(x)$ always 1?

5.* Consider the distribution μ with cdf

$$F(x) := 1 - (1 - \alpha)e^{-x}, \quad x \geq 0,$$

with $0 < \alpha < 1$.

  (a) Show that μ is a mixture of a discrete and an absolutely continuous distribution.

  (b) Give a measure ν and a density f such that μ has density f with respect to ν.

6. Construct a cdf F on the interval $[0, 1]$ in the following way, related to the Cantor set in Example 1.1 and Exercise 1.19. Let

$$D_{n,i} := \left(\frac{3i - 2}{3^{n+1}}, \frac{3i - 1}{3^{n+1}} \right), \quad i = 1, \ldots, 2^n, \quad n \in \mathbb{N}$$

be the ith open interval that is removed in the nth stage of the construction of the Cantor set. For $x \in D_{n,i}$, where $i \in \{1, \ldots, 2^n\}$ and $n \in \mathbb{N}$, define:

$$F(x) := \frac{2i - 1}{2^{n+1}}.$$

We may extend the domain of F to include every point $x \in [0, 1]$ by taking $F(x) := \lim_n F(x_n)$, where (x_n) is any sequence in D that converges to x. This function F is called the *Cantor function*. The function F is increasing and is continuous.

  (a) Draw the graph of F.
  (b) Verify that the derivative of F is almost everywhere equal to 0.
  (c)* Let q be the functional inverse of F (see Exercise 1.24):

$$q(u) := \inf\{x \in [0, 1] : F(x) > u\}, \quad u \in [0, 1].$$

    Show that the range of q is $C \setminus C_0$, where $C_0 := \{\frac{3i-2}{3^{n+1}}, i = 1, \ldots, 2^n, n \in \mathbb{N}\} \cup \{1\}$.
  (d)* Deduce that C has as many elements as the interval $[0, 1]$.

7.* Consider the probability space $(\Omega, \mathcal{H}, \mathbb{P})$, where Ω is the interval $[0, 1)$, $\mathcal{H}$ is the Borel σ-algebra on $[0, 1)$, and $\mathbb{P}$ is the Lebesgue measure (on $[0,1)$). Every $\omega \in [0, 1)$ has a unique binary expansion containing an infinite number of zeros, e.g.,

$$\frac{47}{64} = .101111000000\ldots.$$

Now, for any $\omega \in [0, 1)$ write down the expansion $\omega = .\omega_1 \omega_2 \cdots$, and define

$$X_n(\omega) := \omega_n.$$

(a) Draw the graphs of X_1, X_2, and X_3.

(b) Verify that $X_1, X_2, \ldots$ are $\mathsf{Ber}(1/2)$ distributed random variables.

(c) Verify that the $X_1, X_2, \ldots$ are independent. Hence, we have constructed an analytical model for the coin tossing experiment with a fair coin.

8.* Let $X \sim \mathcal{U}[0, 1]$ and $Y \sim \mathsf{Exp}(1)$ be independent.

(a) Determine the joint pdf of X and Y and draw the graph.

(b) Calculate $\mathbb{P}((X, Y) \in [0, 1] \times [0, 1])$.

(c) Calculate $\mathbb{P}(X + Y < 1)$.

9.* Let $Z_1, \ldots, Z_n$ be iid random variables, each with a standard normal distribution. Determine the joint pdf of the vector $\mathbf{Z} := [Z_1, \ldots, Z_n]^\top$. Let $\mathbf{A}$ be an invertible $n \times n$ matrix. Determine the joint pdf of the random vector $X := [X_1, \ldots, X_n]^\top$ defined by $X := \mathbf{A}Z$.

Show that $X_1, \ldots, X_n$ are iid standard normal only if $\mathbf{A}\mathbf{A}^\top = \mathbf{I}_n$ (identity matrix); in other words, only if $\mathbf{A}$ is an orthogonal matrix. Can you find a geometric interpretation of this?

10.* Let $X \sim \mathsf{Exp}(\lambda)$ and $Y \sim \mathsf{Exp}(\mu)$ be independent.

(a) What distribution does $\min(X, Y)$ have?

(b) Show that
$$\mathbb{P}(X < Y) = \frac{\lambda}{\lambda + \mu}.$$

11.* Let $X \sim \mathcal{U}(-\pi/2, \pi/2)$. What is the pdf of $Y := \tan X$?

12. Prove the following. If $Y := aX + b$ and X has density f_X with respect to the Lebesgue measure on $(\mathbb{R}, \mathcal{B})$, then Y has density f_Y given by
$$f_Y(y) = \frac{1}{|a|} f_X\left(\frac{y - b}{a}\right).$$

(Here, $a \neq 0$.)

13.* Let $X \sim \mathcal{N}(0, 1)$. Prove that $Y := X^2$ has a χ_1^2 distribution.

14. Let $U \sim \mathcal{U}(0, 1)$. Let F be an arbitrary cdf on $\overline{\mathbb{R}}$, and let q denote its quantile (functional inverse):
$$q(u) := \inf\{x \in \mathbb{R} : F(x) > u\}, \quad u \in \mathbb{R}.$$

(a) Let $F(x-) := \lim_{x_n \uparrow x} F(x_n)$ be the left limit of F at x. This exists, since F is increasing. Prove that for any x and u:
$$q(u) \geq x \Leftrightarrow F(x-) \leq u.$$

(b) Show that $X := q(U)$ is a random variable with cdf F. This gives the *inverse-transform method* for simulating random variables from F.

15. Using the definitions of variance and covariance, prove all the properties for the variance and covariance in Table 2.3.

16. Prove, using Fubini's Theorem 1.67, that for any positive random variable X, it holds that

$$\mathbb{E}X^p = \int_0^\infty dx\, px^{p-1}\mathbb{P}(X > x).$$

17. Show that if Y is a random variable taking values in $[-c, c]$ and with $\mathbb{E}Y = 0$, then for $\theta \in \mathbb{R}$,

$$\mathbb{E}e^{\theta Y} \leq e^{\frac{1}{2}\theta^2 c^2}.$$

18. Using MGFs, show that the sum of n independent $\mathsf{Exp}(\lambda)$ distributed random variables has a $\mathsf{Gamma}(n, \lambda)$ distribution.

19.* The *double exponential* distribution has pdf

$$f(x) := \frac{1}{2}e^{-|x|}, \quad x \in \mathbb{R}.$$

Show that its characteristic function ψ is given by

$$\psi(r) = \frac{1}{1 + r^2}, \quad r \in \mathbb{R}.$$

20. Using Exercise 19 and the inversion formula (2.60), prove that the characteristic function of the Cauchy distribution is $e^{-|r|}$, $r \in \mathbb{R}$.

21. Algorithm 2.76 to simulate a $\mathcal{N}(\mathbf{0}, \mathbf{\Sigma})$ requires the Cholesky decomposition of $\mathbf{\Sigma}$. The following alternative method uses instead the *precision matrix* $\mathbf{\Lambda} := \mathbf{\Sigma}^{-1}$. This is useful when $\mathbf{\Lambda}$ is a sparse matrix but $\mathbf{\Sigma}$ is not. Suppose that $\mathbf{DD}^\top$ is the Cholesky factorization of $\mathbf{\Lambda}$. Let Y satisfy $\mathbf{Z} = \mathbf{D}^\top Y$, where $\mathbf{Z}$ is a vector of iid $\mathcal{N}(0, 1)$ random variables. Show that $Y \sim \mathcal{N}(\mathbf{0}, \mathbf{\Sigma})$.

22.* Let N be a Poisson random measure with mean measure μ. Show that its Laplace functional is given by (2.80). Hint: start with functions $f = a\mathbb{1}_A$ with $a \geq 0$ and $A \in \mathcal{E}$ such that $\mu(A) < \infty$. Then, use the defining properties of the Poisson random measure.

23. Consider a compound Poisson process with Lévy measure

$$\nu(dx) := |x|^{-3/2}\, \mathbb{1}_{\{\delta < |x| < \varepsilon\}}\, dx, \quad x \in \mathbb{R}$$

for some $0 < \delta < \varepsilon \leq \infty$. Let $c := \nu(\mathbb{R}) = 4(\delta^{-1/2} - \varepsilon^{-1/2})$.

(a) Show that if $U \sim \mathcal{U}(0, 1)$ and $R \sim \text{Ber}(1/2)$ are independent, then

$$\frac{(2R - 1)\delta}{(1 - U + U\sqrt{\delta/\varepsilon})^2}$$

has distribution v/c.

(b) Implement Algorithm 2.91 and show a typical realization for (a) the case $\delta = 10^{-6}$ and $\varepsilon = \infty$, and (b) the case $\delta = 10^{-6}$ and $\varepsilon = 10^{-5}$.

24. Let (X_t) be a Lévy process. Show that the characteristic function of X_t must be of the form

$$\mathbb{E}e^{i r^\top X_t} = e^{t\,\phi(r)}, \quad r \in \mathbb{R}^d$$

for some complex-valued function ϕ.

25. Show that the $\text{Gamma}(\alpha, c)$, $\text{Poi}(c)$, and $\mathcal{N}(\mu, \sigma^2)$ distributions are infinitely divisible.

CHAPTER 3

CONVERGENCE

The purpose of this chapter is to introduce various modes of convergence in probability and how they are interrelated. We discuss almost sure convergence, convergence in probability, convergence in distribution, and L^p convergence. The notion of uniform integrability connects various modes of convergence. Main applications are the Law of Large Numbers and the Central Limit Theorem.

3.1 Motivation

As a motivating example for convergence, consider the random experiment where we repeatedly toss a biased coin. Suppose the probability of heads is 0.3. We can model this experiment with a Bernoulli process $X_1, X_2, \ldots \sim_{\text{iid}} \mathsf{Ber}(0.3)$, where $\{X_i = 1\}$ is the event that the ith throw is heads. The total number of heads in n throws is the random variable $S_n := X_1 + \cdots + X_n$. The average number of heads in n throws is S_n/n. In Figure 3.1 we see a typical realization $(s_n/n, n = 1, \ldots, 100)$ of the random process $(S_n/n, n = 1, \ldots, 100)$.

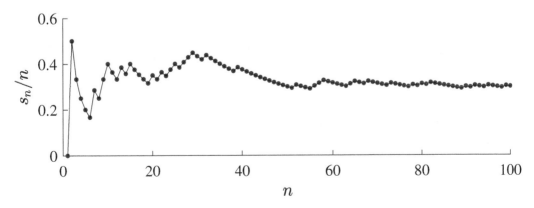

Figure 3.1: Average number of heads in n tosses, against n.

The behavior of the realization in Figure 3.1 is in accordance with our intuition. We would expect the estimate s_n/n for the probability of heads, here 0.3, to be more accurate for large n than for small n. However, it is not true that (s_n/n) converges to 0.3 for *every* realization (s_n). For example, for the realization where only heads occur, we have $\lim s_n/n = 1$, and if only tails occur, $\lim s_n/n = 0$. In fact, any limit between 0 and 1 can be obtained for specific realizations. To explain the intricacies of the above example, we have to look at the convergence behavior of sequences of random variables — in this case the sequence (S_n/n).

To understand the convergence of a sequence of random variables, we need to recall a few things about the convergence of real-valued sequences; see Definition 1.24. For any sequence (x_n) of real numbers,

$$\liminf x_n := \sup_m \inf_{n \geq m} x_m \quad \text{and} \quad \limsup x_n := \inf_m \sup_{n \geq m} x_m$$

are well-defined (possibly infinite). If $\liminf x_n = \limsup x_n$ we say that (x_n) has a limit and we write $\lim x_n$ for it. We say that (x_n) *converges* to $\lim x_n$. If the limit lies in $\mathbb{R}$ then (x_n) is said to be a *convergent* sequence.

Another way to state that (x_n) converges to $x \in \mathbb{R}$ is: for every $\varepsilon > 0$ there is an N_ε such that for all $n > N_\varepsilon$ it holds that $|x_n - x| < \varepsilon$. An equivalent statement is:

$$\sum_n \mathbb{1}_{\{|x_n - x| > \varepsilon\}} < \infty \quad \text{for every } \varepsilon > 0.$$

When we do not know to which limit a sequence converges, we can alternatively use the following Cauchy convergence criteria. The proposition is formulated and proved for real numbers, but the results hold for any convergent sequence (x_n) in a Banach space equipped with a norm $\| \cdot \|$; see Exercise 3. In the proof, we make use of the fundamental property of the real numbers that every bounded sequence (x_n) in $\mathbb{R}$ has a convergent subsequence; this is often referred to as the *Bolzano–Weierstrass theorem*.

Proposition 3.2: Cauchy Criteria for Convergence

The following statements satisfy (1) $\Leftrightarrow$ (2) and (3) $\Rightarrow$ (2):

1. (x_n) converges to some real $x := \lim x_n$.
2. (x_n) is Cauchy convergent, that is, $\lim_{m,n \to \infty} |x_m - x_n| = 0$.
3. $\sum_n \mathbb{1}_{\{|x_{n+1} - x_n| > \varepsilon_n\}} < \infty$ for any (ε_n) such that $c := \sum_n \varepsilon_n < \infty$.

Proof. If (x_n) converges to x, then (x_n) is a Cauchy sequence, since

$$|x_m - x_n| \leq |x_m - x| + |x_n - x|$$

and by letting $m, n \to \infty$, the right-hand side goes to 0. We have thus shown that $(1) \Rightarrow (2)$.

Conversely, suppose that (x_n) is a Cauchy sequence of real numbers; that is, it satisfies $\lim_{m,n\to\infty} |x_m - x_n| = 0$. Every such Cauchy sequence is bounded. To see this, take N large enough such that $|x_m - x_n| < 1$ for $m, n > N$. Then, for $m \geq N$, $|x_m| \leq 1 + |x_N|$, and so for *all* m, we have $|x_m| \leq 1 + \sum_{i=1}^{N} |x_i|$, which is bounded. Therefore, by the Bolzano–Weierstrass theorem, there exists a subsequence (x_{n_k}) that converges to some limit $x \in \mathbb{R}$. Thus, for every $\varepsilon > 0$, there is an N_1 such that $|x_{n_k} - x| < \varepsilon/2$ for all $k \geq N_1$. Also, by the Cauchy property, there is an N_2 such that $|x_m - x_{n_k}| < \varepsilon/2$ for all $m, n_k \geq N_2$. Let k be an integer such that $k \geq N_1$ and $n_k \geq N_2$. Then, for all $m \geq N_2$,

$$|x_m - x| \leq |x_m - x_{n_k}| + |x_{n_k} - x| \leq \varepsilon,$$

which shows that (x_n) converges to x. We have thus shown that $(2) \Rightarrow (1)$.

Finally, suppose that $\sum_k \mathbb{1}_{\{|x_{k+1}-x_k|>\varepsilon_k\}} < \infty$. Then, there is an i such that $|x_{k+1} - x_k| \leq \varepsilon_k$ for all $k \geq i$. So, for any $k > j > i$, we have

$$|x_k - x_j| \leq |x_k - x_{k-1}| + \cdots + |x_{j+1} - x_j| \leq \sum_{m=j}^{k-1} \varepsilon_m \leq c - \sum_{m=1}^{j-1} \varepsilon_m,$$

which goes to 0 as $i \to \infty$. Hence, (x_k) is a Cauchy sequence. This establishes that $(3) \Rightarrow (2)$.   □

All random variables in this chapter are assumed to be *numerical*; that is, they are measurable functions from Ω to $\overline{\mathbb{R}}$. When such random variables are finite, they are said to be *real-valued*.

3.2 Almost Sure Convergence

If a sequence of random variables (X_n) converges pointwise to a random variable X; that is,

$$\lim X_n(\omega) = X(\omega) \quad \text{for all } \omega \in \Omega,$$

we say that (X_n) *converges surely* to X. We have seen in Proposition 1.26 that $\limsup X_n$ and $\liminf X_n$ are again random variables, and if both are equal, then $\lim X_n$ is again a random variable. When the event $\{\lim X_n = X\}$ has probability 1, we say that X_n converges *almost surely* to X.

Definition 3.3: Almost Sure Convergence

The sequence of random variables (X_n) *converges almost surely* (a.s.) to a random variable X if
$$\mathbb{P}(\lim X_n = X) = 1.$$
We denote the almost sure convergence by $X_n \overset{a.s.}{\to} X$.

It will be convenient to characterize almost sure convergence in a different, but equivalent, way:

Theorem 3.4: Almost Sure Convergence

The sequence of random variables (X_n) converges almost surely to a random variable X if and only if for every $\varepsilon > 0$

$$(3.5) \qquad \lim_{m \to \infty} \mathbb{P}(\sup_{n \ge m} |X_n - X| > \varepsilon) = 0.$$

Proof. Consider the event $A_m := \{\sup_{n \ge m} |X_n - X| > \varepsilon\}$, $m \in \mathbb{N}$. Because $A_0 \supseteq A_1 \supseteq \cdots$, we have by the sequential continuity from above (2.5) of the probability measure $\mathbb{P}$ that

$$\lim_{m \to \infty} \mathbb{P}(A_m) = \mathbb{P}\left(\cap_{m=0}^{\infty} A_m\right) = \mathbb{P}(\inf_m \sup_{n \ge m} |X_n - X| > \varepsilon) = \mathbb{P}(\limsup |X_n - X| > \varepsilon).$$

If $X_n \overset{a.s.}{\to} X$, then $\limsup |X_n - X| = \liminf |X_n - X| = 0$ with probability 1, so $\lim \mathbb{P}(\sup_{n \ge m} |X_n - X| > \varepsilon) = 0$ for all $\varepsilon > 0$ and hence (3.5) holds. Conversely, if $\mathbb{P}(\limsup |X_n - X| > \varepsilon) = 0$ for all $\varepsilon > 0$, then $\mathbb{P}(\limsup |X_n - X| = 0) = 1$; and since $\liminf |X_n - X| \ge 0$, the limit of X_n is equal to X with probability 1. □

■ **Example 3.6 (Complete and Almost Sure Convergence)** A sequence of random variables (X_n) is said to *converge completely* to X if for all $\varepsilon > 0$

$$\sum_n \mathbb{P}(|X_n - X| > \varepsilon) < \infty.$$

We write $X_n \overset{cpl.}{\to} X$. Complete convergence implies almost sure convergence. To see this, take any $\varepsilon > 0$ and let $H_n := \{|X_n - X| > \varepsilon\}$. If $X_n \overset{cpl.}{\to} X$, then using the countable subadditivity in Theorem 2.2, we obtain

$$\mathbb{P}(\sup_{n \ge m} |X_n - X| > \varepsilon) = \mathbb{P}\left(\bigcup_{n \ge m} H_n\right) \le \sum_{n \ge m} \mathbb{P}(H_n) \to 0$$

as $m \to \infty$; that is, $X_m \overset{a.s.}{\to} X$. ■

Sometimes we do not know to which limit a sequence converges. To establish if the sequence converges almost surely, we can apply the following Cauchy characterization:

Proposition 3.7: Cauchy Criteria for Almost Sure Convergence

The following statements are equivalent:

1. $X_n \xrightarrow{\text{a.s.}} X$.

2. $\lim_{n \to \infty} \mathbb{P}(\sup_{k \geq 0} |X_{n+k} - X_n| > \varepsilon) = 0$ for any choice of $\varepsilon > 0$.

3. $\lim_{k \to \infty} \mathbb{P}(\sup_{m,n \geq k} |X_m - X_n| > \varepsilon) = 0$ for any choice of $\varepsilon > 0$.

4. $\mathbb{P}\left(\lim_{m,n \to \infty} |X_m - X_n| = 0\right) = 1$; that is, (X_n) is a.s. a Cauchy sequence.

Proof. We first establish that $(1) \Leftrightarrow (4)$. This is really just saying that for almost every $\omega \in \Omega$ the sequence (x_n) with $x_n := X_n(\omega)$ is a Cauchy sequence of real numbers if and only if (x_n) converges to $x := X(\omega)$. This, however, is immediate from Proposition 3.2.

We now show that $(3) \Leftrightarrow (4)$. Define

$$Y_n := \sup_{j,k \geq 0} |X_{n+j} - X_{n+k}|,$$

and note that (Y_n) is monotonically decreasing and that (4) is equivalent to $\mathbb{P}(\lim Y_n = 0) = 1$. Moreover, $\sup_{n \geq m} |Y_n| = Y_m$, which by Theorem 3.4 implies that (3) is equivalent to $Y_n \xrightarrow{\text{a.s.}} 0$. In other words, (3) is equivalent to $\mathbb{P}(\lim Y_n = 0) = 1$, completing the proof.

Finally, we demonstrate that $(2) \Leftrightarrow (3)$. Since (3) is equivalent to $Y_n \xrightarrow{\text{a.s.}} 0$, and

$$\sup_{k \geq 0} |X_{n+k} - X_n| \leq \sup_{j,k \geq 0} |X_{n+j} - X_{n+k}| = Y_n \xrightarrow{\text{a.s.}} 0,$$

then $(3) \Rightarrow (2)$. In addition, $|X_j - X_k| \leq |X_j - X_n| + |X_k - X_n|$ implies that

$$Y_n \leq \sup_{j,k \geq n} \left(|X_j - X_n| + |X_k - X_n|\right) \leq 2 \sup_{k \geq 0} |X_{n+k} - X_n|.$$

Therefore, $\mathbb{P}(Y_n > \varepsilon) \leq \mathbb{P}(\sup_{k \geq 0} |X_{n+k} - X_n| > \varepsilon/2) \to 0$ shows that $Y_n \xrightarrow{\text{a.s.}} 0$, or $(2) \Rightarrow (3)$, completing the proof. $\square$

The following inequalities are useful in many different proofs and applications:

Proposition 3.8: Markov's and Chebyshev's Inequality

For any random variable Y and $\varepsilon > 0$, *Markov's inequality* holds:

$$(3.9) \qquad \mathbb{P}(|Y| > \varepsilon) \leq \frac{\mathbb{E}\,|Y|}{\varepsilon}.$$

As a consequence, we have *Chebyshev's inequality*, where $\mu := \mathbb{E}X$:

$$(3.10) \qquad \mathbb{P}(|X - \mu| > \varepsilon) \leq \frac{\mathbb{V}\text{ar}\,X}{\varepsilon^2}.$$

Proof. We have $\mathbb{E}|Y| \geq \mathbb{E}[|Y|\mathbb{1}_{\{|Y|>\varepsilon\}}] \geq \mathbb{E}[|\varepsilon|\mathbb{1}_{\{|Y|>\varepsilon\}}] = \varepsilon\,\mathbb{P}(|Y| > \varepsilon)$. We obtain Chebyshev's inequality by taking $Y := (X-\mu)^2$ in Markov's inequality and replacing ε with ε^2. $\qquad\qquad\square$

An extension of Chebyshev's inequality (3.10) for the case where X is the sum of (possibly non-identically distributed) independent random variables is given by Kolmogorov's inequality.

Theorem 3.11: Kolmogorov's Inequality

Let $S_n := \sum_{i=1}^n X_i$, where $X_1, X_2, \ldots$ are independent with 0 mean. Then, for every $\varepsilon > 0$:

$$(3.12) \qquad \mathbb{P}(\sup_{k \leq n} |S_k| > \varepsilon) \leq \frac{\mathbb{V}\text{ar}\,S_n}{\varepsilon^2}.$$

A basic proof is outlined in Exercise 4. We provide a more general result in Theorem 5.44, using martingale techniques.

■ **Example 3.13 (Random Series)** Suppose that the random series (that is, sequence of partial sums) (S_n) in Kolmogorov's Theorem 3.11 is such that $c_n := \mathbb{V}\text{ar}\,S_n \leq c < \infty$. We can then show that the series converges almost surely.

Since (c_n) is a monotonically increasing and bounded sequence, it converges. Hence, (c_n) is a Cauchy sequence by Proposition 3.2. Applying Kolmogorov's inequality to $S_{n+k} - S_n = X_{n+1} + \cdots + X_{n+k}$ for $k = 1, \ldots, m$ yields

$$\varepsilon^2\,\mathbb{P}(\sup_{k \leq m} |S_{n+k} - S_n| > \varepsilon) \leq \sum_{j=n+1}^{m+n} \mathbb{V}\text{ar}\,X_j = c_{m+n} - c_n.$$

Taking limits on both sides as $m, n \to \infty$, and using the sequential continuity from above (2.5) yields:

$$\varepsilon^2 \lim_n \mathbb{P}(\sup_k |S_{n+k} - S_n| > \varepsilon) \leq \lim_{m,n \to \infty} (c_{m+n} - c_n) = 0,$$

which by Proposition 3.7 (Item 2) implies that (S_n) converges almost surely to a real-valued random variable S.

As a specific case, let $X_i := (B_i - \frac{1}{2})(\frac{1}{2})^i, i = 1, 2, \ldots$, where $\{B_i\} \sim_{\text{iid}} \text{Ber}(\frac{1}{2})$. Then, $\mathbb{E}X_i = 0$ for all i and $\text{Var } S_n = \frac{1}{12}(1 - 4^{-n}) < \frac{1}{12}$. Hence, we conclude that the random series (S_n) converges almost surely to a random variable

$$S := \sum_{i=1}^{\infty} \left(B_i - \frac{1}{2}\right)\left(\frac{1}{2}\right)^i = \sum_{i=1}^{\infty} B_i \left(\frac{1}{2}\right)^i - \frac{1}{2}$$

and that $Y_n := \sum_{i=1}^{n} B_i(\frac{1}{2})^i$ converges to $S + \frac{1}{2}$.   ∎

The following Borel–Cantelli lemma is useful in many proofs that require almost sure convergence:

Lemma 3.14: Borel–Cantelli

Let (H_n) be a sequence of events. Then, the following hold:

1. $\sum_n \mathbb{P}(H_n) < \infty$ implies that $\mathbb{P}\left(\sum_n \mathbb{1}_{H_n} < \infty\right) = 1$.

2. $\sum_n \mathbb{P}(H_n) = \infty$ implies that $\mathbb{P}\left(\sum_n \mathbb{1}_{H_n} < \infty\right) = 0$, provided that the events (H_n) are pairwise independent.

Proof. Let $\varepsilon > 0$ be an arbitrarily small constant and let $p_k := \mathbb{P}(H_k)$. Define the event

$$A_n := \left\{\sum_{k=1}^{n} \mathbb{1}_{H_k} < \varepsilon^{-1}\right\}$$

and note that, since $A_1 \supseteq A_2 \supseteq \cdots$ is a sequence of decreasing events, $\lim \mathbb{P}(A_n) = \mathbb{P}(\cap_{k=1}^{\infty} A_k)$ by Theorem 2.4. Similarly, $\lim \mathbb{P}(A_n^c) = \mathbb{P}(\cup_{k=1}^{\infty} A_k^c)$ by Theorem 2.2, with $\cup_{k=1}^{\infty} A_k^c = \{\sum_{k=1}^{\infty} \mathbb{1}_{H_k} \geq \varepsilon^{-1}\}$.

Assuming that $\sum_n p_n := c < \infty$, Markov's inequality (3.9) implies that

$$\mathbb{P}(A_n^c) = \mathbb{P}(\sum_{k=1}^{n} \mathbb{1}_{H_k} \geq \varepsilon^{-1}) \leq \varepsilon \mathbb{E}\sum_{k=1}^{n} \mathbb{1}_{H_k} = \varepsilon \sum_{k=1}^{n} p_k.$$

Therefore, taking limits on both sides of the inequality as $n \to \infty$ yields

$$\mathbb{P}(\sum_{k=1}^{\infty} \mathbb{1}_{H_k} \geq \varepsilon^{-1}) \leq \varepsilon c.$$

By taking $\varepsilon \downarrow 0$, we deduce that with probability 0 infinitely many of the events (H_n) occur; or, equivalently, that with probability 1 only finitely many of the events (H_n) occur. This completes the first statement in the Borel–Cantelli lemma.

Now assume that $\sum_n p_n = \infty$; thus, for each $\varepsilon > 0$ we can find an n_ε such that $c_n := \sum_{k=1}^{n} p_k > \varepsilon^{-2}$ for all $n \geq n_\varepsilon$.

Therefore, for all $n \geq n_\varepsilon$ we have that

$$\mathbb{P}(A_n) = \mathbb{P}(\sum_{k=1}^{n} \mathbb{1}_{H_k} < \varepsilon^{-1}) = \mathbb{P}(\sum_{k=1}^{n} (p_k - \mathbb{1}_{H_k}) > c_n - \varepsilon^{-1})$$

$$\leq \mathbb{P}(|\sum_{k=1}^{n} (p_k - \mathbb{1}_{H_k})| > c_n - \varepsilon^{-1})$$

(by Chebyshev's inequality) $\quad \leq \dfrac{\mathbb{V}\mathrm{ar}(\sum_{k=1}^{n} \mathbb{1}_{H_k})}{(c_n - \varepsilon^{-1})^2}$

(by pairwise independence) $\quad = \dfrac{\sum_{k=1}^{n} p_k(1-p_k)}{(c_n - \varepsilon^{-1})^2} \leq \dfrac{1}{c_n(1-\varepsilon)^2}.$

Taking the limit on both sides of the inequality as $n \to \infty$, we obtain $\lim \mathbb{P}(A_n) = \mathbb{P}(\sum_{k=1}^{\infty} \mathbb{1}_{H_k} < \varepsilon^{-1}) \leq \lim \frac{1}{c_n(1-\varepsilon)^2} = 0$. Since $\varepsilon > 0$ was arbitrarily small, we deduce that with probability 1 infinitely many of the events (H_n) occur; that is, $\mathbb{P}(\sum_n \mathbb{1}_{H_n} = \infty) = 1$. $\qquad \square$

3.3   Convergence in Probability

Almost sure convergence of (X_n) to X involves the joint distribution of (X_n) and X. A simpler type of convergence that only involves the distribution of $X_n - X$ is the following:

Definition 3.15: Convergence in Probability

The sequence of random variables (X_n) *converges in probability* to a random variable X if, for all $\varepsilon > 0$,

$$\lim \mathbb{P}(|X_n - X| > \varepsilon) = 0.$$

We denote the convergence in probability by $X_n \xrightarrow{\text{P}} X$.

■ **Example 3.16 (Convergence in Probability Versus Almost Sure Convergence)**
Since the event $\{|X_n - X| > \varepsilon\}$ is contained in $\{\sup_{k \geq n} |X_k - X| > \varepsilon\}$, we can conclude that almost sure convergence implies convergence in probability. However, the converse is not true in general. For instance, consider the sequence $X_1, X_2, \ldots$ of independent random variables with marginal distributions

$$\mathbb{P}(X_n = 1) = 1 - \mathbb{P}(X_n = 0) = 1/n, \quad n = 1, 2, \ldots.$$

Clearly, $X_n \xrightarrow{\mathbb{P}} 0$. However, for $\varepsilon < 1$ and any $n = 1, 2, \ldots$ we have,

$$\mathbb{P}\left(\sup_{n \geq m} |X_n| \leq \varepsilon\right) = \mathbb{P}(X_m \leq \varepsilon, X_{m+1} \leq \varepsilon, \ldots)$$

$$= \mathbb{P}(X_m \leq \varepsilon)\,\mathbb{P}(X_{m+1} \leq \varepsilon) \times \cdots \text{ (using independence)}$$

$$= \lim_{n \to \infty} \prod_{k=m}^{n} \mathbb{P}(X_k \leq \varepsilon) = \lim_{n \to \infty} \prod_{k=m}^{n} \left(1 - \frac{1}{k}\right)$$

$$= \lim_{n \to \infty} \frac{m-1}{m} \times \frac{m}{m+1} \times \cdots \times \frac{n-1}{n} = 0.$$

It follows that $\mathbb{P}(\sup_{n \geq m} |X_n - 0| > \varepsilon) = 1$ for any $0 < \varepsilon < 1$ and all $m \geq 1$. In other words, it is *not* true that $X_n \xrightarrow{a.s.} 0$. ∎

Proposition 3.17: Convergence of a Subsequence

If (X_n) converges to X in probability, there is a subsequence (X_{n_k}) that converges to X almost surely.

Proof. Take any $\varepsilon > 0$ and put $n_0 := 0$. For $k \geq 1$, let n_k be the first time after n_{k-1} such that $\mathbb{P}(|X_{n_k} - X| > \varepsilon) \leq 2^{-k}$. There are such (n_k) because (X_n) converges to X in probability by assumption. Setting $Y_m := X_{n_m}$, $m \geq 0$, we have

$$\mathbb{P}(\sup_{m \geq k} |Y_m - X| > \varepsilon) \leq \sum_{m \geq k} \mathbb{P}(|Y_m - X| > \varepsilon) \leq 2^{1-k}.$$

Taking the limit for $k \to \infty$ shows that (Y_k) converges almost surely to X, by Theorem 3.4. □

Sometimes we do not know the limiting random variable X. The following Cauchy criterion for convergence in probability is then useful:

Proposition 3.18: Cauchy Criterion for Convergence in Probability

The sequence of random variables (X_n) converges in probability if and only if for every $\varepsilon > 0$,

(3.19) $$\lim_{m,n \to \infty} \mathbb{P}(|X_m - X_n| > \varepsilon) = 0.$$

Proof. Necessity is shown as follows. Suppose that $X_n \xrightarrow{\mathbb{P}} X$. For any $\varepsilon > 0$, we have $\{|X_m - X| \leq \varepsilon\} \cap \{|X_n - X| \leq \varepsilon\} \subseteq \{|X_m - X_n| \leq 2\varepsilon\}$, so that

$$\mathbb{P}(|X_m - X_n| > \varepsilon) \leq \mathbb{P}(|X_m - X| > \varepsilon/2) + \mathbb{P}(|X_n - X| > \varepsilon/2),$$

where the right-hand side tends to 0 as $m, n \to \infty$.

For sufficiency, assume that (3.19) holds. Put $n_0 := 0$. For $k = 1, 2, \ldots$ define n_k as the smallest $n > n_{k-1}$ such that $\mathbb{P}(|X_m - X_n| > \varepsilon_k) \le \varepsilon_k$ for all $m, n \ge n_k$, with $\varepsilon_k := 2^{-k}$, so that $\sum_k \varepsilon_k < \infty$. Let $Y_k := X_{n_k}, k \in \mathbb{N}$. The above shows that

$$\mathbb{P}(|Y_{k+1} - Y_k| > \varepsilon_k) \le \varepsilon_k \quad \text{for all } k \in \mathbb{N}.$$

By the first part of the Borel–Cantelli Lemma 3.14, we have almost surely that $\sum_k \mathbb{1}_{\{|Y_{k+1} - Y_k| > \varepsilon_k\}} < \infty$. Define $y_k := Y_k(\omega)$ for an ω for which the above holds. Then, (y_k) is a Cauchy sequence by part three of Proposition 3.2, and hence it converges to some limit. Thus, the random sequence (Y_k) converges almost surely to some limit — call it X. Observe that, for any $\varepsilon > 0$, n, and k,

$$\mathbb{P}(|X_n - X| > \varepsilon) \le \mathbb{P}(|X_n - X_{n_k}| > \varepsilon/2) + \mathbb{P}(|Y_k - X| > \varepsilon/2).$$

Now take the limit for $n, k \to \infty$ and the right-hand side converges to 0 because of the assumption (3.19) and the fact that (Y_k) converges almost surely and hence in probability to X.

□

3.4   Convergence in Distribution

Another important type of convergence is useful when we are interested in estimating expectations or multidimensional integrals via Monte Carlo methodology.

Definition 3.20: Convergence in Distribution

The sequence of random variables (X_n) is said to *converge in distribution* to a random variable X with cdf F provided that:

(3.21) $\lim \mathbb{P}(X_n \le x) = F(x)$ for all x such that $\lim\limits_{a \to x} F(a) = F(x)$.

We denote the convergence in distribution by $X_n \overset{d}{\to} X$.

The generalization to random vectors (or more generally, topological spaces) replaces (3.21) with

(3.22) $\lim \mathbb{P}(X_n \in A) = \mathbb{P}(X \in A)$ for all Borel sets A with $\mathbb{P}(X \in \partial A) = 0$,

where ∂A denotes the boundary of the set A.

A useful tool for demonstrating convergence in distribution is the *characteristic function*; see Section 2.6.2. For a d-dimensional random vector X, it is defined as:

$$\psi_X(r) := \mathbb{E} e^{i r^\top X}, \quad r \in \mathbb{R}^d.$$

■ **Example 3.23 (Characteristic Function of a Gaussian Random Vector)** The density of the multivariate standard normal distribution is given by

$$f_Z(z) := \prod_{k=1}^{d} \frac{1}{\sqrt{2\pi}} e^{-\frac{1}{2}z_k^2} = (2\pi)^{-\frac{d}{2}} e^{-\frac{1}{2}z^\top z}, \quad z \in \mathbb{R}^d,$$

and thus the characteristic function of $Z \sim \mathcal{N}(\mathbf{0}, \mathbf{I}_d)$ is

$$\psi_Z(r) := \mathbb{E}e^{i r^\top Z} = (2\pi)^{-d/2} \int_{\mathbb{R}^d} dz\, e^{i r^\top z - \frac{1}{2}\|z\|^2}$$

$$= e^{-\|r\|^2/2} (2\pi)^{-d/2} \int_{\mathbb{R}^d} dz\, e^{-\frac{1}{2}\|z - i r\|^2} = e^{-\|r\|^2/2}, \quad r \in \mathbb{R}^d.$$

Hence, the characteristic function of the random vector $X = \mu + BZ$ with multivariate normal distribution $\mathcal{N}(\mu, \Sigma)$ is given by

$$\psi_X(r) := \mathbb{E}e^{i r^\top X} = \mathbb{E}e^{i r^\top(\mu + BZ)}$$

$$= e^{i r^\top \mu} \mathbb{E}e^{i (B^\top r)^\top Z} = e^{i r^\top \mu} \psi_Z(B^\top r)$$

$$= e^{i r^\top \mu - \|B^\top r\|^2/2} = e^{i r^\top \mu - r^\top \Sigma r/2}.$$

■

The importance of the characteristic function is mainly derived from the following result, for which a proof can be found, for example, in Billingsley (1995, Sections 26 and 29).

Theorem 3.24: Characteristic Function and Convergence in Distribution

Suppose that $\psi_{X_1}(r), \psi_{X_2}(r), \ldots$ are the characteristic functions of the sequence of d-dimensional random vectors $X_1, X_2, \ldots$ and $\psi_X(r)$ is the characteristic function of X. Then, the following three statements are equivalent:

1. $\lim \psi_{X_n}(r) = \psi_X(r)$ for all $r \in \mathbb{R}^d$.

2. $X_n \xrightarrow{d} X$.

3. $\lim \mathbb{E}h(X_n) = \mathbb{E}h(X)$ for all bounded continuous functions $h : \mathbb{R}^d \to \mathbb{R}$.

The theorem can be slightly extended. It can be shown that any function ψ that is the pointwise limit of characteristic functions is itself a characteristic function of a real-valued random variable/vector if and only if ψ is continuous at 0. This is useful in the case where we do not know the distribution of the limiting random variable X from the outset.

■ **Example 3.25 (Random Series Continued)** Continuing Example 3.13, consider
the random series

$$Y_n := \sum_{k=1}^{n} B_k \left(\frac{1}{2}\right)^k, \qquad n = 1, 2, \ldots,$$

where $B_1, B_2, \ldots \overset{\text{iid}}{\sim} \text{Ber}(\frac{1}{2})$. We already established that (Y_n) converges almost
surely to $Y := \sum_{k=1}^{\infty} B_k (\frac{1}{2})^k$. We now show that $Y_n \overset{\text{d}}{\to} Y \sim \mathcal{U}(0, 1)$. First, note that

$$\mathbb{E} \exp(i r Y_n) = \prod_{k=1}^{n} \mathbb{E} \exp(i r B_k / 2^k) = 2^{-n} \prod_{k=1}^{n} (1 + \exp(i r / 2^k)).$$

Second, from the collapsing product, $(1 - \exp(i r / 2^n)) \prod_{k=1}^{n} (1 + \exp(i r / 2^k)) = 1 - \exp(i r)$, we have

$$\mathbb{E} \exp(i r Y_n) = (1 - \exp(i r)) \frac{1/2^n}{1 - \exp(i r / 2^n)}.$$

It follows that $\lim \mathbb{E} \exp(i r Y_n) = (\exp(i r) - 1)/(i r)$, which we recognize as the
characteristic function of the $\mathcal{U}(0, 1)$ distribution. Because almost sure convergence
implies convergence in distribution (which will be shown in Theorem 3.40), we
conclude that $Y \sim \mathcal{U}(0, 1)$.   ■

3.5 Convergence in L^p Norm

Yet another mode of convergence can be found in L^p spaces; see Section 2.5 for the
definition and properties of L^p spaces.

Definition 3.26: Convergence in L^p Norm

The sequence of random variables (X_n) *converges in L^p norm* (for some
$p \in [1, \infty]$) to a random variable X, if $\|X_n\|_p < \infty$ for all n, $\|X\|_p < \infty$, and

$$\lim \|X_n - X\|_p = 0.$$

We denote the convergence in L^p norm by $X_n \overset{L^p}{\to} X$.

For $p = 2$ this type of convergence is sometimes referred to as convergence in
mean squared error. The following example illustrates that convergence in L^p norm
is qualitatively different from convergence in distribution:

■ **Example 3.27 (Comparison of Modes of Convergence)** Define $X_n := 1 - X$,
where $X \sim \mathcal{U}(0, 1)$; thus, clearly, $X_n \overset{\text{d}}{\to} \mathcal{U}(0, 1)$. However, $\mathbb{E}|X_n - X| \to \mathbb{E}|1 -$

$2X| = 1/2$ and so the sequence does not converge in L^1 norm. In addition, $\mathbb{P}(|X_n - X| > \varepsilon) \to 1 - \varepsilon \neq 0$ and so X_n does not converge in probability either. Thus, in general $X_n \overset{d}{\to} X$ implies neither $X_n \overset{\mathbb{P}}{\to} X$ nor $X_n \overset{L^1}{\to} X$. ∎

Again, when the limiting random variable X is not known, the following Cauchy criterion is useful:

Proposition 3.28: Cauchy Criterion for Convergence in L^p Norm

The sequence of random variables (X_n) converges in L^p norm for $p \in [1, \infty]$ if and only if $\|X_n\|_p < \infty$ for all n and

$$\lim_{m,n \to \infty} \|X_m - X_n\|_p = 0.$$

Proof. If $X_n \overset{L^p}{\to} X$, then $\|X - X_n\|_p \to 0$ for $n \to \infty$, and using Minkowski's inequality in Theorem 2.47 shows that (X_n) is a Cauchy sequence in L^p, because

$$\|X_m - X_n\|_p \leq \|X_m - X\|_p + \|X_n - X\|_p \to 0$$

as $m, n \to \infty$. Showing the converse is not as straightforward and we first consider the case with $1 \leq p < \infty$.

Assume that (X_n) is Cauchy convergent and $\|X_n\|_p < \infty$. For $k = 1, 2, \ldots$ define $n_0 := 0$ and n_k as the smallest $n > n_{k-1}$ such that $\|X_m - X_n\|_p \leq 2^{-2k}$ for all $m, n \geq n_k$. Let $Y_k := X_{n_k}$ and $H_k := \{|Y_{k+1} - Y_k| > 2^{-k}\}$ for all k. It follows from Markov's inequality (3.9) that $\mathbb{P}(H_k) \leq 2^{pk}\|Y_{k+1} - Y_k\|_p^p$, and thus

$$\sum_k \mathbb{P}(H_k) \leq \sum_k 2^{-k} < \infty.$$

Hence, by the first part of Borel–Cantelli's Lemma 3.14, we have almost surely that $\sum_k \mathbb{1}_{H_k} < \infty$. If $y_k := Y_k(\omega)$ for an ω for which the above holds, then (y_k) is a Cauchy sequence by part three of Proposition 3.2 (with $\varepsilon_n := 2^{-n}$). Hence, the random sequence (Y_k) thus converges almost surely to some limit, say X. By Fatou's Lemma 2.35

$$\mathbb{E}|X_n - X|^p = \mathbb{E}\liminf_k |X_n - X_{n_k}|^p \leq \liminf_k \mathbb{E}|X_n - X_{n_k}|^p \to 0,$$

where we take $n \to \infty$ and use the Cauchy convergence assumption. In other words, for a given $\varepsilon > 0$, there exists a large enough N_ε such that $\|X - X_n\|_p < \varepsilon$ for all $n \geq N_\varepsilon$. Since $\|X_n\|_p < \infty$ and $|\|X\|_p - \|X_n\|_p| \leq \|X - X_n\|_p < \varepsilon$ for all $n \geq N_\varepsilon$, this shows that $\|X\|_p < \infty$, and completes the proof for $p \in [1, \infty)$.

For the case $p = \infty$, we have $\|X_n\|_\infty = \inf\{x : \mathbb{P}(|X_n| \le x) = 1\}$. Define $A_{m,n} :=$ $\{|X_m - X_n| > \|X_m - X_n\|_\infty\}$ and $A := \cup_{m,n} A_{m,n}$. Then, $\mathbb{P}(A) \le \sum_{m,n} \mathbb{P}(A_{m,n}) = 0$. Thus, for all $\omega \in A^c$ and any $\varepsilon > 0$ we can find a large enough N_ε such that

$$(3.29) \qquad |X_m(\omega) - X_n(\omega)| \le \|X_m - X_n\|_\infty < \varepsilon, \quad m, n \ge N_\varepsilon,$$

which implies that (x_n), with $x_n := X_n(\omega)$, is a Cauchy sequence in $\mathbb{R}$. The completeness of $\mathbb{R}$ implies that the limit $X(\omega) := \lim_n X_n(\omega)$ exists for each $\omega \in A^c$, and we may set $X(\omega) := 0$ for $\omega \in A$. Taking the limit as $m \to \infty$ in (3.29) yields $\sup_{\omega \in A^c} |X_n(\omega) - X(\omega)| < \varepsilon$ and therefore $\|X_n - X\|_\infty < \varepsilon$ for all $n \ge N_\varepsilon$. Finally, $\|X\|_\infty \le \|X_n - X\|_\infty + \|X_n\|_\infty < \varepsilon + \|X_n\|_\infty$ for all $n \ge N_\varepsilon$ showing that $\|X\|_\infty < \infty$. $\square$

3.5.1 Uniform Integrability

Recall that a numerical random variable X is said to be *integrable* if $\mathbb{E}X$ exists and is a real number; equivalently, if $\mathbb{E}|X| < \infty$. Uniform integrability is a condition on the joint integrability of a collection of random variables. We will see that this concept ties together the notions of L^1 convergence and convergence in probability. This will be particularly relevant for the convergence of martingales; see Chapter 5.

We now give an equivalent definition of integrability, whose proof is left for Exercise 14.

Proposition 3.30: Integrable Random Variable

A real-valued random variable X is integrable if and only if

$$(3.31) \qquad \lim_{b \to \infty} \mathbb{E}|X|\, \mathbb{1}_{\{|X|>b\}} = 0.$$

Uniform integrability extends property (3.31) to an arbitrary *collection* of random variables.

Definition 3.32: Uniform Integrability

A collection $\mathcal{K}$ of random variables is said to be *uniformly integrable* (UI) if

$$\lim_{b \to \infty} \sup_{X \in \mathcal{K}} \mathbb{E}|X|\, \mathbb{1}_{\{|X|>b\}} = 0.$$

■ **Example 3.33 (Sum of two UI sequences)** Let (X_n) and (Y_n) be two UI sequences of random variables. Then, the collection $\{Z_{m,n}\}$, where $Z_{m,n} := X_m + Y_n$, is also UI. To see this, note that $\mathbb{1}_{\{x>b\}}$ is an increasing function for $x \ge 0$, so that

$(x - y)(\mathbb{1}_{\{x>b\}} - \mathbb{1}_{\{y>b\}}) \geq 0$ for $x, y \geq 0$. In other words, after rearrangement:

$$(x + y)(\mathbb{1}_{\{x>b\}} + \mathbb{1}_{\{y>b\}}) \leq 2x\mathbb{1}_{\{x>b\}} + 2y\mathbb{1}_{\{y>b\}}.$$

Using $\mathbb{1}_{\{|x+y|>2b\}} \leq \mathbb{1}_{\{|x|>b\}} + \mathbb{1}_{\{|y|>b\}}$, we obtain the following inequality for any x and y:

$$|x + y|\mathbb{1}_{\{|x+y|>2b\}} \leq (|x| + |y|)(\mathbb{1}_{\{|x|>b\}} + \mathbb{1}_{\{|y|>b\}}) \leq 2|x|\mathbb{1}_{\{|x|>b\}} + 2|y|\mathbb{1}_{\{|y|>b\}}.$$

Therefore, as $b \uparrow \infty$ we have:

$$\sup_{m,n} \mathbb{E}|Z_{m,n}|\mathbb{1}_{\{|Z_{m,n}|>2b\}} \leq 2\sup_m \mathbb{E}|X_m|\mathbb{1}_{\{|X_m|>b\}} + 2\sup_n \mathbb{E}|Y_n|\mathbb{1}_{\{|Y_n|>b\}} \to 0 + 0,$$

which shows that $\{Z_{m,n}\}$ is also UI. ∎

Proposition 3.34: Conditions for Uniform Integrability

Let $\mathcal{K}$ be a collection of random variables.

1. If $\mathcal{K}$ is a finite collection of integrable random variables, then it is UI.

2. If $|X| \leq Y$ for all $X \in \mathcal{K}$ and some integrable Y, then $\mathcal{K}$ is UI.

3. $\mathcal{K}$ is UI if and only if there is an increasing convex function f such that

$$\lim_{x\to\infty} \frac{f(x)}{x} = \infty \quad \text{and} \quad \sup_{X\in\mathcal{K}} \mathbb{E}f(|X|) < \infty.$$

4. $\mathcal{K}$ is UI if $\sup_{X\in\mathcal{K}} \mathbb{E}|X|^{1+\varepsilon} < \infty$ for some $\varepsilon > 0$.

5. $\mathcal{K}$ is UI if and only if

   (a) $\sup_{X\in\mathcal{K}} \mathbb{E}|X| < \infty$, and

   (b) for every $\varepsilon > 0$ there is a $\delta > 0$ such that for every event H:

   (3.35) $\mathbb{P}(H) \leq \delta \Rightarrow \sup_{X\in\mathcal{K}} \mathbb{E}|X|\mathbb{1}_H \leq \varepsilon.$

Proof.

1. Let $\mathcal{K} := \{X_1, \ldots, X_n\}$. Then,

$$\lim_{b\to\infty} \sup_n \mathbb{E}|X_n|\,\mathbb{1}_{\{|X_n|>b\}} = \sup_n \lim_{b\to\infty} \mathbb{E}|X_n|\,\mathbb{1}_{\{|X_n|>b\}} = \sup_n 0 = 0,$$

since (3.31) holds for each integrable random variable.

2. We have

$$\lim_{b\to\infty}\sup_{X\in\mathcal{K}}\mathbb{E}|X|\,\mathbb{1}_{\{|X|>b\}}\le\lim_{b\to\infty}\mathbb{E}Y\,\mathbb{1}_{\{Y>b\}}=0,$$

since Y is integrable.

3. Without loss of generality, we may assume that all $X\in\mathcal{K}$ are positive and that $f\ge 1$. Define $g(x):=x/f(x)$. Then,

$$X\mathbb{1}_{\{X>b\}}=f(X)g(X)\mathbb{1}_{\{X>b\}}\le f(X)\sup_{x>b}g(x).$$

Hence,

$$\sup_{X\in\mathcal{K}}\mathbb{E}X\mathbb{1}_{\{X>b\}}\le\sup_{X\in\mathcal{K}}\mathbb{E}f(X)\sup_{x>b}g(x).$$

The proof of sufficiency is completed by observing that the right-hand side of the above inequality goes to 0 as $b\to\infty$. To show necessity, suppose $\mathcal{K}$ is UI. For every $X\in\mathcal{K}$ we have

$$\mathbb{E}X\mathbb{1}_{\{X>b\}}=\int_0^\infty dy\,\mathbb{P}(X\mathbb{1}_{\{X>b\}}>y)=\int_0^\infty dy\,\mathbb{P}(X>b\vee y)$$

$$\ge\int_b^\infty dy\,\mathbb{P}(X>y).$$

It follows from the UI assumption of $\mathcal{K}$ that

$$\lim_{b\to\infty}\underbrace{\sup_{\mathcal{K}}\int_b^\infty dy\,\mathbb{P}(X>y)}_{=:~h(b)}=0.$$

Thus, there exists a sequence $0=b_0<b_1<\cdots$ increasing to ∞ such that $h(b_n)\le h(0)2^{-n}$ for all $n\in\mathbb{N}$, where $h(0)<\sup_{X\in\mathcal{K}}\mathbb{E}X<\infty$. Now define g as the step function which starts at 1 and increases by 1 at each point b_n, and let $f(x):=\int_0^x dy\,g(y)$ be the area underneath the graph of f in $[0,x]$. Then, f is increasing and convex and $\lim_x f(x)/x=\infty$. It remains to show that $\sup_{\mathcal{K}}\mathbb{E}f(X)<\infty$. This follows from

$$\mathbb{E}f(X)=\sum_{n=0}^\infty\mathbb{E}\int_{b_n}^\infty dy\,\mathbb{1}_{\{X>y\}}\le\sum_{n=0}^\infty h(b_n)\le 2h(0)<\infty$$

for all $X\in\mathcal{K}$.

4. This is a consequence of Point 3, by taking $f(x)=x^{1+\varepsilon}$.

5. To simplify the notation, we can assume that $X \geq 0$. Take any $\varepsilon > 0$ and event H. Let $b \geq 0$. We have

$$(3.36) \quad \sup_{X \in \mathcal{K}} \mathbb{E}X\mathbb{1}_H = \sup_{X \in \mathcal{K}} \mathbb{E}X\mathbb{1}_H(\mathbb{1}_{\{X \leq b\}} + \mathbb{1}_{\{X > b\}}) \leq b\,\mathbb{P}(H) + \overbrace{\sup_{X \in \mathcal{K}} \mathbb{E}X\mathbb{1}_{\{X > b\}}}^{s(b)}.$$

In particular, $\mathbb{E}X \leq b + s(b)$ for all $X \in \mathcal{K}$. If $\mathcal{K}$ is UI, the last term is finite for b large enough. This shows that $s(0) = \sup_{X \in \mathcal{K}} \mathbb{E}X < \infty$. Moreover, if $\mathcal{K}$ is UI, there is a b such that $s(b) < \varepsilon/2$. So if we take $\delta < \varepsilon/(2b)$, then $\sup_{X \in \mathcal{K}} \mathbb{E}X\mathbb{1}_H < \varepsilon$. This shows sufficiency.

To show necessity, suppose that (3.35) holds and $s(0) < \infty$. We must show that for every $\varepsilon > 0$ there exists a b such that $s(b) \leq \varepsilon$. To show this, in (3.35) take $b \geq s(0)/\delta < \infty$ and $H = \{X' > b\}$ for an arbitrary $X' \in \mathcal{K}$. Then, by Markov's inequality (3.9),

$$\mathbb{P}(X' > b) \leq \frac{s(0)}{b} \leq \delta,$$

which in combination with (3.35) implies that $\mathbb{E}X'\mathbb{1}_H = \mathbb{E}X'\mathbb{1}_{\{X' > b\}} \leq \varepsilon$ for $X' \in \mathcal{K}$. Since X' is an arbitrary choice in $\mathcal{K}$, the inequality $\mathbb{E}X\mathbb{1}_{\{X > b\}} \leq \varepsilon$ is true for all $X \in \mathcal{K}$. In other words, $s(b) \leq \varepsilon$, as had to be shown.

$\square$

■ **Example 3.37 (Failure of UI)** When uniform integrability fails to hold, then $X_n \xrightarrow{\mathbb{P}} X$ does not necessarily imply that $\mathbb{E}X_n \to \mathbb{E}X$. For example, consider the sequence of random variables (X_n) with $\mathbb{P}(X_n = n) = 1/n = 1 - \mathbb{P}(X_n = 0)$, $n = 1, 2, \ldots$. Then, $X_n \xrightarrow{\mathbb{P}} 0$, because

$$\mathbb{P}(|X_n - 0| > \varepsilon) = \mathbb{P}(X_n = n) = 1/n \to 0,$$

but $\mathbb{E}X_n = 1 \neq 0$ for all n. This is not surprising, since

$$\mathbb{E}|X_n|\mathbb{1}_{\{|X_n| \geq b\}} = \begin{cases} 1, & b \leq n, \\ 0, & b > n; \end{cases}$$

that is, $\lim_{b \to \infty} \sup_n \mathbb{E}|X_n|\mathbb{1}_{\{|X_n| \geq b\}} = 1$, implying that (X_n) is not UI. ■

Theorem 3.38: L^1 Convergence and Uniform Integrability

A sequence (X_n) of real-valued integrable random variables converges in L^1 if and only if it converges in probability and is uniformly integrable.

Proof. Suppose that (X_n) is UI and converges in probability. Let $\varepsilon > 0$ be an arbitrarily small number. Since (X_n) is UI, then so is the collection $\{Z_{m,n}\}$ with $Z_{m,n} := X_m - X_n$. In other words, for the given ε we can find a large enough b_ε such that $\mathbb{E}|Z_{m,n}| \mathbb{1}_{\{|Z_{m,n}|>b_\varepsilon\}} \leq \varepsilon/3$ for all m, n. Note that the uniform integrability means that b_ε does not depend on m and n. Next, for the given $\varepsilon > 0$ and $\delta := \varepsilon/(3b_\varepsilon)$, Proposition 3.18 tells us that we can find a large enough N_ε such that $\mathbb{P}(|X_m - X_n| > \varepsilon/3) \leq \delta$ for all $m, n \geq N_\varepsilon$. Combining these results for $m, n \geq N_\varepsilon$, we obtain

$$
\begin{aligned}
\mathbb{E}|X_m - X_n| &= \mathbb{E}|Z_{m,n}| \left(\mathbb{1}_{\{|Z_{m,n}|\leq\varepsilon/3\}} + \mathbb{1}_{\{\varepsilon/3<|Z_{m,n}|\leq b_\varepsilon\}} + \mathbb{1}_{\{|Z_{m,n}|>b_\varepsilon\}} \right) \\
&\leq \varepsilon/3 + b_\varepsilon \, \mathbb{P}(|Z_{m,n}| > \varepsilon/3) + \mathbb{E}|Z_{m,n}| \mathbb{1}_{\{|Z_{m,n}|>b_\varepsilon\}} \\
&\leq \varepsilon/3 + b_\varepsilon \, \varepsilon/(3b_\varepsilon) + \varepsilon/3 = \varepsilon,
\end{aligned}
$$

which implies that (X_n) is a Cauchy sequence in L^1. By Proposition 3.28 there exists an integrable X such that $\mathbb{E}|X_n - X| \to 0$. This completes the first part of the proof.

Now, assume that $\mathbb{E}|X_n - X| \to 0$ with X and (X_n) being integrable. From Markov's inequality (3.9) we have that $\mathbb{P}(|X_n - X| > \varepsilon) \leq \varepsilon^{-1}\mathbb{E}|X_n - X| \to 0$, proving convergence in probability. Next, we show that $(X_n - X)$ is UI. For a given $\varepsilon > 0$, we can choose a large enough N_ε so that $\mathbb{E}|X_n - X| < \varepsilon$ for all $n \geq N_\varepsilon$. Therefore, for all b we have

$$
\sup_{n \geq N_\varepsilon} \mathbb{E}|X_n - X| \mathbb{1}_{\{|X_n-X|>b\}} \leq \sup_{n \geq N_\varepsilon} \mathbb{E}|X_n - X| < \varepsilon.
$$

However, since all X_n and X are integrable, the finite sequence $X_1 - X, X_2 - X, \ldots, X_{N_\varepsilon} - X$ is uniformly integrable. Thus, there exists a large enough b_ε such that $\sup_{n \leq N_\varepsilon} \mathbb{E}|X_n - X| \mathbb{1}_{\{|X_n-X|>b\}} < \varepsilon$ for all $b \geq b_\varepsilon$. Combining all the results thus far, we obtain $\sup_n \mathbb{E}|X_n - X| \mathbb{1}_{\{|X_n-X|>b\}} < \varepsilon$ for all $b \geq b_\varepsilon$.

Since $(X_n - X)$ is UI and X is integrable (and hence UI), then the sequence (X_n) is also UI, because it is the sum of two UI sequences. $\square$

■ **Example 3.39 (Squeezing For Random Variables)** Suppose that $X_n \overset{\mathbb{P}}{\to} X$ and that (X_n) is "squeezed" between (W_n) and (Y_n), in the sense that for all n:

$$
W_n \leq X_n \leq Y_n.
$$

If $Y_n \overset{L^1}{\to} Y$ and $W_n \overset{L^1}{\to} W$, then $X_n \overset{L^1}{\to} X$.

To see this, define $T_n := X_n - W_n$ and $V_n := Y_n - W_n \geq T_n \geq 0$. From the L^1 convergence of (Y_n) and (W_n), we have $V_n \overset{L^1}{\to} V := Y - W$. By Theorem 3.38 we then know that (V_n) is UI, and since $|T_n| \leq V_n$ for all n, (T_n) is also UI. Since $X_n \overset{\mathbb{P}}{\to} X$ and $W_n \overset{\mathbb{P}}{\to} W$, we have that $T_n \overset{\mathbb{P}}{\to} T := X - W$. From (T_n) being UI, another application of Theorem 3.38 allows us to conclude that $T_n \overset{L^1}{\to} T$. In other words, $X_n = T_n + W_n \overset{L^1}{\to} T + W = X$, because both $T_n \overset{L^1}{\to} T$ and $W_n \overset{L^1}{\to} W$.   ■

3.6   Relations Between Modes of Convergence

The next theorem shows how the different types of convergence are related to each other. For example, in the diagram below, the notation $\overset{q \geq p}{\Rightarrow}$ means that L^q-norm convergence implies L^p-norm convergence under the assumption that $q \geq p \geq 1$.

Theorem 3.40: Modes of Convergence

The most general relationships among the various modes of convergence for numerical random variables are shown below.

$$\boxed{X_n \overset{\text{cpl.}}{\to} X} \Rightarrow \boxed{X_n \overset{\text{a.s.}}{\to} X}$$
$$\Downarrow$$
$$\boxed{X_n \overset{\text{P}}{\to} X} \Rightarrow \boxed{X_n \overset{\text{d}}{\to} X}$$
$$\Uparrow$$
$$\boxed{X_n \overset{L^q}{\to} X} \overset{q \geq p}{\Rightarrow} \boxed{X_n \overset{L^p}{\to} X}$$

Proof.

1. We show that $X_n \overset{\text{P}}{\to} X \Rightarrow X_n \overset{\text{d}}{\to} X$ by considering the cdfs F_{X_n} and F_X of X_n and X, respectively. We have:

$$F_{X_n}(x) = \mathbb{P}(X_n \leq x) = \mathbb{P}(X_n \leq x, |X_n - X| > \varepsilon) + \mathbb{P}(X_n \leq x, |X_n - X| \leq \varepsilon)$$
$$\leq \mathbb{P}(|X_n - X| > \varepsilon) + \mathbb{P}(X_n \leq x, X \leq X_n + \varepsilon)$$
$$\leq \mathbb{P}(|X_n - X| > \varepsilon) + \mathbb{P}(X \leq x + \varepsilon).$$

Now, in the arguments above we can switch the roles of X_n and X (there is a symmetry) to deduce that: $F_X(x) \leq \mathbb{P}(|X - X_n| > \varepsilon) + \mathbb{P}(X_n \leq x + \varepsilon)$. Therefore, making the switch $x \to x - \varepsilon$ gives $F_X(x - \varepsilon) \leq \mathbb{P}(|X - X_n| > \varepsilon) + F_{X_n}(x)$. Putting it all together gives:

$$F_X(x - \varepsilon) - \mathbb{P}(|X - X_n| > \varepsilon) \leq F_{X_n}(x) \leq \mathbb{P}(|X_n - X| > \varepsilon) + F_X(x + \varepsilon).$$

Taking $n \to \infty$ on both sides yields for any $\varepsilon > 0$:

$$F_X(x - \varepsilon) \leq \lim_{n \to \infty} F_{X_n}(x) \leq F_X(x + \varepsilon).$$

Since F_X is continuous at x by assumption we can take $\varepsilon \downarrow 0$ to conclude that $\lim_{n \to \infty} F_{X_n}(x) = F_X(x)$.

2. We show that $X_n \overset{L^q}{\to} X \Rightarrow X_n \overset{L^p}{\to} X$ for $q \geq p \geq 1$. This is immediate from the monotonicity property of the L^p norm in Theorem 2.47, which shows that $\|X_n - X\|_p \leq \|X_n - X\|_q \to 0$, proving the statement of the theorem.

3. To show $X_n \xrightarrow{L^p} X \Rightarrow X_n \xrightarrow{P} X$, we need only show that $X_n \xrightarrow{L^1} X \Rightarrow X_n \xrightarrow{P} X$, by the monotonicity property of the L^p norm in Theorem 2.47. The result now follows from Theorem 3.38. More directly, we can use Markov's inequality (3.9) and $X_n \xrightarrow{L^1} X$, to conclude that for every $\varepsilon > 0$,

$$\mathbb{P}(|X_n - X| > \varepsilon) \leq \frac{\mathbb{E}|X_n - X|}{\varepsilon} \to 0 \quad \text{as } n \to \infty.$$

4. Finally, $X_n \xrightarrow{\text{cpl.}} X \Rightarrow X_n \xrightarrow{\text{a.s.}} X \Rightarrow X_n \xrightarrow{P} X$ is proved in Examples 3.6 and 3.16.

□

Under certain conditions, the converse statements can be established as follows:

Theorem 3.41: Modes of Convergence under Additional Assumptions

Converse relationships among the various modes of convergence are:

$$\boxed{X_n \xrightarrow{\text{cpl.}} X} \overset{\text{pairwise ind.}}{\Longleftarrow} \boxed{X_n \xrightarrow{\text{a.s.}} X}$$

$$\Uparrow \text{subseq.}$$

$$\boxed{X_n \xrightarrow{P} X} \overset{X \text{ const.}}{\Longleftarrow} \boxed{X_n \xrightarrow{d} X}$$

$$\Downarrow (X_n) \text{ is UI}$$

$$\boxed{X_n \xrightarrow{L^q} X} \overset{(|X_n|^q) \text{ is UI}}{\Longleftarrow} \boxed{X_n \xrightarrow{L^1} X}$$

Proof.

1. We prove that $X_n \xrightarrow{d} c \Rightarrow X_n \xrightarrow{P} c$ for some constant c. To this end, let

$$F(x) := \begin{cases} 1, & x \geq c, \\ 0, & x < c \end{cases}$$

be the cdf of a random variable X such that $\mathbb{P}(X = c) = 1$. Then, $X_n \xrightarrow{d} c$ stands for $\mathbb{P}(X_n \leq x) =: F_n(x) \to F(x)$ for all $x \neq c$. In other words,

$$\lim F_n(x) = \begin{cases} 1, & x > c, \\ 0, & x < c, \end{cases}$$

and we can write:

$$\mathbb{P}(|X_n - c| > \varepsilon) \leq 1 - \mathbb{P}(X_n \leq c + \varepsilon) + \mathbb{P}(X_n < c - \varepsilon) \to 1 - 1 + 0 = 0, \quad n \to \infty,$$

which shows that $X_n \xrightarrow{P} c$ by definition.

2. Theorem 3.38 shows that if $X_n \xrightarrow{\text{P}} X$ and (X_n) is UI, then $X_n \xrightarrow{L^1} X$.

3. We show that $X_n \xrightarrow{L^1} X$ and $(|X_n|^q)$ being UI for $q \geq 1$ implies that $X_n \xrightarrow{L^q} X$. Since $X_n \xrightarrow{L^1} X \Rightarrow X_n \xrightarrow{\text{P}} X$, it follows from Proposition 3.18 that $|X_m - X_n| \xrightarrow{\text{P}} 0$ as $m, n \to \infty$, and therefore $Y_{m,n} := |X_m - X_n|^q \xrightarrow{\text{P}} 0$. Since $u \mapsto |u|^q$ is a convex function, it holds that $(\frac{1}{2}|u| + \frac{1}{2}|v|)^q \leq \frac{1}{2}|u|^q + \frac{1}{2}|v|^q$. Hence,

$$Y_{m,n} \leq (|X_m| + |X_n|)^q \leq 2^{q-1}(|X_m|^q + |X_n|^q).$$

It follows that the collection $\{Y_{m,n}\}$ is UI, because $(|X_m|^q + |X_n|^q)$ is the sum of two UI sequences. Theorem 3.38 then implies that $Y_{m,n} = |X_m - X_n|^q \xrightarrow{L^1} 0$ or, equivalently, that (X_n) is a Cauchy sequence in L^q. Therefore, $X_n \xrightarrow{L^q} X$ by Proposition 3.28.

4. Proposition 3.17 established that if $X_n \xrightarrow{\text{P}} X$, then a subsequence $X_{n_k} \xrightarrow{\text{a.s.}} X$.

5. Finally, we show that if $X_n \xrightarrow{\text{a.s.}} X$ and (X_n) are pairwise independent, then $X_n \xrightarrow{\text{cpl.}} X$. Since $X_n \xrightarrow{\text{a.s.}} X \Rightarrow X_n \xrightarrow{\text{P}} X$, Exercise 30 shows that under these assumptions $\mathbb{P}(X = c) = 1$ for some constant c. Thus, we can define the events

$$H_k := \{|X_k - c| > \varepsilon\}, \quad k = 1, 2, \ldots,$$

and we then need to prove that the pairwise independence of $H_1, H_2, \ldots$ and $\mathbb{P}(\sum_n \mathbb{1}_{H_n} < \infty) = 1$ implies that $\sum_k \mathbb{P}(H_k) =: s < \infty$. To prove this, we argue by contradiction. Suppose that $s = \infty$ and $H_1, H_2, \ldots$ are pairwise independent, then the second part of the Borel–Cantelli Lemma 3.14 tells us that $\mathbb{P}(\sum_n \mathbb{1}_{H_n} < \infty) = 0$, which is in contradiction with $\mathbb{P}(\sum_n \mathbb{1}_{H_n} < \infty) = 1$. Therefore, $s < \infty$, which by definition means that $X_n \xrightarrow{\text{cpl.}} X$.

$\square$

3.7   Law of Large Numbers and Central Limit Theorem

Two main results in probability are the *Law of Large Numbers* and the *Central Limit Theorem*. Both are limit theorems involving sums of independent random variables. In particular, consider a sequence $X_1, X_2, \ldots$ of iid random variables with finite expectation μ and finite variance σ^2. For each n define

$$\overline{X}_n := (X_1 + \cdots + X_n)/n.$$

What can we say about the (random) sequence of averages $\overline{X}_1, \overline{X}_2, \overline{X}_3, \ldots$? By the properties of the expectation and variance we have $\mathbb{E}\,\overline{X}_n = \mu$ and $\mathbb{V}\text{ar}\,\overline{X}_n = \sigma^2/n$. Hence, as n increases, the variance of the (random) average $\overline{X}_n$ goes to 0. This means that, by Definition 3.5, the average $\overline{X}_n$ converges to μ in L^2 norm as $n \to \infty$; that is, $\overline{X}_n \xrightarrow{L^2} \mu$.

In fact, to obtain *convergence in probability* the variance need not be finite — it is sufficient to assume that $\mu := \mathbb{E}X < \infty$.

Theorem 3.42: Weak Law of Large Numbers

If $X_1, X_2, \ldots$ are iid with finite expectation μ, then for all $\varepsilon > 0$

$$\lim_{n \to \infty} \mathbb{P}(|\overline{X}_n - \mu| > \varepsilon) = 0.$$

In other words, $\overline{X}_n \xrightarrow{P} \mu$.

The theorem has a natural generalization for random vectors. Namely, if $\boldsymbol{\mu} := \mathbb{E}\boldsymbol{X} < \infty$, then $\mathbb{P}(\|\overline{\boldsymbol{X}}_n - \boldsymbol{\mu}\| > \varepsilon) \to 0$, where $\| \cdot \|$ is the Euclidean norm. We give a proof in the scalar case and leave the multivariate case for Exercise 37.

Proof. Let $Z_k := X_k - \mu$ for all k, so that $\mathbb{E}Z_k = 0$. We thus need to show that $\overline{Z}_n \xrightarrow{P} 0$. We use the properties of the characteristic function of $Z \sim Z_k$ denoted by ψ_Z. Due to the iid assumption, we have

$$(3.43) \qquad \psi_{\overline{Z}_n}(r) := \mathbb{E}e^{ir\overline{Z}_n} = \mathbb{E}\prod_{k=1}^{n} e^{irZ_k/n} = \prod_{i=1}^{n} \psi_Z(r/n) = [\psi_Z(r/n)]^n.$$

An application of Taylor's theorem (see Proposition 2.61) in the neighborhood of $r = 0$ yields

$$\psi_Z(r/n) = \psi_Z(0) + \frac{r}{n}\psi_Z'(0) + o(1/n).$$

Since $\psi_Z(0) = 1$ and $\psi_Z'(0) = i\,\mathbb{E}Z = 0$, we have:

$$\psi_{\overline{Z}_n}(r) = [\psi_Z(r/n)]^n = [1 + o(1/n)]^n \to 1, \quad n \to \infty.$$

The characteristic function of a random variable that always equals 0 is 1. Therefore, Theorem 3.24 implies that $\overline{Z}_n \xrightarrow{d} 0$. However, according to Theorem 3.41, convergence in distribution to a constant implies convergence in probability. Hence, $\overline{Z}_n \xrightarrow{P} 0$. $\qquad \square$

There is also a stronger version of the Law of Large Numbers, given in Theorem 3.44. It is stated and proved here under the condition that $\mathbb{E}X^2 < \infty$. We will give a martingale-based proof in Section 5.5.2 that does away with this condition. The mildest condition known today is that the variables $X_1, X_2, \ldots$ are pairwise independent and identically distributed with $\mathbb{E}|X_k| < \infty$. The corresponding proof, however, is significantly more difficult; see Exercise 35.

Theorem 3.44: Strong Law of Large Numbers

If $X_1, X_2, \ldots$ are iid with expectation μ and $\mathbb{E}X^2 < \infty$, then for all $\varepsilon > 0$

$$\lim_{m \to \infty} \mathbb{P}\left(\sup_{n \geq m} |\overline{X}_n - \mu| > \varepsilon \right) = 0.$$

In other words, $\overline{X}_n \overset{\text{a.s.}}{\to} \mu$.

Proof. Because X can be written as the difference of its positive and negative part, $X = X^+ - X^-$, we may assume without loss of generality that the $\{X_n\}$ are positive. Next, from the sequence $\overline{X}_1, \overline{X}_2, \overline{X}_3, \ldots$ we can pick the subsequence $\overline{X}_1, \overline{X}_4, \overline{X}_9, \overline{X}_{16}, \ldots =: (\overline{X}_{n^2})$. From Chebyshev's inequality (3.10) and the iid condition, we have

$$\sum_{n=1}^{\infty} \mathbb{P}\left(|\overline{X}_{n^2} - \mu| > \varepsilon \right) \leq \frac{\mathbb{V}\text{ar} X}{\varepsilon^2} \sum_{n=1}^{\infty} \frac{1}{n^2} < \infty.$$

Therefore, $\overline{X}_{n^2} \overset{\text{cpl.}}{\to} \mu$ and from Theorem 3.40 we conclude that $\overline{X}_{n^2} \overset{\text{a.s.}}{\to} \mu$.

For any arbitrary n, we can find a k, say $k = \lfloor \sqrt{n} \rfloor$, such that $k^2 \leq n \leq (k+1)^2$. For such a k and positive $X_1, X_2, \ldots$, it holds that

$$\frac{k^2}{(k+1)^2} \overline{X}_{k^2} \leq \overline{X}_n \leq \overline{X}_{(k+1)^2} \frac{(k+1)^2}{k^2}.$$

Since $\overline{X}_{k^2}$ and $\overline{X}_{(k+1)^2}$ converge almost surely to μ as k (and hence n) goes to infinity, we conclude that $\overline{X}_n \overset{\text{a.s.}}{\to} \mu$. $\square$

The *Central Limit Theorem* describes the approximate distribution of $\overline{X}_n$. Loosely, it states that *the average of a large number of iid random variables approximately has a normal distribution*. Specifically, the random variable $\overline{X}_n$ has a distribution that is approximately normal, with expectation μ and variance σ^2/n.

Theorem 3.45: Central Limit Theorem

If $X_1, X_2, \ldots$ are iid with finite expectation μ and finite variance σ^2, then for all $x \in \mathbb{R}$,

$$\lim_{n \to \infty} \mathbb{P}\left(\frac{\overline{X}_n - \mu}{\sigma/\sqrt{n}} \leq x \right) = \Phi(x),$$

where Φ is the cdf of the standard normal distribution. In other words, $\sqrt{n}(\overline{X}_n - \mu)/\sigma \overset{\text{d}}{\to} \mathcal{N}(0,1)$.

Proof. Let $Z_k := (X_k - \mu)/\sigma$ for all k, so that $\mathbb{E}Z_k = 0$ and $\mathbb{E}Z_k^2 = 1$. We thus need to show that $\sqrt{n}\,\overline{Z}_n \overset{d}{\to} \mathcal{N}(0, 1)$. We again use the properties of the characteristic function. Let ψ_Z be the characteristic function of a generic $Z \sim Z_k$. A similar calculation to the one in (3.43) yields:

$$\psi_{\sqrt{n}\,\overline{Z}_n}(r) = \mathbb{E}e^{ir\sqrt{n}\,\overline{Z}_n} = [\psi_Z(r/\sqrt{n})]^n.$$

An application of Taylor's theorem (see Proposition 2.61) in the neighborhood of $r = 0$ yields

$$\psi_Z(r/\sqrt{n}) = 1 + \frac{r}{\sqrt{n}}\,\psi_Z'(0) + \frac{r^2}{2n}\,\psi_Z''(0) + o(r^2/n).$$

Since $\psi_Z'(0) = \mathbb{E}\frac{d}{dr}e^{irZ}\big|_{r=0} = i\,\mathbb{E}Z = 0$ and $\psi_Z''(0) = i^2\,\mathbb{E}Z^2 = -1$, we have:

$$\psi_{\sqrt{n}\,\overline{Z}_n}(r) = [\psi_Z(r/\sqrt{n})]^n = \left[1 - \frac{r^2}{2n} + o(1/n)\right]^n \to e^{-r^2/2}, \quad n \to \infty.$$

From Example 3.23, we recognize $e^{-r^2/2}$ as the characteristic function of the standard normal distribution. Thus, from Theorem 3.24 we conclude that $\sqrt{n}\,\overline{Z}_n \overset{d}{\to} \mathcal{N}(0, 1)$.

$\square$

■ **Example 3.46 (CLT and Sums of Random Variables)** Figure 3.47 shows the Central Limit Theorem in action. The left part shows the pdfs of $S_n = X_1 + \cdots + X_n$ for $n = 1, \ldots, 4$, where $X_1, \ldots, X_4 \sim_{iid} \mathcal{U}(0, 1)$. The right part shows the same for the case where $X_1, \ldots, X_4 \sim_{iid} \mathsf{Exp}(1)$, so that $S_n \sim \mathsf{Gamma}(n, 1)$. In both cases, we clearly see convergence to a bell-shaped curve, characteristic of the normal distribution.

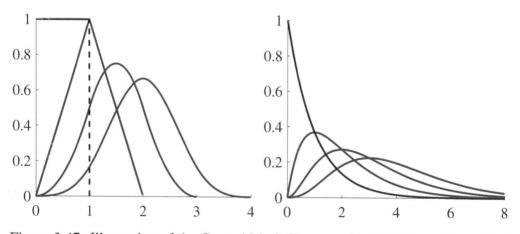

Figure 3.47: Illustration of the Central Limit Theorem for (left) the uniform distribution and (right) the exponential distribution.

Exercises

1. Let (x_k) and (y_k) be two sequences of real numbers such that $\lim_{n\to\infty} x_n/y_n = l$. Prove the *Stolz–Cesàro limit*:

$$\text{(3.48)} \qquad \lim_{n\to\infty} \frac{x_1 + \cdots + x_n}{y_1 + \cdots + y_n} = l$$

under either one of the following conditions:

(a) $\lim_{n\to\infty} \sum_{k=1}^n y_k = \infty$ and $y_k > 0$ for all $k \geq n_1$ and some integer n_1;

(b) $\lim_{n\to\infty} \sum_{k=1}^n x_k = \lim_{n\to\infty} \sum_{k=1}^n y_k = 0$ and $y_k < 0$ for all $k \geq n_1$ and some integer n_1.

2. Suppose that $x_n \to x$ as $n \to \infty$, and that $f : \mathbb{R} \to \mathbb{R}$ is a continuous function. Further, let (w_n) be positive weights such that $\sum_{k=1}^n w_k \to \infty$.

(a) Use the Stolz–Cesàro limit (3.48) to prove the convergence of the *Cesàro average* as $n \to \infty$:

$$\text{(3.49)} \qquad \frac{w_1 f(x_1) + \cdots + w_n f(x_n)}{w_1 + \cdots + w_n} \to f(x).$$

(b) Suppose that $0 < b_1 \leq b_2 \leq b_3 \leq \cdots$ with $b_n \uparrow \infty$ and $\lim_{n\to\infty} \sum_{k=1}^n a_k = x < \infty$. Use the Cesàro average (3.49) to prove the *Kronecker lemma*

$$\text{(3.50)} \qquad \lim_{n\to\infty} \frac{1}{b_n} \sum_{k=1}^n b_k a_k = 0.$$

3. Modify the proof of Proposition 3.2 to demonstrate that the results also hold for any convergent sequence (x_n) in a Banach space equipped with a norm $\|\cdot\|$.

4.* To prove Kolmogorov's inequality (3.12) from first principles, we define for a fixed $a > 0$ and a fixed $n \geq 1$ the random variable

$$N := \inf\{k \geq 1 : |S_k| > a\}.$$

(a) Show that $\{N \leq n\}$ is the same event as $\{\max_{k \leq n} |S_k| > a\}$.

(b) For $k < n$, the random variable $S_k \mathbb{1}_{\{N=k\}}$ depends only on $X_1, \ldots, X_k$, and the random variable $S_n - S_k$ depends only on $X_{k+1}, \ldots, X_n$. Prove this.

(c) Show that $\mathbb{E} S_k (S_n - S_k) \mathbb{1}_{\{N=k\}} = 0$ for all $k \leq n$.

(d) Using $S_n^2 \geq S_k^2 + 2S_k(S_n - S_k)$, show that

$$\mathbb{E}[S_n^2 \mathbb{1}_{\{N=k\}}] \geq a^2 \, \mathbb{P}(N = k).$$

(e) Summing the previous inequality over $k \le n$, finish the proof of Kolmogorov's inequality.

5. Let $X_1, X_2, \ldots$ be a sequence of random variables (not necessarily independent), with

$$X_n := \begin{cases} n^2 - 1 & \text{with probability } 1/n^2 \\ -1 & \text{with probability } 1 - 1/n^2. \end{cases}$$

Let S_n be the sum of the first n of the $\{X_i\}$.

(a) Show that $\mathbb{E}X_n = 0$ for all $n = 1, 2, \ldots$.

(b) Prove that, almost surely,

$$\sum_{n=1}^{\infty} \mathbb{1}_{\{X_n > -1\}} < \infty.$$

(c) Prove that (S_n/n) converges to -1 almost surely.

6. Suppose $X_2, X_3, \ldots$ is a sequence of pairwise independent random variables with

$$\mathbb{P}(X_k = \pm k) := \frac{1}{2k \ln k}, \quad \mathbb{P}(X_k = 0) := 1 - \frac{1}{k \ln k}, \quad k \ge 2.$$

Show that $\overline{X}_n \xrightarrow{\mathbb{P}} 0$, but not almost surely (that is, it is not true that $\overline{X}_n \xrightarrow{\text{a.s.}} 0$).

7. Let $X_1, X_2, \ldots$ be independent random variables with means $\mu_1, \mu_2, \ldots$.

(a) If $\sum_{k=1}^{\infty} k^{-1} \operatorname{Var} X_k < \infty$, show that $Z_n := \frac{1}{n} \sum_{k=1}^{n} (X_k - \mu_k) \xrightarrow{\text{cpl.}} 0$.

(b) If $\sum_{k=1}^{\infty} b_k^{-2} \operatorname{Var} X_k < \infty$, for some $0 < b_1 \le b_2 \le \cdots$ and $b_n \uparrow \infty$, show that $S_n := \sum_{k=1}^{n} (X_k - \mu_k)/b_k$ converges almost surely. Hence, using the Kronecker lemma (3.50), deduce that $Z_n := \frac{1}{b_n} \sum_{k=1}^{n} (X_k - \mu_k) \xrightarrow{\text{a.s.}} 0$.

8. Let M be the space of all real-valued random variables that are defined on a probability space $(\Omega, \mathcal{H}, \mathbb{P})$, where random variables that are almost surely equal are considered to be one and the same. On M we can introduce a metric d, by defining

$$d(X, Y) := \mathbb{E}(|X - Y| \wedge 1).$$

(a) Prove that d is indeed a metric (see Definition B.1 for the properties of a metric).

(b) Show that a sequence (X_n) converges to X in probability if and only if the sequence $(d(X_n, X))$ converges to 0 as $n \to \infty$. Hint: show that for $\varepsilon \in (0, 1)$,

$$\varepsilon \mathbb{1}_{\{z > \varepsilon\}} \le z \wedge 1 \le \varepsilon + \mathbb{1}_{\{z > \varepsilon\}}$$

for all $z \ge 0$.

9. Let $X, X_1, X_2, \ldots$ be random variables that take values in a metric space E with metric r. We say that (X_n) *converges in probability* to X if for every $\varepsilon > 0$.

$$(3.51) \qquad \lim \mathbb{P}(r(X_n, X) > \varepsilon) = 0.$$

On the space of random variables taking values in E, define the metric

$$d(X, Y) := \mathbb{E}(r(X, Y) \wedge 1),$$

where X and Y are identified if $\mathbb{P}(X = Y) = 1$. Show that (3.51) holds if and only if $\lim d(X_n, X) = 0$.

10. Below $X_1, X_2, \ldots$ are assumed to be independent random variables. Use Theorem 3.24 to prove the following results:

(a) If $X_n \sim \mathsf{Bin}(n, \lambda/n)$, then, as $n \to \infty$, X_n converges in distribution to a $\mathsf{Poi}(\lambda)$ random variable.

(b)* If $X_n \sim \mathsf{Geom}(\lambda/n)$, then, as $n \to \infty$, $\frac{1}{n}X_n$ converges in distribution to an $\mathsf{Exp}(\lambda)$ random variable.

(c) If $X_n \sim \mathcal{U}(0, 1)$ and $M_n = \max\{X_1, X_2, \ldots, X_n\}$, then, as $n \to \infty$, $n(1 - M_n)$ converges in distribution to an $\mathsf{Exp}(1)$ random variable. Can you also find a proof without using characteristic functions?

11.* Let $X_1, X_2, \ldots$ be an iid sequence of random variables. Define $S_0 := 0$ and $S_n := \sum_{i=1}^{n} X_i$ for $n = 1, 2, \ldots$. Suppose that each X_i has a Cauchy distribution; i.e., it has probability density function f (with respect to the Lebesgue measure) given by

$$f(x) := \frac{1}{\pi} \frac{1}{1 + x^2}, \qquad x \in \mathbb{R}.$$

(a) Using characteristic functions, show that (S_n/n) converges in distribution. Identify the limiting distribution.

(b) Find the limiting distribution of $M_n := \pi \max\{X_1, \ldots, X_n\}/n$ as $n \to \infty$.

12. Let $X_1, X_2, \ldots$ be iid random variables with pdf

$$f(x) := \frac{1 - \cos(x)}{\pi x^2}, \qquad x \in \mathbb{R}.$$

Define $S_n := \sum_{i=1}^{n} X_i$ for $n = 1, 2, \ldots$.

(a) Show that the characteristic function of each X_i is:

$$\psi(r) := \begin{cases} 1 - |r| & \text{if } |r| \leq 1, \\ 0 & \text{if } |r| > 1. \end{cases}$$

(b) Prove that (S_n/n) converges in distribution to a Cauchy random variable as $n \to \infty$.

(c) Show that
$$\lim_{n \to \infty} \mathbb{P}(S_n \le n) = \frac{3}{4}.$$

13. Suppose that $X_1, X_2, \ldots$ are iid with $\mathbb{E}X^p =: \mu_p < \infty$ for $p = 1, 2, 3, 4$, and characteristic function ψ_X. Using the fact that $\psi_X^{(k)}(0) = i^k \, \mathbb{E}X^k$, find an explicit formula for $\mathbb{E}(\overline{X}_n - \mu_1)^4$ in terms of $\mu_1, \mu_2, \mu_3, \mu_4$, and n.

14. Prove Proposition 3.30 using the Dominated Convergence Theorem 2.36.

15. Suppose that the random variables $X, Y, (X_n, n \in \mathbb{N})$, and $(Y_n, n \in \mathbb{N})$ satisfy the following conditions:

C1. $X_n \overset{a.s.}{\to} X$.

C2. $|X_n| \le Y_n$.

C3. $Y_n \overset{a.s.}{\to} Y$ and $\mathbb{E}Y_n \to \mathbb{E}Y < \infty$.

Using Fatou's Lemma 2.35 on $Y_n \pm X_n$, prove that $\mathbb{E}X_n \to \mathbb{E}X$. We refer to this result as the *extended dominated convergence theorem*.

16. Suppose that $X_n \overset{L^1}{\to} 0$ and $Y_n \overset{L^1}{\to} 0$, but that the sum $X_n + Y_n =: Z$ is independent of n. Show that Z must be almost surely 0.

17. Suppose $X_n \overset{a.s.}{\to} X$, where $\|X\|_p < \infty$ and $\|X_n\|_p < \infty$ for all n and $p \in [1, \infty)$. Prove that $\|X_n - X\|_p \to 0$ if and only if $\|X_n\|_p \to \|X\|_p$.

18. Let $\xi_1, \xi_2, \ldots$ be a sequence of iid zero-mean random variables. Let (α_n) be a sequence of positive constants such that $\sum_{n=1}^{\infty} \alpha_n = 1$. Define the sequence of random variables $X_0, X_1, X_2, \ldots$ via the recursion
$$X_0 := 0, \quad X_n := (1 - \alpha_n)X_{n-1} + \alpha_n \xi_n, \quad n = 1, 2, \ldots$$

(a) Show that the sequence (ξ_n) is uniformly integrable.

(b) Use Proposition 3.34 to show that (X_n) is uniformly integrable. Hint: For a convex function $f : \mathbb{R} \to \mathbb{R}$ and $\alpha \in (0, 1)$, $f(\alpha x + (1 - \alpha)y) \le \alpha f(x) + (1 - \alpha)f(y)$.

19. Let (X_n) be a sequence of uncorrelated random variables (i.e., $\mathbb{C}\text{ov}(X_i, X_j) = 0$ for $i \ne j$) with expectation 0 and variance 1. Prove that for any bounded random variable Y, we have $\lim \mathbb{E}[X_n Y] = 0$. Hint: With $\alpha_n := \mathbb{E}[X_n Y]$, first consider
$$\mathbb{E}\left(Y - \sum_{k=1}^{n} \alpha_k X_k\right)^2.$$

20. Show that a collection $\mathcal{K}$ of random variables X is UI if

$$\sup_{X \in \mathcal{K}} \mathbb{P}(|X| \geq x) \leq \mathbb{P}(|Y| \geq x)$$

for all $x \in \mathbb{R}$, where Y is an integrable random variable ($\mathbb{E}|Y| < \infty$).

21.* Show that a collection of random variables $\mathcal{K}$ is UI, provided there exists a function f such that $f(x)/x \uparrow \infty$ as $x \uparrow \infty$ and $\sup_{X \in \mathcal{K}} \mathbb{E}f(|X|) = c_1 < \infty$.

22. Suppose C_1 and C_2 are two uniformly integrable families of random variables. Define $C := \{X + Y : X \in C_1, Y \in C_2\}$. Show that C is uniformly integrable.

23. If $X_n \xrightarrow{d} X$ and $(X_m - Y_n) \xrightarrow{\mathbb{P}} 0$ as $m, n \to \infty$, show that $Y_n \xrightarrow{d} X$.

24. If $g : \mathbb{R} \to \mathbb{R}$ is a continuous function, show that:

(a) $X_n \xrightarrow{d} X \Rightarrow g(X_n) \xrightarrow{d} g(X)$.
(b) $X_n \xrightarrow{\mathbb{P}} X \Rightarrow g(X_n) \xrightarrow{\mathbb{P}} g(X)$.
(c) $X_n \xrightarrow{a.s} X \Rightarrow g(X_n) \xrightarrow{a.s} g(X)$.

25. Suppose that (X_n) is UI and $X_n \xrightarrow{d} X$. Show that $\mathbb{E}|X_n| \to \mathbb{E}|X| < \infty$.

26. Suppose $X_n \xrightarrow{d} X$ and $\sup_n \mathbb{E}|X_n|^\alpha < \infty$ for $\alpha > 1$. Show that $\mathbb{E}|X_n|^\beta \to \mathbb{E}|X|^\beta$ for any $\beta \in [1, \alpha)$.

27. Consider $Y_n := 1 - \cos(2\pi n U)$, where $U \sim \mathcal{U}(0, x)$ for $x \in [0, 1]$. Verify Fatou's Lemma 2.35 by showing that

$$\mathbb{E} \liminf_{n} Y_n = 0 \leq 1 = \liminf_{n} \mathbb{E}Y_n.$$

You may use Dirichlet's approximation theorem, which states that for any real numbers r and $0 < \varepsilon \leq 1$, we can find integers n and k such that $n \in [1, \varepsilon^{-1}]$ and $|nr - k| < \varepsilon$.

28. Show that if $\sup_n X_n \leq Y$ and $\mathbb{E}Y < \infty$, then the *reverse Fatou* inequality holds:

$$\limsup_{n} \mathbb{E}X_n \leq \mathbb{E} \limsup_{n} X_n.$$

29. Consider the following two statements:

(a) $x^r \mathbb{P}(|X| > x) \to 0$ for $r > 0$, as $x \uparrow \infty$.
(b) $\mathbb{E}|X|^s < \infty$ for $s > 0$.

Show that (a) implies (b), provided that $s < r$. Then, show that (b) implies (a), provided that $s = r$.

30. If $X_n \xrightarrow{\text{P}} X$ and $X_1, X_2, \ldots$ are pairwise independent, show that X is almost surely constant.

31.* Let $g(x, y)$ be a continuous real-valued function of x and y. Suppose that $X_n \xrightarrow{\text{d}} X$ and $Y_n \xrightarrow{\text{P}} c$ for some finite constant c. Then,

$$g(X_n, Y_n) \xrightarrow{\text{d}} g(X, c).$$

This is *Slutsky's theorem*. We can prove it as follows. Let

$$\psi_{X_n, Y_n}(\boldsymbol{r}) := \mathbb{E}\, e^{i(r_1 X_n + r_2 Y_n)} \quad \text{and} \quad \psi_{X,c}(\boldsymbol{r}) := e^{ir_2 c}\, \mathbb{E} e^{ir_1 X}, \quad \forall \boldsymbol{r} \in \mathbb{R}^2$$

be the characteristic functions of $\boldsymbol{Z}_n := [X_n, Y_n]^\top$ and $\boldsymbol{Z} := [X, c]^\top$. We wish to show that ψ_{X_n, Y_n} converges to $\psi_{X,c}$ which, by Theorem 3.24, implies that $\boldsymbol{Z}_n \xrightarrow{\text{d}} \boldsymbol{Z}$.

(a) First, show that

(3.52) $\quad |\psi_{X_n, Y_n}(\boldsymbol{r}) - \psi_{X,c}(\boldsymbol{r})| \le |\psi_{X_n}(r_1) - \psi_X(r_1)| + \mathbb{E}\,|e^{ir_2(Y_n - c)} - 1|.$

(b) Second, show that for any $\varepsilon > 0$,

$$\mathbb{E}|e^{ir_2(Y_n - c)} - 1| \le 2\mathbb{P}[|Y_n - c| > \varepsilon] + |r_2|\varepsilon.$$

(c) Third, using (a) and (b) prove that $\boldsymbol{Z}_n \xrightarrow{\text{d}} \boldsymbol{Z}$ and consequently, that $g(X_n, Y_n) \xrightarrow{\text{d}} g(X, c)$.

32. Suppose that $Z_n \le X_n \le Y_n$ for all n, $X_n \xrightarrow{\text{d}} X$, and $Y_n \xrightarrow{L^1} Y$, $Z_n \xrightarrow{L^1} Z$. Show that $\mathbb{E}|X_n| \to \mathbb{E}|X| < \infty$.

33. Suppose that $f : [0, 1] \to \mathbb{R}$ is a continuous function on $[0, 1]$ and define the n-degree polynomial function:

$$f_n(t) := \sum_{k=0}^{n} f(k/n) \binom{n}{k} t^k (1 - t)^{n-k}, \quad t \in [0, 1].$$

(a) Show that $f_n(t) = \mathbb{E} f(\overline{Y}_n)$, where $\overline{Y}_n$ is the average of n iid Bernoulli variables with success probability t.

(b) Use the *Bolzano–Weierstrass theorem* to prove that there exists a constant $c < \infty$ such that $\sup_{t \in [0,1]} |f(t)| \le c$.

(c) Show that $f_n(t) \to f(t)$ for every $t \in [0, 1]$.

(d) Use the *Heine–Cantor theorem* (see Example B.6) to prove that

$$\sup_{t \in [0,1]} |f_n(t) - f(t)| \to 0.$$

This proves the *Weierstrass approximation theorem* — any continuous function on $[0, 1]$ can be approximated arbitrarily well (in the supremum norm) by a polynomial function.

34. Suppose that $X_1, X_2, \ldots$ are iid with finite mean $\mu := \mathbb{E}X < \infty$, and let $Y_k := X_k \mathbb{1}_{\{|X_k| \le k\}}$ be a truncated version of X_k for all k.

(a) Show that $\overline{Y}_n - \overline{X}_n \xrightarrow{\text{a.s.}} 0$.

(b) Using the identity $\mathbb{E}Y_k^2 = \int_0^\infty dt\, 2t \mathbb{1}_{\{t \le k\}} \mathbb{P}(|X_k| > t)$, show that

$$\sum_{k=1}^{\infty} \frac{\mathbb{E}Y_k^2}{k^2} \le 4\mu.$$

(c) Define the variable $S_n := \sum_{k=1}^{n} \frac{Y_k - \mathbb{E}Y_k}{k}$. Use the result in Example 3.13 to prove that S_n converges with probability 1.

(d) Use the Kronecker lemma (3.50) to show that $\overline{Y}_n - \mathbb{E}\overline{Y}_n \xrightarrow{\text{a.s.}} 0$. Hence, deduce that $\overline{X}_n \xrightarrow{\text{a.s.}} \mu$.

35. Let $X_1, X_2, \ldots$ be a sequence of positive pairwise independent and identically distributed random variables with mean $\mathbb{E}X =: \mu < \infty$. The following exercises culminate in a proof that $\overline{X}_n \xrightarrow{\text{a.s.}} \mu$. We may drop the positivity requirement by considering the positive and negative parts of the random variables separately.

(a) Let $Y_n := X_n \mathbb{1}_{\{X_n < n\}}$ be a truncated version of each X_n. Show that $\overline{X}_n - \overline{Y}_n \xrightarrow{\text{a.s.}} 0$.

(b) Use the fact that $\sum_{j=1}^{k} \mathbb{1}_{\{j-1 \le Y_k < j\}} = 1$ and $\{X_k < k\} \Rightarrow \{X_k = Y_k\}$ to show that

$$\mathbb{E}Y_k^2 \le \sum_{j=1}^{k} j^2 \mathbb{P}(j - 1 \le X_1 < j) \quad \text{and} \quad \sum_{j=1}^{\infty} j\, \mathbb{P}(j - 1 \le X_1 < j) \le 1 + \mu.$$

(c) Let $\beta_k := \lceil \alpha^k \rceil$ for $\alpha > 1$ be a subsequence of $k = 1, 2, \ldots$. Show that

$$\sum_{n: \beta_n \ge k} \frac{1}{\beta_n^2} \le \frac{c_\alpha}{k^2},$$

where $c_\alpha < \infty$ is some constant depending on α.

(d) By changing the order of summation, deduce the following bounds:

$$\sum_{n=1}^{\infty} \frac{1}{\beta_n^2} \sum_{k=1}^{\beta_n} \mathbb{E}Y_k^2 \le c_\alpha \sum_{k \ge 1} \frac{\mathbb{E}Y_k^2}{k^2} \le 2c_\alpha(1 + \mu).$$

(e) Using the previous result to deduce that $\overline{Y}_{\beta_k} - \mathbb{E}\overline{Y}_{\beta_k} \xrightarrow{\text{a.s.}} 0$.

(f) Use the Cesàro average in (3.49) to show that $\mathbb{E}\overline{Y}_{\beta_k} \to \mu$. Hence, deduce that $\overline{Y}_{\beta_k} \xrightarrow{\text{a.s.}} \mu$.

(g) Show that for $\beta_k \leq n \leq \beta_{k+1}$, we have:

$$\overline{Y}_{\beta_k} \frac{\beta_k}{\beta_{k+1}} \leq \overline{Y}_n \leq \frac{\beta_{k+1}}{\beta_k} \overline{Y}_{\beta_{k+1}}.$$

Hence, using all of the previous results deduce that $\overline{X}_n \overset{\text{a.s.}}{\rightarrow} \mu$.

36.* Suppose $X_1, X_2, \ldots$ are independent random variables with

$$\mathbb{P}(X_k = k) = \mathbb{P}(X_k = -k) := \frac{1}{2k^2},$$
$$\mathbb{P}(X_k = 1) = \mathbb{P}(X_k = -1) := \frac{1}{2}\left(1 - \frac{1}{k^2}\right).$$

Show that, while $\mathbb{E}\overline{X}_n = 0$ and $\mathbb{V}\text{ar}(\sqrt{n}\,\overline{X}_n) \rightarrow 2$, it is *not* true that $\sqrt{n}(\overline{X}_n - 0)/\sqrt{2} \overset{\text{d}}{\rightarrow} \mathcal{N}(0,1)$.

37. Prove that if $X_1, X_2, X_3, \ldots$ are iid d-dimensional random vectors with finite expectation $\mu := \mathbb{E}X < \infty$ (or equivalently $\|\mu\| < \infty$), then for all $\varepsilon \in (0,1)$ we have $\mathbb{P}(\|\overline{X}_n - \mu\| > \varepsilon) \rightarrow 0$ as $n \rightarrow \infty$.

CHAPTER 4

CONDITIONING

The concept of *conditioning* in the theory of probability and stochastic processes aims to model in a precise way the (additional) *information* that we have about a random experiment. In this chapter, we define the concepts of conditional expectation and conditional probability and explore their properties. A main application is in the study of Markov processes.

4.1 A Basic Example

Let $(\Omega, \mathcal{H}, \mathbb{P})$ be a probability space for a random experiment. Think of the random experiment of drawing a uniform point in $\Omega := (0, 1]$ and let $\mathcal{H}$ be the Borel σ-algebra on $(0, 1]$ and $\mathbb{P}$ the (restriction of the) Lebesgue measure to $(0, 1]$. Let X be a numerical random variable that describes a measurement on this experiment. In particular, suppose that X represents the squared value of the point drawn; so $X(\omega) = \omega^2, \omega \in \Omega$.

What would be your "best guess" for X if you had to guess a number between 0 and 1? It would have to be the expectation $\mathbb{E}X = \int_0^1 dx\, x^2 = 1/3$.

Now, let $\mathcal{F}$ be the σ-algebra generated by the sets $(0, 1/3], (1/3, 2/3], (2/3, 1]$. Suppose we have extra information that $\omega \in (2/3, 1]$. What is now your best guess for X? It would be $3 \int_{2/3}^1 dx\, x^2 = 57/81$. Similarly, if we knew that $\omega \in (0, 1/3]$ our guess would be $3/81$, and knowing $\omega \in (1/3, 2/3]$ would give the guess $21/81$.

In terms of σ-algebras of information, our initial guess of 1/3 was based on the trivial σ-algebra $\{\Omega, \emptyset\}$; i.e., no information where the outcome of the experiment will lie other than that it lies somewhere in $(0, 1]$. However, we can "refine" our initial guess of 1/3 based on the information in the σ-algebra $\mathcal{F}$, to give the set of guesses $\{3/81, 21/81, 57/81\}$. Or, equivalently, we can define a function $\widehat{X}$ via:

$$\widehat{X}(\omega) := \frac{3}{81} \mathbb{1}_{(0,1/3]}(\omega) + \frac{21}{81} \mathbb{1}_{(1/3,2/3]}(\omega) + \frac{57}{81} \mathbb{1}_{(2/3,1]}(\omega).$$

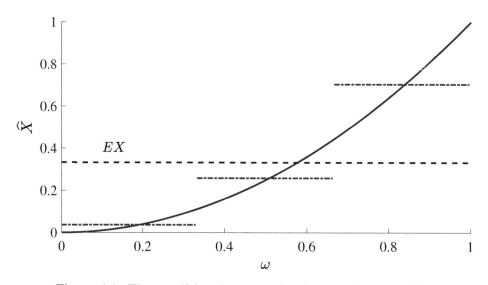

Figure 4.1: The conditional expectation is a random variable.

This will be our definition for the conditional expectation of X given $\mathcal{F}$ in this case. Written $\mathbb{E}_{\mathcal{F}} X$. Note that $\widehat{X}$ is $\mathcal{F}$-measurable.

4.2   Conditional Expectation

In elementary probability, we define the conditional expectation after defining conditional probability. In advanced probability, it is more convenient to reverse the order, as we often want to condition on σ-algebras. In what follows, $(\Omega, \mathcal{H}, \mathbb{P})$ is a probability space, X a random variable, and $\mathcal{F}$ a sub-σ-algebra of $\mathcal{H}$; that is a σ-algebra whose every member (i.e., set) belongs to $\mathcal{H}$ as well. Recall the notation $V \in \mathcal{F}_{+}$ to mean that $V \geq 0$ and is $\mathcal{F}$-measurable.

Definition 4.2: Conditional Expectation

The conditional expectation of X given $\mathcal{F}$, denoted $\mathbb{E}_{\mathcal{F}} X$, is defined in two steps:

1. For $X \geq 0$, $\mathbb{E}_{\mathcal{F}} X$ is defined as any random variable $\widehat{X}$ with
   (a) $\widehat{X} \in \mathcal{F}_{+}$.
   (b) $\mathbb{E} V X = \mathbb{E} V \widehat{X}$ for every $V \in \mathcal{F}_{+}$ *(projection property)*.
2. For arbitrary X, if $\mathbb{E} X$ exists (possibly $\pm\infty$), $\mathbb{E}_{\mathcal{F}} X := \mathbb{E}_{\mathcal{F}} X^{+} - \mathbb{E}_{\mathcal{F}} X^{-}$. Otherwise, the conditional expectation is left undefined.

Note that the conditional expectation is defined *almost surely*. That is, there are in general many functions $\widehat{X}$ that satisfy the conditions in the definition above. However, fortunately all these *versions* are equal to each other up to a set of probability 0. This is a consequence of the following theorem:

Theorem 4.3: Existence and Uniqueness of the Conditional Expectation

Let $X \geq 0$ be a random variable and let $\mathcal{F}$ be a sub-σ-algebra of $\mathcal{H}$. Then, $\mathbb{E}_{\mathcal{F}} X$ exists and is unique up to equivalence.

Proof. This is an application of the Radon–Nikodym Theorem 1.59. For each event $H \in \mathcal{F}$, define

$$P(H) := \mathbb{P}(H) \quad \text{and} \quad Q(H) := \mathbb{E}\mathbb{1}_H X = \int_H X \, d\mathbb{P}.$$

Both P and Q define measures on $(\Omega, \mathcal{F})$ and $Q \ll P$ (i.e., Q is absolutely continuous with respect to P). Hence, by Theorem 1.59 there exists up to equivalence a unique density $\widehat{X} = dQ/dP$ such that

$$\int V \, dQ = \int V\widehat{X} \, dP$$

for every $V \in \mathcal{F}_+$. That is, $\mathbb{E}VX = \mathbb{E}V\widehat{X}$. $\square$

The properties of conditional expectation, such as monotonicity and linearity, derive directly from the properties of the expectation in Theorem 2.34. To illustrate, suppose that $X \leq Y$. Let $\widehat{X}$ and $\widehat{Y}$ be versions of the conditional expectation of X and Y with respect to $\mathcal{F}$. Take any $V \in \mathcal{F}_+$. Then, $\mathbb{E}VX \leq \mathbb{E}VY$, by the monotonicity of the (ordinary) expectation. Hence, by the definition of conditional expectation, we have $\mathbb{E}V\widehat{X} \leq \mathbb{E}V\widehat{Y}$; i.e., $\mathbb{E}V(\widehat{X} - \widehat{Y}) \leq 0$ for all $V \in \mathcal{F}_+$. This can only be true if $\widehat{X} \leq \widehat{Y}$ almost surely; that is, $\mathbb{E}_{\mathcal{F}} X \leq \mathbb{E}_{\mathcal{F}} Y$. In a similar way, we can show that the other properties of the ordinary expectation in Theorem 2.34 also hold for the conditional expectation.

Two important new properties emerge as well, as given in the following theorem:

Theorem 4.4: Additional Properties of the Conditional Expectation

Let $\mathcal{F}$ and $\mathcal{G}$ be sub-σ-algebras of $\mathcal{H}$. Let W and X be random variables such that $\mathbb{E}X$ and $\mathbb{E}WX$ exist. Then, the following hold:

1. *(Taking out what is known)*: If $W \in \mathcal{F}$, then $\mathbb{E}_{\mathcal{F}} WX = W\mathbb{E}_{\mathcal{F}} X$.

2. *(Repeated conditioning)*: If $\mathcal{F} \subseteq \mathcal{G}$, then

$$(4.5) \qquad \mathbb{E}_{\mathcal{F}}\mathbb{E}_{\mathcal{G}} X = \mathbb{E}_{\mathcal{G}}\mathbb{E}_{\mathcal{F}} X = \mathbb{E}_{\mathcal{F}} X.$$

Proof. It suffices to consider only positive W and X.

1. Suppose $W \in \mathcal{F}_+$. Let $\widehat{X} := \mathbb{E}_{\mathcal{F}} X$. By definition, $\widehat{X} \in \mathcal{F}_+$ and for all $V \in \mathcal{F}_+$:

$$\mathbb{E} V(WX) = \mathbb{E} V W \widehat{X} = \mathbb{E} V(W\widehat{X}).$$

   In other words, $W\widehat{X} = W \mathbb{E}_{\mathcal{F}} X$ is a version of $\mathbb{E}_{\mathcal{F}}(WX)$.

2. Let $\mathcal{F} \subseteq \mathcal{G}$. Define $W := \mathbb{E}_{\mathcal{F}} X$. We have $W \in \mathcal{F}_+$ and hence also $W \in \mathcal{G}_+$. So by part 1 (taking out what is known) we have $\mathbb{E}_{\mathcal{G}} W = W$; that is, $\mathbb{E}_{\mathcal{G}} \mathbb{E}_{\mathcal{F}} X = \mathbb{E}_{\mathcal{F}} X$.

   To prove the first equality in (4.5), define $Y := \mathbb{E}_{\mathcal{G}} X$ and $\widehat{X} := \mathbb{E}_{\mathcal{F}} X \in \mathcal{F}_+$. Take any $V \in \mathcal{F}_+$ (and hence $V \in \mathcal{G}_+$ as well). By definition of $\widehat{X}$, we have $\mathbb{E} V \widehat{X} = \mathbb{E} V X$ and by the definition of Y, we have $\mathbb{E} V Y = \mathbb{E} V X$. Hence, $\mathbb{E} V \widehat{X} = \mathbb{E} V Y$; that is, $\mathbb{E}_{\mathcal{F}} \mathbb{E}_{\mathcal{G}} X = \mathbb{E}_{\mathcal{F}} X$.

$\square$

There is an amusing way to remember the repeated conditioning: The σ-algebras $\mathcal{F}$ and $\mathcal{G}$ represent the information about the measurement on a random experiment by a fool and a genius. The fool has no use for the genius' information: $\mathbb{E}_{\mathcal{F}} \mathbb{E}_{\mathcal{G}} X = \mathbb{E}_{\mathcal{F}} X$, and, sadly, the genius cannot further improve on the poor information of the fool: $\mathbb{E}_{\mathcal{G}} \mathbb{E}_{\mathcal{F}} X = \mathbb{E}_{\mathcal{F}} X$.

In many applications of conditioning, we wish to take the expectation of a random variable X conditional on one or more random variables or, in general, on a stochastic process $Y := \{Y_t, t \in \mathbb{T}\}$, in which case $\mathcal{F} = \sigma Y = \sigma\{Y_t, t \in \mathbb{T}\}$. In this case we often write

$$\mathbb{E}_{\mathcal{F}} X = \mathbb{E}[X \mid Y_t, t \in \mathbb{T}] = \mathbb{E}[X \mid Y].$$

When conditioning on a random variable Y, we can use the fact (see Theorem 2.65) that any random variable in σY, in particular $\mathbb{E}[X \mid Y]$, must be a deterministic measurable function of Y; that is, $\mathbb{E}[X \mid Y] = h(Y)$ for some measurable function h. This is in accordance with our use of conditional expectations in elementary probability: We first determine $h(y) := \mathbb{E}[X \mid Y = y]$ and then use $\mathbb{E}[X \mid Y] = h(Y)$ in computations. Here are some examples.

■ **Example 4.6 (Bernoulli Process)** Let $X_1, X_2, \ldots$ be a Bernoulli process with success parameter p. Let, $S_1 := X_1$ and $S_n := X_1 + \cdots + X_n$ for $n = 2, 3 \ldots$. Given the event $\{X_1 = x_1, \ldots, X_n = x_n\}$, the only values that S_{n+1} can take are $s_n := x_1 + \cdots + x_n$ and $s_n + 1$, with probability $1 - p$ and p, respectively. Consequently,

$$(4.7) \qquad \mathbb{E}[S_{n+1} \mid X_1, \ldots, X_n] = S_n (1 - p) + (S_n + 1) p = S_n + p.$$

In a similar way, we can derive that

$$(4.8) \qquad \mathbb{E}[S_{n+1} \mid S_1, \ldots, S_n] = S_n + p.$$

Intuitively, we know that (4.7) and (4.8) should be the same because the amount of information that $X_1, \ldots, X_n$ and $S_1, \ldots, S_n$ carry is exactly the same. In particular, if we know the outcomes of $X_1, \ldots, X_n$ then we can deduce the outcomes of $S_1, \ldots, S_n$ and vice versa. ∎

■ **Example 4.9 (Poisson Count of Exponential Interarrivals)** Suppose that $A_1, A_2, \ldots$ is an iid sequence of $\mathsf{Exp}(\lambda)$ random variables. Think of A_i as the time interval between the arrival of the ith and $(i-1)$st customer to a post office; thus, $T_1 := A_1, T_2 := A_1 + A_2$, and so on, are the actual arrival times. Let N_t be the number of customers that arrive during the interval $[0, t]$. Note that N_t is a random variable with values in $\mathbb{N}$. What is the distribution of N_t?

Let us solve this using generating functions and conditioning. First, consider the probability generating function of N_t; in particular, let $g(t) := \mathbb{E}\, s^{N_t}$. By conditioning on T_1, the time the first customer arrives, we have

$$g(t) = \mathbb{E}\, s^{N_t} = \mathbb{E}\, \underbrace{\mathbb{E}[s^{N_t} \mid T_1]}_{h(T_1)} = \int_0^\infty dx\, \underbrace{\mathbb{E}[s^{N_t} \mid T_1 = x]}_{h(x)}\, \lambda\, e^{-\lambda x},$$

where we have defined the function h implicitly above. For $x > t$, we have $h(x) = \mathbb{E}[s^{N_t} \mid T_1 = x] = s^0 = 1$, because in that case there are no arrivals in $[0, t]$. For $x \le t$, the conditional distribution of N_t given $T_1 = x$ is the same as the distribution of the number of arrivals in $(x, t]$ plus 1, which in turn has the same distribution as $N_{t-x} + 1$. Hence, in this case $h(x) = s\,\mathbb{E}\, s^{N_{t-x}} = s\, g(t-x)$. Combining these cases, we have

$$g(t) = \int_t^\infty dx\, \lambda\, e^{-\lambda x} + \int_0^t dx\, s\, g(t-x)\, \lambda\, e^{-\lambda x}$$

$$= e^{-\lambda t} + e^{-\lambda t} \int_0^t dy\, s\, g(y)\, \lambda\, e^{\lambda y}.$$

Differentiating g with respect to t thus gives

$$g'(t) = -\lambda\, e^{-\lambda t} - \lambda\, e^{-\lambda t} \int_0^t dy\, s\, g(y)\, \lambda\, e^{\lambda y} + e^{-\lambda t} g(t)\, \lambda\, e^{\lambda t} s$$

$$= -\lambda g(t) + s\lambda g(t) = -\lambda(1-s)\, g(t).$$

Since $g(1) = 1$, by definition, it follows that

$$g(t) = e^{-\lambda(1-s)t}.$$

In other words, $N_t \sim \mathsf{Poi}(\lambda t)$. ∎

For random variables in an L^2 Hilbert space (see Section 2.5), the conditional expectation can be defined as a projection, in which case the "projection property" in Definition 4.2 becomes evident. Recall that the inner product of X and Y in L^2 is $\langle X, Y \rangle = \mathbb{E}XY$, and hence $\|X - Y\|_2^2 = \mathbb{E}(X - Y)^2$.

Theorem 4.10: Conditional Expectation in L^2

For $X \in L^2$, the conditional expectation $\mathbb{E}_{\mathcal{F}} X$ is the orthogonal projection of X onto the subspace $\mathcal{V}$ of $\mathcal{F}$-measurable random variables in L^2.

Proof. Let $\widehat{X}$ be the orthogonal projection of X onto the subspace $\mathcal{V}$. That is,

$$\widehat{X} := \underset{V \in \mathcal{V}}{\operatorname{argmin}} \|X - V\|_2.$$

Then, X can be written as $X = \widehat{X} + X^\perp$, where $X^\perp := X - \widehat{X}$ is perpendicular to any $V \in \mathcal{V}$, meaning that

$$\langle V, X^\perp \rangle = 0.$$

But this is just another way of saying that $\mathbb{E}V(X - \widehat{X}) = 0$. This shows that the projection property in Definition 4.2 holds. Since $\widehat{X} \in \mathcal{V}$, we have that $\widehat{X} = \mathbb{E}_{\mathcal{F}} X$. □

4.3 Conditional Probability and Distribution

A basic treatment of conditional probability starts with the following definition:

Definition 4.11: Conditional Probability for Events

Let A and B be two events with $\mathbb{P}(B) > 0$. The *conditional probability* that A occurs given that B occurs is denoted by

(4.12)
$$\mathbb{P}(A \mid B) := \frac{\mathbb{P}(A \cap B)}{\mathbb{P}(B)}.$$

Definition 4.11 is sufficient to derive a slew of useful results, summarized in Theorem 4.13. The proofs are elementary; see Exercise 1.

> ### Theorem 4.13: Properties Involving Conditional Probabilities
>
> Let $(\Omega, \mathcal{H}, \mathbb{P})$ be a probability space and $A, A_1, A_2, \ldots, A_n, B_1, B_2, \ldots$ events such that $\mathbb{P}(A_1 \cap \cdots \cap A_n) > 0$ and $\{B_i\}$ form a partition of Ω, with $\mathbb{P}(B_i) > 0$ for all i. Then, the following hold:
>
> 1. *Law of total probability:* $\mathbb{P}(A) = \sum_i \mathbb{P}(A \mid B_i) \, \mathbb{P}(B_i)$.
>
> 2. *Bayes' rule:* $\mathbb{P}(B_j \mid A) = \dfrac{\mathbb{P}(B_j) \, \mathbb{P}(A|B_j)}{\sum_i \mathbb{P}(B_i) \, \mathbb{P}(A|B_i)}$.
>
> 3. *Chain or product rule:*
>
> $$\mathbb{P}\left(\cap_{i=1}^{n} A_i \right) = \mathbb{P}(A_1) \, \mathbb{P}(A_2 \mid A_1) \cdots \mathbb{P}(A_n \mid A_1 \cap \cdots \cap A_{n-1}).$$

■ **Example 4.14** (**Memoryless Property**) The single most important property of the exponential distribution is that it is *memoryless*. By this we mean the following. Let $X \sim \mathsf{Exp}(\lambda)$. Then, for any $x, y > 0$,

$$\mathbb{P}(X > x + y \mid X > x) = \frac{\mathbb{P}(X > x + y, X > x)}{\mathbb{P}(X > x)} = \frac{\mathbb{P}(X > x + y)}{\mathbb{P}(X > x)}$$

$$= \frac{e^{-\lambda(x+y)}}{e^{-\lambda x}} = e^{-\lambda y} = \mathbb{P}(X > y).$$

For example, if X denotes the lifetime of a machine, then, knowing that the machine at time x is still operating, the *remaining* lifetime of the machine is the same as that of a brand new machine.

We can view the exponential distribution as a continuous version of the geometric distribution. The latter also has the memoryless property: in a coin flip experiment, knowing that the first n tosses yielded tails, does not give you extra information about the remaining number of tosses that you have to throw until heads appears; the coin does not have a "memory". ■

How do we define the conditional probability of an event A given the information contained in a σ-algebra $\mathcal{F}$, denoted $\mathbb{P}(A \mid \mathcal{F})$ or $\mathbb{P}_{\mathcal{F}}(A)$? The answer is both simple and complicated. It is simple, because we can define

(4.15) $$\mathbb{P}_{\mathcal{F}}(A) := \mathbb{E}_{\mathcal{F}} \mathbb{1}_A.$$

When $\mathcal{F}$ is the σ-algebra generated by an event B, i.e., $\mathcal{F} = \{B, B^c, \Omega, \emptyset\}$, and $0 < \mathbb{P}(B) < 1$, then $\mathbb{P}_{\mathcal{F}}(A)$ is the random variable

$$\mathbb{P}(A \mid B) \mathbb{1}_B + \mathbb{P}(A \mid B^c) \mathbb{1}_{B^c},$$

using Definition 4.11. This agrees with our intuitive notion of a conditional probability: if we know that the outcome lies in B, then our best guess for the probability that it (also) lies in A is $\mathbb{P}(A \cap B)/\mathbb{P}(B)$. If on the other hand, we know the outcome does not lie in B, then our best guess is $\mathbb{P}(A \cap B^c)/\mathbb{P}(B^c)$.

Things get complicated when conditioning on events that have 0 probability. We can still use (4.15), as this is well-defined. However, the definition specifies the conditional probability *up to equivalence*. Hence, in general there may be many versions of $\mathbb{P}_{\mathcal{F}}(A) = \mathbb{E}_{\mathcal{F}} \mathbb{1}_A$ that are equal to each other with probability 1. Take one such version and call it $Q(\cdot, A)$. By definition, it is an $\mathcal{F}$-measurable random variable taking values in $[0, 1]$; denote its value at ω by $Q(\omega, A)$. We can take $Q(\cdot, \Omega) = 1$ and $Q(\cdot, \emptyset) = 0$. It appears that Q is a *probability transition kernel* from $(\Omega, \mathcal{F})$ into $(\Omega, \mathcal{H})$, as defined in Definition 1.63; in particular, a mapping from $\Omega \times \mathcal{H}$ to $[0, 1]$ such that:

1. $Q(\cdot, A)$ is $\mathcal{F}$-measurable for every event $A \in \mathcal{H}$.
2. $Q(\omega, \cdot)$ is a probability measure on $(\Omega, \mathcal{H})$ for every $\omega \in \Omega$.

Unfortunately, most versions Q are *not* transition kernels, due to the "up to equivalence" definition. Nevertheless, and fortunately, most measurable spaces $(\Omega, \mathcal{H})$ of practical interest admit a version of the conditional probability that are *regular*, in the sense that Q is a proper probability transition kernel. We will prove this after we discuss the concepts of conditional distribution and standard measurable space.

Definition 4.16: Conditional Distribution

Let Y be a random variable on the probability space $(\Omega, \mathcal{H}, \mathbb{P})$ taking values in the measurable space $(E, \mathcal{E})$. Let $\mathcal{F}$ be a sub-σ-algebra of $\mathcal{H}$. The *conditional distribution of Y given $\mathcal{F}$* is any probability transition kernel L from $(\Omega, \mathcal{F})$ to $(E, \mathcal{E})$ such that

$$\mathbb{P}_{\mathcal{F}}(Y \in B) = L(\cdot, B), \quad B \in \mathcal{E}.$$

A measurable space $(E, \mathcal{E})$ is said to be *standard*, if there exists a Borel subset $R \subseteq [0, 1]$ and a bijection $g : E \to R$ with inverse $h : R \to E$ such that g is $\mathcal{E}/\mathcal{R}$-measurable and h is $\mathcal{R}/\mathcal{E}$-measurable, where $\mathcal{R}$ is the restriction of the Borel σ-algebra to R.

Theorem 4.17: Existence of the Conditional Distribution

For every standard measurable space $(E, \mathcal{E})$ there exists a version of the conditional distribution of Y given $\mathcal{F}$.

Proof. Let g, h, and R be as defined above. If Y takes values in $(E, \mathcal{E})$, then the random variable $\widetilde{Y} := g \circ Y$ is real-valued; it takes values in $(R, \mathcal{R})$. Suppose $\widetilde{Y}$ has conditional distribution $\widetilde{L}$. Define,

$$L(\omega, A) := \widetilde{L}(\omega, h^{-1}A), \quad \omega \in \Omega, \; A \in \mathcal{E}.$$

Then, $L(\cdot, A)$ is a positive random variable for every $A \in \mathcal{E}$; and for every $H \in \mathcal{F}_+$,

$$\mathbb{E}\mathbb{1}_H \mathbb{1}_{\{Y \in A\}} = \mathbb{E}\mathbb{1}_H \mathbb{1}_{\{\widetilde{Y} \in h^{-1}A\}} = \mathbb{E}\mathbb{1}_H \widetilde{L}(\cdot, h^{-1}A) = \mathbb{E}\mathbb{1}_H L(\cdot, A).$$

Consequently, L is a transition probability kernel from $(\Omega, \mathcal{F})$ to $(E, \mathcal{E})$, with

$$L(\cdot, A) = \mathbb{E}_{\mathcal{F}} \mathbb{1}_{\{Y \in A\}} = \mathbb{P}_{\mathcal{F}}(Y \in A), \quad A \in \mathcal{E}.$$

In other words, L is a conditional distribution of Y given $\mathcal{F}$.

Thus, to prove the theorem, it remains to show the existence of $\widetilde{L}$; that is, to prove the theorem for the case $E := \overline{\mathbb{R}}$ and $\mathcal{E} := \overline{\mathcal{B}}$. To this end, for each rational number $q \in \mathbb{Q}$ define

$$C_q := \mathbb{P}_{\mathcal{F}}(Y \le q)$$

and

$$\Omega_{q<r} := \{C_q < C_r\}, \quad q, r \in \mathbb{Q}, \quad q < r.$$

Obviously, $\Omega_{q<r} \in \mathcal{F}$ and $\mathbb{P}(\Omega_{q<r}) = 1$. Let Ω_0 be the intersection of all the $\Omega_{q<r}$. Then, $\Omega_0 \in \mathcal{F}$ and $\mathbb{P}(\Omega_0) = 1$. For a fixed $\omega \in \Omega_0$ consider the function $C_q(\omega) : \mathbb{Q} \to [0, 1]$. It is increasing and so for each $t \in \mathbb{R}$ the limit $\overline{C}_t(\omega)$ exists, resulting in a cdf $\overline{C}_t(\omega), t \in \mathbb{R}$. Corresponding to this cdf is a unique probability measure $\overline{L}_\omega$ on $(\overline{\mathbb{R}}, \overline{\mathcal{B}})$. Define

$$L(\omega, B) := \mathbb{1}_{\Omega_0}(\omega) \overline{L}_\omega(B) + \mathbb{1}_{\Omega \setminus \Omega_0}(\omega) \delta_0(B), \quad \omega \in \Omega, \; B \in \overline{\mathcal{B}},$$

where δ_0 is the Dirac measure at 0. We show that L is a probability transition kernel from $(\Omega, \mathcal{F})$ to $(\overline{\mathbb{R}}, \overline{\mathcal{B}})$. Obviously, for each $\omega \in \Omega$, $L(\omega, \cdot)$ is a probability measure on $(\overline{\mathbb{R}}, \overline{\mathcal{B}})$. To show that $L(\cdot, B)$ is $\overline{\mathcal{B}}/\mathcal{F}$-measurable, we invoke the Monotone Class Theorem. Let

$$\mathcal{D} := \{B \in \overline{\mathcal{B}} : L(\cdot, B) \in \mathcal{F}_+\}.$$

This is a monotone class (check this yourself). Hence, by the Monotone Class Theorem 1.12, $\mathcal{D} = \overline{\mathcal{B}}$ if intervals of the form $[-\infty, t]$ belong to $\mathcal{D}$. Let (r_n) be a sequence of rationals strictly decreasing to t. Then,

$$L(\omega, [-\infty, t]) = \mathbb{1}_{\Omega_0}(\omega) \lim_n C_{r_n}(\omega) + \mathbb{1}_{\Omega \setminus \Omega_0}(\omega) \delta_0([-\infty, t]).$$

Both $\mathbb{1}_{\Omega_0}$ and $\mathbb{1}_{\Omega \setminus \Omega_0}$ are in $\mathcal{F}_+$, as are $C_{r_n} \in \mathcal{F}_+$ for all n, and hence also their monotone limit $\overline{C}_t$. It follows that $L(\cdot, [-\infty, t])$ is also in $\mathcal{F}_+$. Consequently, $L(\cdot, B) \in \mathcal{F}_+$ for any $B \in \overline{\mathcal{B}}$. Thus, L is a genuine probability transition kernel.

The only thing that remains to be verified is that L is the conditional distribution of Y given $\mathcal{F}$; in other words, that

$$\mathbb{E}\mathbb{1}_H \mathbb{1}_{\{Y \in B\}} = \mathbb{E}\mathbb{1}_H L(\cdot, B), \quad B \in \overline{\mathcal{B}}, \; H \in \mathcal{F}_+.$$

By the same monotone class argument as above, it suffices to check this for $B :=$ $[-\infty, t]$, in which case we have indeed

$$\mathbb{E}\,\mathbb{1}_H \mathbb{1}_{\{Y \le t\}} = \lim_n \mathbb{E}\,\mathbb{1}_H \mathbb{1}_{\{Y \le r_n\}} = \lim_n \mathbb{E}\,\mathbb{1}_H C_{r_n}$$
$$= \mathbb{E}\,\mathbb{1}_{H \cap \Omega_0} L(\cdot, [-\infty, t]) = \mathbb{E}\,\mathbb{1}_H L(\cdot, [-\infty, t]),$$

where the third equality follows from the facts that $\mathbb{1}_H = \mathbb{1}_{H \cap \Omega_0}$ with probability 1 and that $C_{r_n}(\omega)$ converges pointwise to $L(\omega, [-\infty, t])$ for $\omega \in \Omega_0$. $\qquad\square$

We are now in the position to prove that *regular* conditional probabilities exist.

Theorem 4.18: Regular Version of a Conditional Probability

Let $(\Omega, \mathcal{H})$ be a standard measurable space and $\mathcal{F}$ a sub-σ-algebra of $\mathcal{H}$. Then, there *exists* a regular version of the conditional probability $\mathbb{P}_\mathcal{F}$.

Proof. Consider the random variable $Y(\omega) := \omega$, $\omega \in \Omega$. Theorem 4.17, with $E := \Omega$ and $\mathcal{E} := \mathcal{H}$, guarantees the existence of the conditional distribution of Y given $\mathcal{F}$ in the form of a probability transition kernel L. In particular,

$$\mathbb{P}_\mathcal{F}(B) = \mathbb{P}_\mathcal{F}(Y \in B) = L(\cdot, B), \quad B \in \mathcal{H}.$$

This shows that L is a regular version of the conditional probability. $\qquad\square$

We now consider the situation where we have two random variables X and Y, taking values in measurable spaces $(D, \mathcal{D})$ and $(E, \mathcal{E})$, respectively. We can think of the random point (X, Y) as the result of two random experiments: first draw X according to some distribution μ on $(D, \mathcal{D})$ and then, given $X = x$, draw Y according to the distribution $K(x, \mathrm{d}y)$ on $(E, \mathcal{E})$ for some probability transition kernel K. The joint distribution π of X and Y is then given (see (1.65)) by

$$\pi(A \times B) = \int_A \mu(\mathrm{d}x) K(x, B), \quad A \in \mathcal{D}, B \in \mathcal{E}$$

or in terms of integrals:

$$\pi f = \int \pi(\mathrm{d}x, \mathrm{d}y) f(x, y) = \int_D \mu(\mathrm{d}x) \int_E K(x, \mathrm{d}y) f(x, y), \quad f \in \mathcal{D} \otimes \mathcal{E}.$$

Recall from (1.66) that we can also write

(4.19) $\pi(\mathrm{d}x, \mathrm{d}y) = \mu(\mathrm{d}x) K(x, \mathrm{d}y).$

Using μ and K in this way is a convenient method for constructing joint distributions on product spaces. Moreover, the probability kernel $(\omega, B) \mapsto K(X(\omega), B)$ is the obvious candidate for the conditional distribution of Y given X (more precisely, given σX), as confirmed by the following theorem:

Theorem 4.20: Conditional Distribution

If π is of the form (4.19), then the kernel L defined by

$$L(\omega, B) := K(X(\omega), B), \quad \omega \in \Omega, B \in \mathcal{E}$$

is a version of the conditional distribution of Y given X.

Proof. Since K is a kernel, L is one as well, and hence L has the two defining properties of a kernel. Moreover, for every $B \in \mathcal{E}$ the random variable $L(\cdot, B)$ is a $\mathcal{D}$-measurable function of X and hence it belongs to σX. It remains to verify that for all $g \in \mathcal{D}_+$ and $B \in \mathcal{E}$:

$$\mathbb{E}\, g(X)\mathbb{1}_{\{Y \in B\}} = \mathbb{E}\, g(X)L(\cdot, B).$$

But this follows from

$$\mathbb{E}\, g(X)\mathbb{1}_{\{Y \in B\}} = \int_D \mu(\mathrm{d}x)\, g(x) \int_B K(x, \mathrm{d}y)$$

$$= \int_D \mu(\mathrm{d}x)g(x)K(x, B) = \mathbb{E}\, g(X)K(X, B).$$

$\square$

Conversely, if we are given a general joint probability measure π, we can "disintegrate" it into the form (4.19) by finding the marginal distribution of X and the conditional distribution of Y given X. Here is the precise statement:

Theorem 4.21: Disintegration

Let π be a probability measure on the product space $(D \times E, \mathcal{D} \otimes \mathcal{E})$, where $(E, \mathcal{E})$ is a standard measurable space. Then, there exists a probability measure μ on $(D, \mathcal{D})$ and a transition kernel K from $(D, \mathcal{D})$ to $(E, \mathcal{E})$ such that

$$\pi(\mathrm{d}x, \mathrm{d}y) = \mu(\mathrm{d}x)\, K(x, \mathrm{d}y), \quad x \in D, y \in E.$$

Proof. This is in essence a corollary of Theorem 4.17. On the probability space $(D \times E, \mathcal{D} \otimes \mathcal{E}, \pi)$ define random variables X and Y via $X(\omega) := x$ and $Y(\omega) := y$, where $\omega = (x, y)$. Let μ be the distribution of X; that is $\mu(A) = \pi(A \times E)$, $A \in \mathcal{D}$. Since Y takes values in a standard measurable space, by Theorem 4.17

there exists a regular version L of the conditional distribution of Y given X. A random variable V that takes values in the product space belongs to $(\sigma X)_+$ if and only if $V((x, y)) = v(x)$ for some function $v \in \mathcal{D}_+$. It follows that $L(\omega, B)$ must be of the form $K(X(\omega), B)$. Writing $\mathbb{E}$ for the integration under the probability measure π, we have by the projection property in Definition 4.2:

$$\pi(A \times B) = \underbrace{\mathbb{E}\, \mathbb{1}_A(X)\, \mathbb{1}_B(Y) = \mathbb{E}\, \mathbb{1}_A(X)\, K(X, B)}_{\text{projection property}} = \int_A \mu(\mathrm{d}x)\, K(x, B).$$

As this holds for any $A \in \mathcal{D}$ and $B \in \mathcal{E}$, the theorem is proved. $\square$

In Theorem 4.21, we viewed π as the distribution of a random vector (X, Y), μ as the distribution of X, and $K(X, \cdot)$ as the conditional distribution of Y given X. When (X, Y) has a density p with respect to some product measure $\mu_0 \otimes \nu_0$ on $(D \times E, \mathcal{D} \otimes \mathcal{E})$, i.e., $\pi(\mathrm{d}x, \mathrm{d}y) = p(x, y)\, \mu_0(\mathrm{d}x)\, \nu_0(\mathrm{d}y)$, then the distribution μ of X has density

$$m(x) := \int_E \nu_0(\mathrm{d}y)\, p(x, y), \quad x \in D$$

with respect to μ_0. Moreover, K is the kernel

$$K(x, \mathrm{d}y) := k(x, y)\, \nu_0(\mathrm{d}y)$$

with

(4.22)
$$k(x, y) := \begin{cases} p(x, y)/m(x) & \text{if } m(x) > 0, \\ \int \mu_0(\mathrm{d}x')\, p(x', y) & \text{if } m(x) = 0. \end{cases}$$

The function $y \mapsto k(x, y)$ is called the *conditional density (with respect to ν_0) of Y given $X = x$.* This includes the case where the distribution of (X, Y) is discrete and the case where it is absolutely continuous with respect to the Lebesgue measure.

Note that for x where $m(x) = 0$, $k(x, \cdot)$ is simply the marginal density of Y. This ensures that $k(x, y)$ is defined for every $(x, y) \in D \times E$.

■ **Example 4.23 (Density of a Product of Random Variables)** Let (X, Y) have a density p with respect to the Lebesgue measure on $(\mathbb{R}^2, \mathcal{B}^2)$. What is the density of the product $Z := XY$?

Let $k(x, \cdot)$ be the conditional density of Y given $X = x$ and let m be the density of X. The conditional density of Z at z, given $X = x$, is the same as the conditional density of xY given $X = x$, which is $k(x, \frac{z}{x})/|x|$, using the transformation rule (Theorem 2.38). The density of (X, Z) at (x, z) is thus $m(x)k(x, \frac{z}{x})/|x| = p(x, \frac{z}{x})/|x|$. By integrating out x, we obtain that Z has density f, given by

$$f(z) := \int_{\mathbb{R}} \mathrm{d}x\, \frac{p(x, \frac{z}{x})}{|x|}.$$

An alternative approach to derive this is to apply Theorem 2.38 to the transformation $Z_1 = X_1 X_2$ and $Z_2 = X_1$, writing (X_1, X_2) for (X, Y). Let f_{Z_1, Z_2} be the density of (Z_1, Z_2). The matrix of partial derivatives for the mapping $x \mapsto z$ is

$$\frac{\partial z}{\partial x} = \begin{bmatrix} x_2 & x_1 \\ 1 & 0 \end{bmatrix},$$

which gives the Jacobian $|x_1|$ $(= |z_2|)$. It follows that for $(z_1, z_2) \in \mathbb{R}^2$:

$$f_{Z_1, Z_2}(z_1, z_2) = \frac{p(z_2, z_1/z_2)}{|z_2|}.$$

The probability density function of Z_1 is obtained by integrating out z_2. ∎

We finally mention the important notion of conditional independence.

Definition 4.24: Conditional Independence

Let $\mathcal{F}, \mathcal{F}_1, \ldots, \mathcal{F}_n$ be sub-σ-algebras of $\mathcal{H}$. The $\{\mathcal{F}_i\}$ are said to be *conditionally independent* given $\mathcal{F}$ if

$$\mathbb{E}_{\mathcal{F}} V_1 \cdots V_n = \mathbb{E}_{\mathcal{F}} V_1 \cdots \mathbb{E}_{\mathcal{F}} V_n$$

for all positive random variables $V_1, \ldots, V_n$ in $\mathcal{F}_1, \ldots, \mathcal{F}_n$, respectively.

For example, if $\mathcal{F} = \sigma X$, $\mathcal{F}_1 = \sigma Y$, and $\mathcal{F}_2 = \sigma Z$, the meaning is that, as far as predicting the value of any function of Y is concerned, the extra knowledge provided by Z loses all its significance once the value of X is known. The following proposition puts this in mathematical terms:

Proposition 4.25: Conditional Independence

$\mathcal{F}_1$ and $\mathcal{F}_2$ are conditionally independent given $\mathcal{F}$ if and only if

$$\mathbb{E}_{\mathcal{F} \vee \mathcal{F}_1} V_2 = \mathbb{E}_{\mathcal{F}} V_2 \quad \text{for all positive } V_2 \in \mathcal{F}_2.$$

Proof. By Definition 4.24, $\mathcal{F}_1$ and $\mathcal{F}_2$ are conditionally independent given $\mathcal{F}$ if and only if for all positive $V_1 \in \mathcal{F}_1$ and $V_2 \in \mathcal{F}_2$,

$$\mathbb{E}_{\mathcal{F}}[V_1 V_2] = (\mathbb{E}_{\mathcal{F}} V_1)\,(\mathbb{E}_{\mathcal{F}} V_2) = \mathbb{E}_{\mathcal{F}}[V_1 \mathbb{E}_{\mathcal{F}} V_2],$$

where the second equality follows from the "taking out what is known" property of the conditional expectation (see Theorem 4.4). Using the definition of conditional expectation (see Definition 4.2), the above equation holds if and only if

$$\mathbb{E}[V V_1 V_2] = \mathbb{E}[V V_1 \mathbb{E}_{\mathcal{F}} V_2]$$

for all positive $V \in \mathcal{F}$. Since random variables of the form VV_1 generate the σ-algebra $\mathcal{F} \vee \mathcal{F}_1$, this is equivalent to

$$\mathbb{E}_{\mathcal{F} \vee \mathcal{F}_1} V_2 = \mathbb{E}_{\mathcal{F}} V_2.$$

□

4.4 Existence of Probability Spaces

Consider a random experiment modelled via a random variable X that takes values in some measurable space $(E, \mathcal{E})$ with distribution μ. Given $(E, \mathcal{E})$ and μ, does there indeed exist a probability space $(\Omega, \mathcal{H}, \mathbb{P})$ and random variable X such that X has distribution μ? The answer is affirmative: take $\Omega := E$, $\mathbb{P} := \mu$ and $X(\omega) := \omega$, $\omega \in \Omega$.

What if we specify the random experiment in terms of the following "chained" simulation experiment?

1. Draw X_0 from μ, which is a probability measure on $(E_0, \mathcal{E}_0)$.

2. Given $X_0 = x_0$, draw X_1 from $K_1(x_0, \cdot)$, where K_1 is a probability transition kernel from $(E_0, \mathcal{E}_0)$ to $(E_1, \mathcal{E}_1)$.

3. Given $(X_0, X_1) = (x_0, x_1)$, draw X_2 from $K_2((x_0, x_1), \cdot)$, where K_2 is a probability transition kernel from $(E_0 \times E_1, \mathcal{E}_0 \otimes \mathcal{E}_1)$ to $(E_2, \mathcal{E}_2)$.

4. And so on.

Does there exist a probability space $(\Omega, \mathcal{H}, \mathbb{P})$ and a random process $(X_0, X_1, \ldots)$ that models this experiment? Motivated by our "trivial" construction at the beginning of this section, we take Ω as the space of all possible sequences $(x_0, x_1, \ldots)$, with each x_n in E_n. Thus, $\Omega := E_1 \times E_2 \times \cdots$. On Ω we put the product σ-algebra, so $\mathcal{H} := E_0 \otimes E_1 \otimes \cdots$. We define the random variables $X_0, X_1, \ldots$ as coordinate functions: for $\omega := (x_0, x_1, \ldots)$, let $X_n(\omega) := x_n$, $n \in \mathbb{N}$.

The main thing now is to construct a probability measure $\mathbb{P}$ on $(\Omega, \mathcal{H})$ that matches our experiment. Let

$$\boldsymbol{X}_n := (X_0, X_1, \ldots, X_n)$$

be the random vector obtained after the nth step of the simulation experiment. The probability measure $\mathbb{P}$ should be such that the distribution of $\boldsymbol{X}_0 = X_0$ is μ, and the distribution of $\boldsymbol{X}_n$ is

$$(4.26) \qquad \pi_n(\mathrm{d}\boldsymbol{x}_n) := \mu(\mathrm{d}x_0)K_1(x_0, \mathrm{d}x_1) \cdots K_n(\boldsymbol{x}_{n-1}, \mathrm{d}x_n).$$

Let $\mathcal{F}_n := \sigma X_n$. Every set in $\mathcal{F}_n$ is of the form

(4.27) $\qquad \{X_n \in B\} = B \times E_{n+1} \times \cdots, \qquad B \in \mathcal{E}_0 \otimes \cdots \otimes \mathcal{E}_n.$

Thus, $\mathbb{P}$ must be such that

(4.28) $\mathbb{P}(B_0 \times \cdots \times B_n \times E_{n+1} \times \cdots) = \pi_n(B_0 \times \cdots \times B_n), \qquad B_i \in \mathcal{E}_i, i = 0, \ldots, n.$

The following theorem shows that such a $\mathbb{P}$ exists and is unique:

Theorem 4.29: Ionescu–Tulcea's Existence Theorem

There exists a unique probability measure $\mathbb{P}$ on $(\Omega, \mathcal{H})$ such that (4.28) holds.

Proof. The goal is to use Carathéordory's extension Theorem 1.40. First, note that the union $\mathcal{A} := \cup_n \mathcal{F}_n$ is an *algebra* that generates $\mathcal{H}$. Consider the mapping $\mathbb{P}_0 : \mathcal{A} \to [0, 1]$ defined by (4.28). This mapping is finitely additive on $\mathcal{A}$, as each π_n is a probability measure. If we can show that this mapping $\mathbb{P}_0$ is *countably additive* on $\mathcal{A}$, then Carathéordory's extension theorem applies to conclude that $\mathbb{P}_0$ can be extended to a probability measure $\mathbb{P}$ on the σ-algebra $\mathcal{H}$. Since finite additivity already holds, showing that $\mathbb{P}_0$ is countably additive is equivalent (see Exercise 1.11) to showing that for every sequence of sets (H_k) in $\mathcal{A}$ it holds that

$$(H_k) \downarrow \emptyset \quad \text{implies} \quad \mathbb{P}_0(H_k) \downarrow 0.$$

Take $x := (x_0, x_1, \ldots) \in \Omega$ and $H \in \mathcal{A}$. Note that each $H \in \mathcal{A}$ is of the form (4.27) for some $n \in \mathbb{N}$. For this n and H, define

$$Q_m(x_m, H) := \mathbb{1}_H(x) = \mathbb{1}_B(x_n), \quad m \geq n,$$

where x_m and x_n are the truncated versions of x. Then, define recursively for $m = n - 1, \ldots, 0$:

(4.30) $\qquad Q_m(x_m, H) := \int_{E_{m+1}} K_{m+1}(x_m, dx_{m+1}) \, Q_{m+1}(x_{m+1}, H).$

We can think of $Q_m(x_m, H)$ as the probability that the outcome $x \in \Omega$ lies in H, after specifying the first $m + 1$ coordinates $x_0, \ldots, x_m$. In fact, (4.30) holds for *all* $m \in \mathbb{N}$. This is by definition true for $m < n$, but it holds also for $m \geq n$. For example,

$$Q_n(x_n, H) = \int_{E_{n+1}} K_{n+1}(x_n, dx_{n+1}) \, Q_{n+1}(x_{n+1}, H) = \int_{E_{n+1}} K_{n+1}(x_n, dx_{n+1}) \mathbb{1}_H(x)$$
$$= K_{n+1}(x_n, H) = \mathbb{1}_H(x).$$

Moreover, by expanding all the recursive integrals, we have

$$\int_{E_0} \mu_0(dx_0) Q_0(x_0, H) = \int_{E_0} \int_{E_1} \mu_0(dx_0) K_1(x_0, dx_1) Q_1(\boldsymbol{x}_1, H) = \cdots$$

$$= \int_{E_0 \times \cdots \times E_n} \mu_0(dx_0) K_1(x_0, dx_1) K_2(\boldsymbol{x}_1, dx_2) \cdots K_n(\boldsymbol{x}_{n-1}, dx_n) \mathbb{1}_H(\boldsymbol{x}) = \mathbb{P}_0(H).$$

Let $(H_k) \downarrow \emptyset$. Then, $(\mathbb{P}_0(H_k))$ is a decreasing sequence, as $\mathbb{P}_0$ is a finitely additive set function on $\mathcal{A}$. Suppose that this last sequence does *not* converge to 0. We want to show that this leads to a contradiction where $\cap_k H_k$ is not empty.

The function $f_k : x_0 \mapsto Q_0(x_0, H_k)$ is bounded by 1 and so, by the Bounded Convergence Theorem 2.36,

$$\lim \mathbb{P}_0(H_k) = \lim \int_{E_0} \mu_0(dx_0) Q_0(x_0, H_k) = \lim \mu_0 f_k$$

$$= \mu_0 \lim f_k = \int_{E_0} \mu_0(dx_0) \lim Q_0(x_0, H_k).$$

Since $\lim \mathbb{P}_0(H_k) > 0$ by our assumption (which we hope to prove wrong), there must exist $x_0^* \in E_0$ such that $\lim Q_0(x_0^*, H_k) > 0$. Similarly, because

$$Q_0(x_0, H_k) = \int_{E_1} K_1(x_0, dx_1) \, Q_1(\boldsymbol{x}_1, H_k),$$

we deduce, following the same line of reasoning, that there must be a x_1^* such that $\lim Q_1((x_0^*, x_1^*), H_k) > 0$, and by induction we conclude that there is a $\boldsymbol{x}^* \in \Omega$ such that $\lim \mathbb{1}_{H_k}(\boldsymbol{x}^*) > 0$; that is, $\lim \mathbb{1}_{H_k}(\boldsymbol{x}^*) = 1$. That means that $\boldsymbol{x}^* \in \cap_k H_k$, which contradicts the assumption that $H_k \downarrow \emptyset$.   □

Theorem 4.29 pertains only to stochastic processes with countable index sets. The next theorem extends this to arbitrary index sets. Below I, J, and K are index sets, with $I \subseteq J \subseteq K$ — mimicking their alphabetic order. Let proj_{JI} denote the natural projection from E^J to E^I.

Theorem 4.31: Kolmogorov's Extension Theorem

Let $(E, \mathcal{E})$ be a standard measurable space and K an index set. For each finite subset J of K, let π_J be a probability measure on the product space $(E^J, \mathcal{E}^J)$. If the (π_J) are *consistent* in the sense that

$$\pi_I = \pi_J \circ \text{proj}_{JI}^{-1}, \quad I \subseteq J, \quad |J| < \infty,$$

then there exists a unique probability measure $\mathbb{P}$ on $(\Omega, \mathcal{H})$ and a stochastic process X taking values in $(E, \mathcal{E})^K$ such that, with $X_J := (X_j)_{j \in J}$:

$$\mathbb{P}(X_J \in A) = \pi_J(A), \quad A \in \mathcal{E}^J, \quad |J| < \infty.$$

Proof. As usual, we take $\Omega := E^K$, $\mathcal{H} := \mathcal{E}^K$, and let X_t by the tth coordinate variable, i.e., $X_t(\omega) := \omega(t)$ for all t and ω. Obviously, $\mathcal{H} = \sigma\{X_t, t \in K\}$ and each X_t is $\mathcal{H}/\mathcal{E}$-measurable.

Take any countably infinite set of indexes $J := \{t_0, t_1, \ldots\}$. Let $J_n := \{t_0, \ldots, t_n\}$. By the disintegration Theorem 4.21, π_{J_n} is of the form (4.26), and so Theorem 4.29 guarantees the existence of a unique probability measure P_J on $(E, \mathcal{E})^J$ such that for all n:

$$\pi_{J_n} = P_J \circ \text{proj}_{JJ_n}^{-1}.$$

Moreover, for any countably infinite subset of indexes $I \subseteq J$, it holds that $P_I = P_J \circ \text{proj}_{JI}^{-1}$.

Now consider any $H \in \mathcal{E}^K$. By Theorem 2.66, the indicator $\mathbb{1}_H$ must be a function of X_j, $j \in J$ for some countably infinite set of indexes J. In other words, $H = \{X_J \in A\}$ for some $A \in \mathcal{E}^J$. Now define

$$\mathbb{P}(H) := P_J(A).$$

The consistency requirements ensure that this definition is without ambiguity. It remains to show that $\mathbb{P}$ is a probability measure. Countable additivity is shown as follows. Take (H_n) disjoint with union H, with $H_n = \{X_{J_n} \in A_n\}$ for some J_n and $A_n \in \mathcal{E}^{J_n}$. We may assume that the J_n are all the same, by replacing J_n with $J := \cup_n J_n$. Then, $H = \cup_n H_n = \cup_n\{X_J \in A_n\} = \{X_J \in A\}$, with $A := \cup_n A_n$, the (A_n) being disjoint. Hence, by countable additivity of P_J:

$$\mathbb{P}(H) = P_J(A) = \sum_n P_J(A_n) = \sum_n \mathbb{P}(H_n).$$

Since $\mathbb{P}(\Omega) = 1$, $\mathbb{P}$ is a probability measure. Uniqueness follows from the fact that events $\{X_J \in A\}$ form a p-system that generates $\mathcal{H}$. $\square$

■ **Example 4.32 (Existence of Gaussian Processes)** We show that for any given covariance function γ on $\mathbb{R}_+ \times \mathbb{R}_+$, there exists a probability space $(\Omega, \mathcal{H}, \mathbb{P})$ and a zero-mean Gaussian process $X := (X_t, t \geq 0)$ that has this covariance function.

Let $\Omega := \mathbb{R}^{\mathbb{R}_+}$ be the set of all mappings from $\mathbb{R}_+$ to $\mathbb{R}$, let $\mathcal{H} := \mathcal{B}^{\mathbb{R}_+}$ be the Borel σ-algebra thereon, and let (X_t) be the coordinate mappings; that is, $X_t(\omega) := \omega(t)$ for $\omega \in \Omega$. For each finite subset $J \subset \mathbb{R}_+$, let π_J be the $|J|$-dimensional Gaussian distribution on $\mathbb{R}^J$ with mean vector $\mathbf{0}$ and covariance matrix induced by γ. To apply Kolmogorov's extension Theorem 4.31, we need to verify that the (π_J) form a consistent family, but this is immediate by the properties of the multidimensional Gaussian distribution. Thus, there exists a probability measure $\mathbb{P}$ on $(\Omega, \mathcal{H})$ such that X is a zero-mean Gaussian process with covariance function γ. ■

The above example holds in particular for the covariance function $\gamma_{s,t} = s \wedge t$ of the Wiener process. However, Kolmogorov's extension theorem does not say anything about the path properties of the Gaussian process. Fortunately, path continuity can be established without too much difficulty using the concept of Hölder continuous modifications.

> **Definition 4.33: Hölder Continuity**
>
> A function $f : \mathbb{R}_+ \to \mathbb{R}$ is said to be *Hölder continuous* of order $\alpha > 0$ on $B \subseteq \mathbb{R}_+$ if there is a constant k such that
>
> $$|f(t) - f(s)| \le k\,|t - s|^\alpha, \quad s, t, \in B.$$

A Hölder continuous function is evidently continuous. It is even *uniformly* continuous; that is, for every $\varepsilon > 0$ there is a $\delta > 0$ such that for every x it holds that if y satisfies $|y - x| < \delta$, then $\lfloor f(y) - f(x)| < \varepsilon$.

A stochastic process $\widetilde{X} := \{\widetilde{X}_t, t \in \mathbb{T}\}$ is said to be a *modification* of $X := \{X_t, t \in \mathbb{T}\}$ if $\mathbb{P}(\widetilde{X}_t = X_t) = 1$ for all $t \in \mathbb{T}$. The following theorem gives sufficient moment conditions for a stochastic process to have a modification that is Hölder continuous:

> **Theorem 4.34: Hölder Continuous Modification**
>
> Let $X := (X_t, t \in [0, 1])$ be a stochastic process with state space $\mathbb{R}$. If there exist constants c, p, q in $(0, \infty)$ such that
>
> (4.35) $$\mathbb{E}\,|X_t - X_s|^p \le c\,|t - s|^{1+q}, \quad s, t \in [0, 1],$$
>
> then for every $\alpha \in (0, q/p)$ there is a modification $\widetilde{X}$ of X that is almost surely Hölder continuous of order α on $[0, 1]$.

Proof. Let D be the set of *dyadic numbers*:

(4.36) $$D := \cup_{n=0}^\infty D_n \quad \text{with} \quad D_n := \left\{ \frac{k}{2^n}, k = 0, 1, \ldots, 2^n \right\}.$$

Define, for $\alpha \in [0, q/p]$, the random variable K by

$$K := \sup_{\substack{s,t \in D \\ s \ne t}} \frac{|X_t - X_s|}{|t - s|^\alpha}.$$

Then, by the definition of K,

(4.37) $$|X_t - X_s| \le K\,|t - s|^\alpha, \quad s, t \in D.$$

In particular, if $K(\omega) < \infty$, then the path $X(\omega)$ is Hölder continuous of order α on D. We want to show that K is almost surely finite. We do this by proving that $\mathbb{E} K^p < \infty$. Consider

$$M_n := \max_{\substack{s,t \in D_n \\ t-s=2^{-n}}} |X_t - X_s|.$$

Note that the maximum is taken here over 2^n variables. We now want to bound $\mathbb{E} M_n^p$ using the elementary bound for the expectation of the maximum of positive random variables $\{U_i\}$,

$$\mathbb{E} \max_i U_i \leq \mathbb{E} \sum_i U_i = \sum_i \mathbb{E} U_i,$$

and the assumption (4.35). This gives:

$$\|M_n\|_p^p = \mathbb{E} M_n^p \leq 2^n c \, (2^{-n})^{1+q} = c \, 2^{-nq}.$$

Next, take $s, t \in D$ with $s < t$. Let $s_n := \inf D_n \cap [s, 1]$, so that $s_n \downarrow s$. Similarly, let $t_n := \sup D_n \cap [0, t]$, so that $t_n \uparrow t$. Since $s, t \in D$, we have $s_n = s$ and $t_n = t$ for all n large enough. Thus, for every m, we can write

$$X_t - X_s = \sum_{n \geq m} (X_{t_{n+1}} - X_{t_n}) + X_{t_m} - X_{s_m} + \sum_{n \geq m} (X_{s_n} - X_{s_{n+1}}).$$

Now consider an arbitrary pair (s, t) in D with $0 < t - s \leq 2^{-m}$. Then, $t_m - s_m$ is either 0 or 2^{-m}. Hence,

$$|X_t - X_s| \leq \sum_{n \geq m} M_{n+1} + M_m + \sum_{n \geq m} M_{n+1} \leq 2 \sum_{n \geq m} M_n.$$

By the definition of K, we have

$$K = \sup_{\substack{s,t \in D \\ s \neq t}} \frac{|X_t - X_s|}{|t - s|^\alpha} \leq \sup_m \sup_{\substack{s,t \in D \\ 2^{-m-1} < |t-s| \leq 2^{-m}}} \frac{|X_t - X_s|}{|t - s|^\alpha} \leq \sup_m (2^{m+1})^\alpha \, 2 \sum_{n \geq m} M_n$$

$$= \sup_m 2^{1+\alpha} \sum_{n \geq m} (2^m)^\alpha M_n \leq 2^{1+\alpha} \sum_{n \geq 0} 2^{n\alpha} M_n.$$

We now prove that $\mathbb{E} K^p < \infty$ for all $p \geq 0$. For $p \geq 1$, we have

$$(\mathbb{E} K^p)^{1/p} = \|K\|_p \leq 2^{1+\alpha} \sum_{n \geq 0} 2^{n\alpha} \|M_n\|_p \leq 2^{1+\alpha} \sum_{n \geq 0} 2^{n\alpha} c^{1/p} \, 2^{-nq/p} < \infty,$$

since $\alpha < q/p < 1$. For $p < 1$, we use the fact that $(\sum_n \beta_n)^p \leq \sum_n \beta_n^p$ for positive $\{\beta_n\}$. Hence,

$$\mathbb{E} K^p \leq 2^{p(1+\alpha)} \sum_{n \geq 0} 2^{n\alpha p} c \, 2^{-nq} < \infty,$$

since $\alpha < q/p$.

Since $\mathbb{E} K^p < \infty$, the event $\Omega_0 := \{K < \infty\}$ is almost sure. Outside Ω_0 define $\widetilde{X}(\omega) := 0$. For $\omega \in \Omega_0$, $X(\omega)$ satisfies (4.37) and is thus Hölder continuous of order α on D. For $\omega \in \Omega_0$ put

$$\widetilde{X}_t(\omega) := \lim_{\substack{s \to t \\ s \in D}} X_s(\omega), \quad t \in [0, 1].$$

Then, for those ω, the path $\widetilde{X}(\omega)$ is Hölder continuous of order α on $[0,1]$, and the same holds trivially for $\omega \notin \Omega_0$. Finally, for each $t \in [0, 1]$ we have $X_t = \widetilde{X}_t$ almost surely, so $\widetilde{X}$ is the sought modification. $\qquad\qquad\square$

■ **Example 4.38** (**Existence of the Wiener Process**) For the Gaussian process $(X_t, t \in [0, 1])$ in Example 4.32 with $\gamma_{s,t} := s \wedge t$, the conditions of Theorem 4.34 apply with, for instance, $p := 4$, $q := 1$, and $c := 3$. Thus, there exists a continuous modification $(W_t, t \in [0, 1])$ of $(X_t, t \in [0, 1])$. Doing the same for $(X_t, t \in [n, n + 1])$, we obtain a process $W := (W_t, t \geq 0)$ that is continuous and has the same finite-dimensional distribution as X. This is the Wiener process whose existence had to be shown. ■

4.5 Markov Property

One of the most studied stochastic processes is the Markov process. In a way it can be viewed as the probabilistic analogue of a difference or differential equation, in the sense that, conditional on the past history, future increments depend only on the present state. The concepts of conditional probability and expectation are thus essential to the analysis.

Throughout this section, $\mathbb{T}$ is some subset of $\mathbb{R}$ and $X := (X_t, t \in \mathbb{T})$ is a stochastic process on some probability space $(\Omega, \mathcal{H}, \mathbb{P})$, taking values in some measurable space $(E, \mathcal{E})$.

Definition 4.39: Adaptedness

A stochastic process $X := (X_t, t \in \mathbb{T})$ taking values in some measurable space $(E, \mathcal{E})$ is said to be *adapted* to a filtration $\mathcal{F} := (\mathcal{F}_t, t \in \mathbb{T})$ if for all $t \in \mathbb{T}$, X_t is $\mathcal{F}_t/\mathcal{E}$-measurable.

In what follows, we assume that the process X is adapted to some filtration $\mathcal{F} := (\mathcal{F}_t, t \in \mathbb{T})$ — for example, the natural filtration of X, in which case $\mathcal{F}_t := \sigma(X_u, u \leq t, u \in \mathbb{T})$. Let $\mathcal{G}_t := \sigma(X_u, u \geq t, u \in \mathbb{T})$ be the *future* σ-algebra of X for time t.

Definition 4.40: Markov Property

X is said to be *Markovian* relative to $\mathcal{F}$ if for every t, the past $\mathcal{F}_t$ and the future $\mathcal{G}_t$ are conditionally independent, given the present state X_t.

This means, by Proposition 4.25, that for every t and positive $V \in \mathcal{G}_t$, it holds that $\mathbb{E}[V \mid \mathcal{F}_t] = \mathbb{E}[V \mid X_t]$, which in turn can be shown (e.g., see Çinlar (2011, Section 6.1)) to be equivalent to the requirement that for every time $u > t$ and $f \in \mathcal{E}_+$,

(4.41) $$\mathbb{E}[f(X_u) \mid \mathcal{F}_t] = \mathbb{E}[f(X_u) \mid X_t].$$

Markov processes are often associated with a family of probability transition kernels; see Definition 1.63 for the meaning of the latter.

Definition 4.42: Markovian Transition Function

A *Markovian transition function* on $(E, \mathcal{E})$ is a family $(P_{t,u}, t, u \in \mathbb{T}, t < u)$ of probability transition kernels on $(E, \mathcal{E})$ that satisfy the *Chapman–Kolmogorov equations*:

(4.43) $$P_{s,t} P_{t,u} = P_{s,u}, \quad s < t \le u.$$

A Markovian process X is said to admit $(P_{t,u})$ as a transition function if

$$\mathbb{E}[f(X_u) \mid X_t] = (P_{t,u} f)(X_t), \quad t < u, f \in \mathcal{E}_+.$$

For $f = \mathbb{1}_A$, $A \in \mathcal{E}$, this means

$$\mathbb{P}(X_u \in A \mid X_t = x) = P_{t,u}(x, A),$$

giving an intuitive meaning to the transition functions. Define $P_t := P_{0,t}$. If $P_{t,u} = P_{u-t}$ for all $t \le u$, the associated Markovian process X is said to be *time-homogeneous* and P_t satisfies the *semi-group* property

$$P_{t+s} = P_s P_t, \quad s, t \ge 0.$$

Markov processes can be classified according to the nature (discrete or continuous) of their state space E and time set $\mathbb{T}$. We call a Markovian process a *Markov chain*[1] if $\mathbb{T}$ is discrete.

The Markov process *par excellence* in continuous time is the Lévy process; this includes the Wiener, Poisson, compound Poisson, and pure-jump Lévy processes

---

[1]There is no consensus on this nomenclature. Some call a Markovian process a Markov chain if E is discrete.

discussed in Section 2.8. If $(X_t, t \geq 0)$ is a d-dimensional Lévy process, then, writing $X_u = X_t + (X_u - X_t)$ for $0 \leq t < u$, we have that X_u is the sum of $X_t \in \mathcal{F}_t$ and a random increment that is independent of $\mathcal{F}_t$. Hence, (4.41) holds, and by combining this with the stationarity of increments, we conclude that a Lévy process is a time-homogeneous Markov process. Indeed, if X_t has a probability distribution π_t on $\mathbb{R}^d$, then it admits the transition function

$$P_t(\boldsymbol{x}, A) = \pi_t(A - \boldsymbol{x}), \quad \boldsymbol{x} \in \mathbb{R}^d, \ A \in \mathcal{B}^d,$$

where $A - \boldsymbol{x}$ denotes the set $\{\boldsymbol{y} - \boldsymbol{x} : \boldsymbol{y} \in A\}$. For example, for a Poisson process with rate c, we have

$$P_t(x, A) = \sum_{y \in A - x} \frac{e^{-ct}(ct)^y}{y!}.$$

In Section 4.5.2 we generalize the pure-jump Lévy processes to pure-jump Markov processes.

4.5.1   Time-homogeneous Markov Chains

We defined a Markov chain as a Markovian process with a discrete, i.e., countable, time set. Without loss of generality, we may take $\mathbb{T} = \mathbb{N}$ for the time set. The state space E may be countable or uncountable. Let μ a probability distribution on E and let K be a probability transition kernel on $(E, \mathcal{E})$. Consider the following random experiment:

1. Draw X_0 from μ.
2. Given $X_0 = x_0$, draw X_1 from $K(x_0, \cdot)$.
3. Given $(X_0, X_1) = (x_0, x_1)$, draw X_2 from $K(x_1, \cdot)$.
4. Given $(X_0, X_1, X_2) = (x_0, x_1, x_2)$, draw X_3 from $K(x_2, \cdot)$.
5. And so on.

The first part of Section 4.4 proves the existence of the resulting stochastic process $(X_n, n \in \mathbb{N})$, which, by construction, is a Markov chain. Starting with an initial distribution μ, each X_{n+1} is drawn from the conditional distribution of X_{n+1} given X_n, which is $K(X_n, \cdot)$. The Markov chain admits the Markovian transition function

$$P_{m,n} = K^{n-m}, \quad m \leq n, \ m, n \in \mathbb{N},$$

where K^{n-m} is the $(n - m)$-fold product of K. Thus, X is time-homogeneous and

$$\mathbb{P}(X_n \in A \mid X_0 = x) = K^n(x, A).$$

In typical applications, the conditional distribution of X_{n+1} given X_n can be specified in two common ways as follows:

- An explicit expression for $K(x, dy)$ is known, and it is easy to sample from $K(x, \cdot)$ for all $x \in E$.

- The process $(X_n, n \in \mathbb{N})$ satisfies a recurrence relation

$$(4.44) \qquad\qquad X_{n+1} := g(X_n, U_n), \quad n \in \mathbb{N},$$

where g is an easily evaluated function and U_n is an easily generated random variable whose distribution may depend on X_n.

An important instance of the first case occurs when the Markov chain X has a discrete state space E. Its distribution is then completely specified by the distribution of X_0 (the initial distribution) and the matrix of one-step transition probabilities $\mathbf{P} = [p_{ij}]$, where

$$p_{ij} := \mathbb{P}(X_{n+1} = j \mid X_n = i), \quad i, j \in E.$$

The conditional distribution of X_{n+1} given $X_n = i$ is therefore the discrete distribution given by the ith row of $\mathbf{P}$. This leads to the following algorithm:

■ **Algorithm 4.45 (Simulating a Markov Chain on a Discrete State Space)**

1. Draw X_0 from the initial distribution. Set $n := 0$.

2. Draw X_{n+1} from the discrete distribution corresponding to the X_nth row of $\mathbf{P}$.

3. Set $n := n + 1$ and go to Step 2.

■ **Example 4.46 (A Markov Chain Maze)** At time $n = 0$ a robot is placed in compartment 3 of the maze in Figure 4.47. At each time $n = 1, 2, \ldots$ the robot chooses one of the adjacent compartments with equal probability. Let X_n be the robot's compartment at time n. Then, (X_n) is a time-homogeneous Markov chain with transition matrix $\mathbf{P}$ given in Figure 4.47.

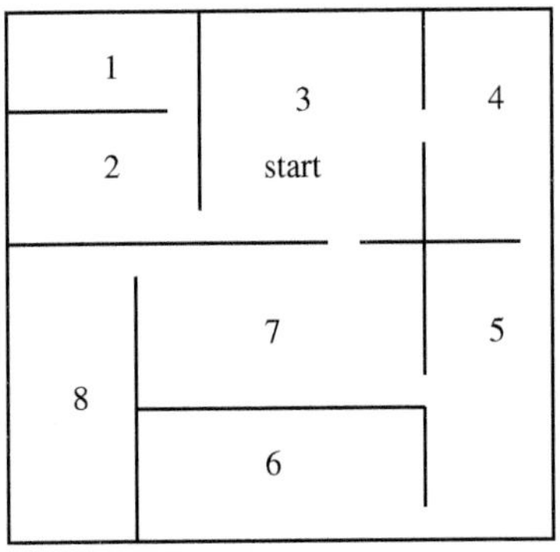

$$\mathbf{P} = \begin{pmatrix} 0 & 1 & 0 & 0 & 0 & 0 & 0 & 0 \\ \frac{1}{2} & 0 & \frac{1}{2} & 0 & 0 & 0 & 0 & 0 \\ 0 & \frac{1}{3} & 0 & \frac{1}{3} & 0 & 0 & \frac{1}{3} & 0 \\ 0 & 0 & \frac{1}{2} & 0 & \frac{1}{2} & 0 & 0 & 0 \\ 0 & 0 & 0 & \frac{1}{3} & 0 & \frac{1}{3} & \frac{1}{3} & 0 \\ 0 & 0 & 0 & 0 & 1 & 0 & 0 & 0 \\ 0 & 0 & \frac{1}{3} & 0 & \frac{1}{3} & 0 & 0 & \frac{1}{3} \\ 0 & 0 & 0 & 0 & 0 & 0 & 1 & 0 \end{pmatrix}.$$

Figure 4.47: A maze and the corresponding transition matrix.

The following MATLAB program implements Algorithm 4.45. The first 101 states of the process are given in Figure 4.48.

```
nmax = 101; a = 0.5; b = 1/3;
P = [0, 1, 0, 0, 0, 0, 0, 0; a, 0, a, 0, 0, 0, 0, 0;
     0, b, 0, b, 0, 0, b, 0; 0, 0, a, 0, a, 0, 0, 0;
     0, 0, 0, b, 0, b, b, 0; 0, 0, 0, 0, 1, 0, 0, 0;
     0, 0, b, 0, b, 0, 0, b; 0, 0, 0, 0, 0, 0, 1, 0 ]
x = zeros(1,nmax); x(1)= 3;
for n=1:nmax-1
    x(n+1) = min(find(cumsum(P(x(n),:))> rand));
end
hold on, plot(0:nmax-1,x,'.'), plot(0:nmax-1,x), hold off
```

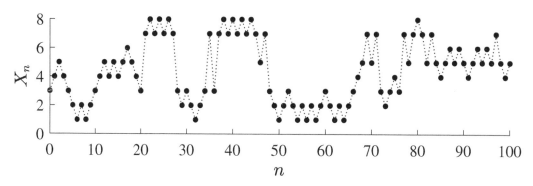

Figure 4.48: Realization of the maze process.

∎

Next is a typical example of a Markov chain that is specified by a recurrence relation such as (4.44).

∎ **Example 4.49 (Random Walk)** In a random walk process $(X_n, n \in \mathbb{N})$, the recurrence is

$$X_{n+1} := X_n + U_n , \quad n \in \mathbb{N},$$

where $X_0 := 0$ and $U_0, U_1, \ldots$ is a sequence of iid random variables. The following MATLAB program simulates a random walk where the $\{U_i\}$ are standard normal. A typical path is given in Figure 4.50.

```
n=200;
U = randn(n,1);
X = cumsum(U);
plot(1:n,X)
```

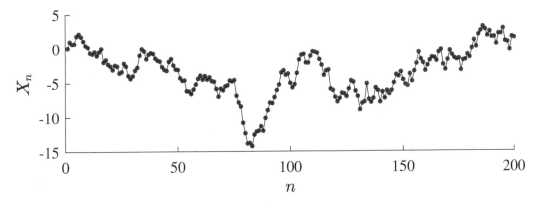

Figure 4.50: Realization of a random walk with standard normal increments.

■

4.5.2 Markov Jump Processes

The easiest way to create a Markov jump process is to "subordinate" a Markov chain, in the following sense. Let $X := (X_n, n \in \mathbb{N})$ be a Markov chain with probability transition kernel K, as in Section 4.5.1, and let $(N_t, t \ge 0)$ be a homogeneous Poisson process with rate $0 < c < \infty$, independent of X. Then, the process $Y := (Y_t, t \ge 0)$ defined by

(4.51) $$Y_t := X_{N_t}, \quad t \ge 0,$$

is a pure-jump Markov process. We say that (N_t) is *subordinated* to X. Think of N_t as the time on a clock which moves 1 unit forward at each jump time of the Poisson process. The transition function of Y is given by

$$P_t(x, A) = \sum_{n=0}^{\infty} \frac{e^{-ct}(ct)^n}{n!} K^n(x, A),$$

where K^n is the nth power of the Markov kernel K.

■ **Example 4.52 (Markov Jump Process)** Let X be a Markov chain with state space $\{1, 2, 3\}$ and one-step transition matrix $\mathbf{P} = [p_{ij}]$, given by

$$\mathbf{P} := \begin{bmatrix} 1/3 & 1/3 & 1/3 \\ 1/2 & 1/2 & 0 \\ 3/4 & 0 & 1/4 \end{bmatrix},$$

and let $(N_t, t \ge 0)$ be a Poisson process with rate $c := 1$. Figure 4.53 shows a typical realization of the Markov jump process $(Y_t, t \ge 0) := (X_{N_t}, t \ge 0)$.

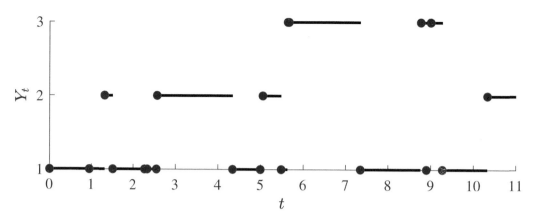

Figure 4.53: Realization of a Markov jump process.

When the process is in state i, the holding or *sojourn* time S_i at i is distributed according to the random sum $\sum_{i=1}^{M} V_i$, where the $\{V_i\}$ are iid $\mathsf{Exp}(c)$ distributed and $M \sim \mathsf{Geom}(1 - p_{ii})$, independently. Consequently, $S_i \sim \mathsf{Exp}(c(1 - p_{ii}))$. When the process leaves state i at the end of the sojourn time S_i, it jumps to state $j \neq i$ with probability $p_{ij}/(1 - p_{ii})$.   ∎

A time-homogeneous Markov jump process $(X_t, t \geq 0)$ taking values in a countable set $E := \{1, 2, \ldots\}$ is often defined via its *Q-matrix*,

$$(4.54) \qquad \mathbf{Q} := \begin{pmatrix} -q_1 & q_{12} & q_{13} & \cdots \\ q_{21} & -q_2 & q_{23} & \cdots \\ q_{31} & q_{32} & -q_3 & \cdots \\ \vdots & \vdots & \vdots & \ddots \end{pmatrix},$$

where q_{ij} is the *transition rate* from i to j:

$$q_{ij} := \lim_{h \downarrow 0} \frac{\mathbb{P}(X_{t+h} = j \mid X_t = i)}{h}, \quad i \neq j, \quad i, j \in E$$

and q_i is the *holding rate* in i:

$$q_i := \lim_{h \downarrow 0} \frac{1 - \mathbb{P}(X_{t+h} = i \mid X_t = i)}{h}, \quad i \in E.$$

In Example 4.52, we have $q_i = c(1 - p_{ii})$ and $q_{ij} = q_i\, p_{ij}/(1 - p_{ii}) = c p_{ij}$.

Subordination, as in (4.51), provides the simplest way to create and study Markov jump processes, but the most general way uses stochastic integrals with respect to Poisson random measures, in a similar way as in the construction of pure-jump Lévy processes in Section 2.8.4.

In contrast to the Lévy process construction, we will employ a homogeneous Poisson random measure N on $\mathbb{R}_+ \times \mathbb{R}_+$, rather than a Poisson random measure on

$\mathbb{R}_+ \times \mathbb{R}^d$ with mean measure Leb $\otimes \nu$. In particular, the mean measure of N is the Lebesgue measure on $\mathbb{R}_+ \times \mathbb{R}_+$. We construct a d-dimensional Markov jump process $X := (X_t, t \geq 0)$ based on N by imposing that any atom (t, u) of N creates a jump of X at time t of magnitude $j(X_t, u)$, where $j : \mathbb{R}^d \times \mathbb{R}_+ \to \mathbb{R}^d$ is a *jump* function. Specifically,

$$(4.55) \qquad X_t := \int_{[0,t] \times \mathbb{R}_+} N(\mathrm{d}s, \mathrm{d}u) \, j(X_{s-}, u), \quad t \geq 0,$$

where $X_{s-} := \lim_{r \uparrow s} X_r$. This is a stochastic integral equation, which is pathwise defined; i.e., for every realization of N. Define the corresponding *Lévy kernel* by

$$(4.56) \qquad L(x, B) := \mathrm{Leb}\{u \geq 0 : j(x, u) \neq 0, x + j(x, u) \in B\}, \quad B \in \mathcal{B}^d.$$

The Lévy kernel is a transition kernel from $(\mathbb{R}^d, \mathcal{B}^d)$ into $(\mathbb{R}^d, \mathcal{B}^d)$. Think of it as a generalization of a Lévy measure that depends on x. For a pure-jump Lévy process X with Lévy measure ν, we have $j(x, u) = j(u)$, and $L(x, B) = \nu(B - x)$. Heuristically, $L(x, B)$ represents the *rate* at which X jumps from x into B, in the sense that in a small time interval $[t, t + \delta]$ this probability is approximately $\delta L(x, B)$. In particular, $k(x) := L(x, \mathbb{R}^d)$ is the rate at which X leaves state x — also called the *killing rate*. This rate can be finite or infinite. If $k(x) < \infty$, then we can write $L(x, B) = k(x) K(x, B)$, where K is a probability transition kernel, and $K(x, B)$ denotes the probability that the killed particle at state x is reborn in set $B \in \mathcal{B}^d$.

The behavior of the process is similar to what was discussed in Example 4.52. When at state x, the process stays there an exponential amount of time with parameter $k(x)$ and then jumps to a new state y with probability $K(x, \mathrm{d}y)$, independent of the sojourn time in x. If the process hits a state x for which $k(x) = 0$, it will stay there forever. Such a state is called an *absorbing state*.

Similar to the construction of pure-jump Lévy processes, some restriction must be placed on the size of the jumps. The least restrictive condition is

$$\int_0^\infty \mathrm{d}u \, (\|j(x, u)\| \wedge 1) < \infty \quad \text{for all } x.$$

4.5.3 Infinitesimal Generator

We conclude this chapter with the derivation of the infinitesimal generator of the Markov jump process X defined by (4.55). Suppose X has transition function P_t. This can be viewed as an operator which acts on a function f via

$$(P_t f)(x) = \mathbb{E}^x f(X_t),$$

where $\mathbb{E}^x$ denotes the expectation under which the process starts from state x. By the right-continuity of the paths, as $t \downarrow 0$, P_t tends to the identity operator I, with

$If = f$. It makes sense to consider the operator $P_t - I$ for small t, scaled by t. This gives the *infinitesimal generator* (or simply *generator*) G of the Markov process:

$$(Gf)(x) := \lim_{t\downarrow0} \frac{(P_t f)(x) - f(x)}{t} = \lim_{t\downarrow0} \mathbb{E}^x \frac{f(X_t) - f(x)}{t},$$

provided the limit exists. The following specifies the generator:

Theorem 4.57: Generator of a Markov Jump Process

The infinitesimal generator of a Markov jump process with Lévy kernel L is given by

$$(Gf)(x) = \int_{\mathbb{R}^d} L(x, dy)[f(y) - f(x)], \quad f \in \mathcal{B}^d(\text{bounded}).$$

The proof is given after we have introduced some new concepts and a useful result on Poisson integration. Let N be a Poisson random measure on $\mathbb{R}_+ \times E$ with mean measure μ. The σ-algebra on E is $\mathcal{E}$. We think of an atom (t, x) of N as representing the time t where a "mark" x enters. Define $\mathcal{F} := (\mathcal{F}_t, t \geq 0)$ as the "natural" filtration associated with N, in the sense that

$$\mathcal{F}_t := \sigma(N((a, b] \times B) : 0 \leq a < b \leq t, B \in \mathcal{E}), \quad t \in \mathbb{R}_+.$$

Thus, $\mathcal{F}_t$ contains all the information about N up to time t, specifically with regard to the positions of its atoms that arrive before or at time t.

Suppose $Y := (Y(t), t \geq 0)$ is some real-valued stochastic process associated with N. The process Y is said to be $\mathcal{F}$-predictable if for any interval $(a, b]$ the information about $(Y(t), t \in (a, b])$ is already contained in $\mathcal{F}_a$. More precisely, Y is $\mathcal{F}$-predictable if it is $\mathcal{F}^p/\overline{\mathcal{B}}$-measurable[2], where $\mathcal{F}^p$ is the σ-algebra on $\overline{\mathbb{R}}_+ \times \Omega$ defined by

(4.58) $\quad \mathcal{F}^p := \sigma((a, b] \times H : a, b \in \mathbb{R}_+, H \in \mathcal{F}_a) \vee \sigma(\{0\} \times H : H \in \mathcal{F}_0),$

where $\vee$ signifies that we take the smallest σ-algebra that contains the union — as the union of two σ-algebras is not usually a σ-algebra itself.

The following proposition says that, under suitable predictability conditions, the expected value of the stochastic integral

$$NR := \int N(dt, dz)R(t, z)$$

only depends on the mean measure of N:

---
[2]Recall that this is the same as saying that the process lies in $\mathcal{F}^p$.

Proposition 4.59: Poisson Integrals

Let N be a Poisson random measure on $\mathbb{R}_+ \times E$ with mean measure μ satisfying $\mu(\{0\} \times E) = 0$. Let $R := (R(t,z), t \geq 0, z \in E)$ be a positive stochastic process in $\mathcal{F}^p \otimes \mathcal{E}$. Then,

$$(4.60) \qquad\qquad \mathbb{E}\, N R = \mathbb{E}\, \mu R.$$

Proof. The collection of all functions $R : \Omega \times \mathbb{R}_+ \times E \to \overline{\mathbb{R}}_+$ that are $(\mathcal{F}^p \otimes E)$-measurable and satisfy (4.60) forms a monotone class, and so by the Monotone Class Theorem 1.33 it suffices to show (4.60) only for indicators of sets $H \times A \times B$, where either

1. $H \in \mathcal{F}_a$, $A = (a, b]$, and $B \in \mathcal{E}$, or

2. $H \in \mathcal{F}_0$, $A = \{0\}$, and $B \in \mathcal{E}$.

In the first case, we have

$$\mathbb{E}\, N R = \mathbb{E}\, \mathbb{1}_H N(A \times B) = \mathbb{E}\, \mathbb{1}_H \mathbb{E}_a N(A \times B)$$
$$= \mathbb{E}\, \mathbb{1}_H \mathbb{E}\, N(A \times B) = \mathbb{E}\, \mathbb{1}_H \mu(A \times B) = \mathbb{E}\, \mu R,$$

where $\mathbb{E}_a$ denotes the conditional expectation given $\mathcal{F}_a$. In the second case, $\mu(A \times E) = 0$, which implies $N(A \times E) = 0$ almost surely, so that both sides of (4.60) are 0. □

Proof of Theorem 4.57. Let g be a bounded positive Borel function on $\mathbb{R}^d \times \mathbb{R}^d$. Consider the stochastic process

$$R(s, u) := \mathbb{1}_{[0,t]}(s)\, g\left(X_{s-}, X_{s-} + j(X_{s-}, u)\right) \mathbb{1}_{\{j(X_{s-}, u) \neq 0\}}, \quad s, u \geq 0.$$

Verify that it satisfies the predictability conditions of Proposition 4.59. Thus, we have

$$\mathbb{E}^x \sum_{s \leq t} g(X_{s-}, X_s) \mathbb{1}_{\{X_{s-} \neq X_s\}} = \mathbb{E}^x \int_{\mathbb{R}_+ \times \mathbb{R}_+} N(ds, du) R(s, u)$$
$$= \mathbb{E}^x \int_{\mathbb{R}_+ \times \mathbb{R}_+} ds\, du\, R(s, u).$$

Writing out $R(s, u)$ in the expected integral on the right-hand side gives

$$\mathbb{E}^x \int_{[0,t]} ds \int_{\mathbb{R}_+} du\, g\left(X_{s-}, X_{s-} + j(X_{s-}, u)\right) \mathbb{1}_{\{j(X_{s-}, u) \neq 0\}}$$
$$= \mathbb{E}^x \int_0^t ds \int_{\mathbb{R}^d} L(X_{s-}, y) g(X_{s-}, y),$$

where the equality follows from the definition of L in (4.56). In the last integral we may replace $s-$ with s, as $X_s = X_{s-}$ for all but a countable set of times s. Now, as a special case, take $g(x, y) := f(y) - f(x)$ and apply the previous to find

$$\mathbb{E}^x\big(f(X_t) - f(X_0)\big) = \mathbb{E}^x \sum_{s \le t} \big(f(X_s) - f(X_{s-})\big)$$

$$= \mathbb{E}^x \int_0^t \mathrm{d}s \int_{\mathbb{R}^d} L(X_s, y)\big(f(y) - f(X_s)\big)$$

$$= \mathbb{E}^x \int_0^t \mathrm{d}s\, Gf(X_s) = \int_0^t \mathrm{d}s\, \mathbb{E}^x Gf(X_s),$$

assuming we can swap the integral and expectation (by Fubini). Consequently,

$$\lim_{t \downarrow 0} \frac{(P_t f)(x) - f(x)}{t} = \frac{\mathrm{d}}{\mathrm{d}t} P_t f(x)\Big|_{t=0} = \mathbb{E}^x Gf(X_0) = Gf(x).$$

$\square$

Exercises

1. Prove Theorem 4.13, using the properties of the probability measure $\mathbb{P}$ and the definition of a conditional probability given an event in (4.12).

2. Let $X_1, X_2, \ldots$ be a Bernoulli process with success parameter p. Let, $S_n := X_1 + \cdots + X_n$ and $Z_n := S_n - np$, for $n = 1, 2, \ldots$. Find the following:

(a) $\mathbb{E}[S_{n+1} \mid X_1, \ldots, X_n]$

(b) $\mathbb{E}[S_{n+1} \mid S_1, \ldots, S_n]$

(c) $\mathbb{E}[Z_{n+1} \mid X_1, \ldots, X_n]$

(d) $\mathbb{E}[Z_{n+1} \mid Z_1, \ldots, Z_n]$

3. Let X be a positive random variable and Y a random variable that takes values in a countable set D. Show that $\mathbb{E}_{\sigma Y} X = h(Y)$, where h is defined by

$$h(y) := \mathbb{E}[X \mid Y = y] = \int_0^\infty \mathrm{d}x\, \mathbb{P}(X > x \mid Y = y), \quad y \in D.$$

4.* If $X \sim \mathrm{Poi}(\lambda)$ and $Y \sim \mathrm{Poi}(\mu)$ are independent, show that, conditional on $X + Y = z$, the distribution of X is $\mathrm{Bin}(z, \lambda/(\mu + \lambda))$.

5. Suppose that (X, Y) has a density $f_{X,Y}$ with respect to the Lebesgue measure on $(\mathbb{R}^2, \mathcal{B}^2)$. The (marginal) densities of X and Y with respect to the Lebesgue

measure of $(\mathbb{R}, \mathcal{B})$ are denoted by f_X and f_Y, respectively. Show that $Z = X + Y$ has density f_Z given by

$$(4.61) \qquad f_Z(z) := \int_{\mathbb{R}} \mathrm{d}x \, k(x, z - x) f_X(x), \quad z \in \mathbb{R},$$

where $k(x, \cdot)$ is the conditional density of Y given $X = x$. When X and Y are positive and independent, show that

$$f_Z(z) := \int_0^z \mathrm{d}x \, f_Y(z - x) f_X(x), \quad z \in \mathbb{R}.$$

6. Let (X_1, X_2) have density f_X with respect to the Lebesgue measure on $(\mathbb{R}^2, \mathcal{B}^2)$. Using conditioning and/or transformation arguments, express the pdf of X_1/X_2 in terms of f_X.

7.* Let $X_1, X_2, \ldots$ be an iid sequence of random variables. Define $S_0 := 0$ and $S_n := \sum_{i=1}^n X_i$ for $n = 1, 2, \ldots$. Suppose that each X_i has a $\mathsf{Ber}(p)$ distribution. For $k \in \mathbb{N}$, let

$$N_k := \inf\{n : S_n = k\}$$

be the first time that the process $\{S_n, n \in \mathbb{N}\}$ hits level k. By conditioning on S_{j-1}, derive $\mathbb{P}(N_k = j), j = k, k+1, \ldots$.

8. Let $X_1, X_2, \ldots$ be an iid sequence of random variables. Define $S_0 := 0$ and $S_n := \sum_{i=1}^n X_i$ for $n = 1, 2, \ldots$. Suppose that each X_i takes the value 1 or -1, with probability p and $q = 1 - p$, respectively. Let $N := \inf\{n : S_n = 1\}$. By conditioning on X_1, determine the probability generating function $z \mapsto \mathbb{E}z^N$ of N. From this, find $\mathbb{P}(N = n), n \in \mathbb{N}$. Hint: you may use Newton's formula: $(1 + t)^\alpha = \sum_{i=0}^\infty \binom{\alpha}{i} t^i$, where $\alpha \in \mathbb{R}$ and $\binom{\alpha}{i}$ is the generalized binomial coefficient; i.e., $\binom{\alpha}{i} := \alpha(\alpha - 1) \cdots (\alpha - i + 1)/i!$ for $i = 1, 2, \ldots$ and $\binom{\alpha}{0} := 1$.

9. Let Y and Z be independent, with $Y \sim \mathsf{Gamma}(a, c)$ and $Z \sim \mathsf{Gamma}(b, c)$. Define $X := Y + Z$. Find the kernel K such that the conditional distribution of Y given X is $K(X, \cdot)$. For the case where $a = b = 1$, show that $K(x, \mathrm{d}y) = \frac{1}{x}\mathrm{d}y, y \in (0, x)$.

10. Let $X \sim \mathsf{Gamma}(n, p/(1 - p))$ for some $p \in (0, 1)$ and integer n. Suppose that, conditionally on $X = x$, the random variable Z has a $\mathsf{Poi}(x)$ distribution. What is the distribution of Z? Hint: use the MGF Tables 2.4 and 2.5 and the repeated conditioning property of conditional expectations.

11.* Using MGFs and repeated conditioning, show that $Z := (X_1 + X_2 X_3)/\sqrt{1 + X_3^2}$, where X_1, X_2, X_3 are iid $\mathcal{N}(0, 1)$ distributed, has a standard normal distribution.

12. Let $Z := X/Y$, where $X \sim \mathcal{U}[-\frac{1}{2}, \frac{1}{2}]$ and $Y \sim \mathcal{U}[0, 1]$ independently. What is the distribution of Z conditional on $X^2 + Y^2 \leq 1$?

13. Let $Z := X/(X+Y)$, where $X \sim \text{Beta}(\alpha, 1)$ and $Y \sim \text{Beta}(\beta, 1)$ independently. What is the conditional distribution of Z given $X + Y \leq 1$?

14. Let $U_1, \ldots, U_n \sim_{\text{iid}} \mathcal{U}[0,1]$. Sorting these in ascending order, gives the *order statistics* $U_{(1)} < \cdots < U_{(n)}$. Show that $U_{(k)} \sim \text{Beta}(k, n-k+1)$.

15. Suppose that $U_1, U_2, \ldots \sim_{\text{iid}} \mathcal{U}[0,1]$ and $N \sim \text{Poi}(\lambda)$, independently. Denote the order statistics of $U_1, \ldots, U_N$ as $U_{(1)} < U_{(2)} < \cdots < U_{(N)}$ and let r be a fixed integer. Conditional on $N \geq r$, find the distribution of $U_{(r)}$.

16.* Suppose that $\{X_t, t \in \mathbb{Z}\}$ is a Gaussian process. Show that the following two statements are equivalent:

(a) *(Markov Property):* $\mathbb{E}[f(X_{t+1}) \mid X_j, \forall j \leq t] = \mathbb{E}[f(X_{t+1}) \mid X_t]$ for any positive measurable function f.

(b) $\mathbb{E}[X_{t+1} \mid X_j, \forall j \leq t] = \mathbb{E}[X_{t+1} \mid X_t]$ for all $t \in \mathbb{Z}$.

17. Suppose that $\{X_t, t \in \mathbb{Z}\}$ is a Gaussian process satisfying:

(a) $\mathbb{E}[X_{t+1} \mid X_j, j \leq t] = \mathbb{E}[X_{t+1} \mid X_t]$.

(b) *(Stationarity):* For all integers t, s, there holds $\mathbb{E}X_t = \mu$ and $\text{Cov}(X_t, X_s) = \varrho(t-s)$ for some even function $\varrho(x) = \varrho(-x)$ with $\varrho(0) = 1$.

Show that ϱ must satisfy the recursion $\varrho(t) = \varrho(1)\varrho(t-1)$ for $t = 1, 2, 3, \ldots$.

18. Show that a Markov jump process that is defined by a Q-matrix (4.54) has the form (4.51) provided that $\sup_i q_i < \infty$.

CHAPTER **5**

MARTINGALES

Martingales form a mainstay of modern probability. In this chapter, we introduce (sub)martingales and show their use. Stopping times, filtrations, and uniform integrability play important roles in the analysis. The key results are Doob's stopping theorem and the martingale convergence theorem. Example applications include proofs for the Law of Large Numbers and the Radon–Nikodym theorem.

Martingales are real-valued stochastic processes in continuous or discrete time that model "fair" betting games, in which at any time the expected future profit is always 0, irrespective of any information about the past of the game. Many proofs in probability theory are facilitated by the use of martingales. In the analysis of martingales, it will be useful to consider the process at certain random times, called *stopping times* (also referred to as *optional times*). We introduce these next.

5.1 Stopping Times

In what follows, $(\Omega, \mathcal{H}, \mathbb{P})$ is a probability space, $\mathbb{T}$ a subset of $\mathbb{R}$, and $\mathcal{F} := (\mathcal{F}_t, t \in \mathbb{T})$ a filtration of sub σ-algebras of $\mathcal{H}$. We add a point of infinity to $\mathbb{T}$ to define $\overline{\mathbb{T}} := \mathbb{T} \cup \{\infty\}$. Recall that each $\mathcal{F}_t$ is a sub σ-algebra of $\mathcal{H}$ that models all the information on a random experiment (described by the probability space) that is available at time t. If X describes our measurements on the random experiment, then the information we have available at time t is given by $\sigma(X_s, s \le t)$. However, $\mathcal{F}_t$ could have more information than $\sigma(X_s, s \le t)$; that is, $\mathcal{F}$ could be "finer" than the natural filtration of X.

> **Definition 5.1: Stopping Time**
>
> A random time $T : \Omega \to \overline{\mathbb{T}}$ is called a *stopping time* of $\mathcal{F}$ if
>
> $$\{T \le t\} \in \mathcal{F}_t \quad \text{for each } t \in \mathbb{T}.$$

Consider the stochastic process $(Z_t) := (\mathbb{1}_{\{T \le t\}}, t \in \mathbb{T})$. Imagine that T is the time when a "catastrophe" happens in the random experiment, at which time the process (Z_t) jumps from 0 to 1. If $\mathcal{F}_t$ is the available information at time t, then T is a stopping time if and only if we can tell, at time t, using only the information available in $\mathcal{F}_t$, whether the catastrophe has occurred yet or not. In other words, for a stopping time T we are able to construct an alarm system that uses only the information in $\mathcal{F}$ and that for every $\omega \in \Omega$ sounds exactly at the time of catastrophe $T(\omega)$.

■ **Example 5.2 (Hitting Time for a Symmetric Random Walk)** Let $\{B_n, n = 1, 2, \ldots\}$ be a collection of iid $\mathsf{Ber}(1/2)$ random variables. Consider the symmetric random walk on the integers $X := (X_n, n \in \mathbb{N})$, defined by $X_{n+1} := X_n + 2B_n - 1$ for $n = 1, 2, \ldots$ and $X_0 := 0$; see also Example 2.31 and Figure 2.32. Define the hitting time of state 1 by

$$T := \inf\{n : X_n = 1\}.$$

Then, T is a stopping time with respect to the natural filtration of X. At each time n we are able to tell, based on the history of X up to and including time n, whether the process has hit state 1 at or before time n or not.

An example of a random time that is *not* a stopping time is:

$$R := \min\{n : X_n = \max_{0 \le i \le 100} X_i\},$$

that is, the first time that the maximum of X up to time 100 is reached. For example, to assess whether the event $\{R \le 50\}$ occurs or not, we have to take into consideration the whole history of X up to time 100. ■

Let $\mathcal{F}$ be a filtration on $\mathbb{T}$. Because we typically extend $\mathbb{T}$ to $\overline{\mathbb{T}}$ by including the point ∞, it makes sense to extend $\mathcal{F}$ to a filtration with index set $\overline{\mathbb{T}}$. The way to do this is to define

$$\mathcal{F}_\infty := \bigvee_{t \in \mathbb{T}} \mathcal{F}_t.$$

That is, the σ-algebra generated by the union of all the $\mathcal{F}_t$. Then, $(\mathcal{F}_t, t \in \overline{\mathbb{T}})$ is a filtration on $\overline{\mathbb{T}}$. We may use the *same* notation, $\mathcal{F}$, for this extended filtration, as it carries the same information as the original $\mathcal{F}$. The advantage, however, is that we can extend any adapted process $X := (X_t, t \in \mathbb{T})$ to a process $(X_t, t \in \overline{\mathbb{T}})$ by appending a random variable X_∞ to X. Moreover, every stopping time T of the extended $\mathcal{F}$ is a stopping time of the original $\mathcal{F}$, and vice versa.

Definition 5.3: Past until T

Let $\mathcal{F}$ be a filtration on $\overline{\mathbb{T}}$, and let T be a stopping time of it. The σ-algebra of *past information until T* is defined as

$$\mathcal{F}_T := \{H \in \mathcal{H} : H \cap \{T \le t\} \in \mathcal{F}_t \text{ for every } t \in \overline{\mathbb{T}}\}.$$

We can think of $\mathcal{F}_T$ as the information on a random experiment that is available at a random stopping time T. If T is constant, $T = t$, then $\mathcal{F}_T$ is simply $\mathcal{F}_t$. As usual, we will use the notation $\mathcal{F}_T$ also for the collection of numerical random variables that are $\mathcal{F}_T/\overline{\mathcal{B}}$-measurable.

Theorem 5.4: $\mathcal{F}_T$ Random Variables

A random variable V belongs to $\mathcal{F}_T$ if and only if

$$X_t := V \mathbb{1}_{\{T \le t\}} \in \mathcal{F}_t \text{ for every } t \in \overline{\mathbb{T}}.$$

Proof. Without loss of generality we may assume that $V \ge 0$. For all $v \ge 0$ and $t \in \mathbb{T}$, we have:

$$\{V > v\} \cap \{T \le t\} = \{X_t > v\}.$$

Hence, $\{V > v\} \in \mathcal{F}_T$ for all $v \ge 0$ if and only if $X_t \in \mathcal{F}_t$ for every $t \in \overline{\mathbb{T}}$. $\square$

Intuitively, $V \in \mathcal{F}_T$ means that for every ω, the value $V(\omega)$ can be discovered at time $T(\omega)$.

5.2 Martingales

We have already mentioned that a martingale can be thought of as a process that models a fair game. We now make this more precise. Let $(X_t, t \in \mathbb{T})$ be a stochastic process that describes the profit X_t at each time t in betting game that is ruled by chance. Suppose that the information available to us at time t is given by the σ-algebra $\mathcal{F}_t$. If for any $s < t$ the expected incremental profit $X_t - X_s$ is 0 *given $\mathcal{F}_s$*, then the game is "fair", in the sense that we cannot win, whatever betting strategy we devise. The process is then a martingale. As martingale analysis heavily relies on conditional expectations given a filtration $\mathcal{F} = (\mathcal{F}_t, t \in \mathbb{T})$, we will use the following notation convention:

$$\mathbb{E}_s V = \mathbb{E}_{\mathcal{F}_s} V = \mathbb{E}[V \mid \mathcal{F}_s],$$

the first notation being the most economical and elegant. If we do not specify the filtration, the natural filtration is assumed.

Definition 5.5: (Sub/Super)Martingale

Let $X := (X_t, t \in \mathbb{T})$ be a real-valued stochastic process that is adapted to a filtration $\mathcal{F}$, and for which each X_t is integrable. The process X is called an $\mathcal{F}$-(sub/super)martingale if for all $t > s$:

$$\begin{aligned} (5.6) \qquad & \mathbb{E}_s[X_t - X_s] \geq 0 \quad \text{(submartingale)}, \\ & \mathbb{E}_s[X_t - X_s] = 0 \quad \text{(martingale)}, \\ & \mathbb{E}_s[X_t - X_s] \leq 0 \quad \text{(supermartingale)}. \end{aligned}$$

When specifying (sub/super)martingales, we will often drop the quantifier "$\mathcal{F}$-" if the corresponding filtration is obvious; for example, when it is the natural filtration of the process.

Using the betting game analogy, for submartingales the expected incremental profit is positive and for supermartingales it is negative. Think of a *sub* martingale as a process that starts *low* and has the tendency to *increase*. You want your betting game to be a submartingale! It suffices to only consider submartingales and martingales for further analysis, as the negative of a supermartingale is a submartingale. Note that in the definition we may replace $\mathbb{E}_s[X_t - X_s]$ with $\mathbb{E}_s X_t - X_s$, as $\mathbb{E}_s X_s = X_s$, by the "taking out what is known" property of the conditional expectation; see Theorem 4.4. From the same theorem we will frequently use the "repeated conditioning" property of the conditional expectation. In particular, taking the expectation of $\mathbb{E}_s X_t - X_s$ shows that a martingale has constant expectations:

$$\mathbb{E}X_t = \mathbb{E}X_s.$$

Here are some more easy properties of (sub)martingales.

Theorem 5.7: Properties of (Sub)Martingales

1. When $\mathbb{T} = \mathbb{N}$, the martingale equality $\mathbb{E}_s[X_t - X_s] = 0$ holds if and only if $\mathbb{E}_k[X_{k+1} - X_k] = 0, k \in \mathbb{N}$.

2. If X and Y are $\mathcal{F}$-submartingales, then so is $aX + bY$ for $a, b \geq 0$.

3. If X and Y are $\mathcal{F}$-submartingales, then so is $X \vee Y$.

4. Let f be a convex function on $\mathbb{R}$ and X an $\mathcal{F}$-martingale. If $f(X_t)$ is integrable for all t, then $f(X) := (f(X_t))$ is an $\mathcal{F}$-submartingale.

Proof.

1. Necessity is obvious. For sufficiency, suppose that $\mathbb{E}_k[X_{k+1} - X_k] = 0$ for all $k \in \mathbb{N}$. Take any $m, n \in \mathbb{N}$ with $n > m$. Then,

$$\mathbb{E}_m[X_n - X_m] = \mathbb{E}_m[(X_n - X_{n-1}) + \cdots + (X_{m+1} - X_m)]$$
$$= \mathbb{E}_m\mathbb{E}_{n-1}[X_n - X_{n-1}] + \cdots + \mathbb{E}_m\mathbb{E}_m[X_{m+1} - X_m] = 0,$$

where we have used the repeated conditioning rule $\mathbb{E}_m\mathbb{E}_k = \mathbb{E}_m$ for $k \geq m$, as well as the linearity of the conditional expectation.

2. This follows from the linearity of the conditional expectation and the submartingale properties of X and Y: for any $t > s$, $\mathbb{E}_s[aX_t + bY_t] = a\,\mathbb{E}_sX_t + b\,\mathbb{E}_sY_t \geq aX_s + bY_s$.

3. This is immediate from $\mathbb{E}_s[X_t \vee Y_t] \geq (\mathbb{E}_sX_t) \vee (\mathbb{E}_sY_t) \geq X_s \vee Y_s$.

4. This is a consequence of Jensen's inequality (Lemma 2.45). Namely, for $t > s$, we have

$$\mathbb{E}_s f(X_t) \geq f(\mathbb{E}_sX_t) = f(X_s). \qquad \square$$

■ **Example 5.8 (Random Walk)** Consider the random walk $X := (X_n, n \in \mathbb{N})$ defined by $X_0 := 0$ and

$$X_{n+1} := X_n + U_n, \quad n \in \mathbb{N},$$

where $\{U_n, n \in \mathbb{N}\}$ is a collection of iid random variables. If $\mathbb{E}U_n = 0$, then X is a martingale (with respect to its natural filtration). Namely, each $\mathbb{E}X_n = 0$, so every X_n is integrable, and

$$\mathbb{E}_nX_{n+1} = \mathbb{E}_n[X_n + U_n] = X_n + \mathbb{E}_nU_n = X_n.$$

More generally, if $\mathbb{E}U_n = c$ is finite, then $(X_n - nc, n \in \mathbb{N})$ is a martingale.

■

■ **Example 5.9 (Sum of Bernoullis)** Let $B_1, B_2, \ldots$ be a Bernoulli process with success parameter p, and define $S_n := B_1 + \cdots + B_n$ and $Z_n := S_n - np$, for $n = 1, 2, \ldots$, with $S_0 := 0$. From the previous example, we see that (Z_n) is a martingale. However, there are more martingales that can be constructed from the Bernoulli process. For example, with $q := 1 - p$, define

$$M_n := \frac{(q/p)^{S_n}}{(2q)^n}, \quad n \in \mathbb{N}.$$

Then, M_n is obviously integrable, and

$$\mathbb{E}_nM_{n+1} = M_n\,\mathbb{E}_n\frac{(q/p)^{B_{n+1}}}{2q} = M_n\frac{(q + (q/p)p)}{2q} = M_n,$$

so that (M_n) is a martingale.

■

■ **Example 5.10 (Poisson Process)** Let $(N_t, t \geq 0)$ be a Poisson (counting) process with rate c. Then,

$$M_t := N_t - ct, \quad t \geq 0,$$

is a martingale. Adaptedness and integrability are evident, and the martingale property follows from the stationarity and independence of the increments of the Poisson process:

$$\mathbb{E}_s[M_t - M_s] = \mathbb{E}_s[N_t - N_s] - c(t - s) = \mathbb{E}[N_t - N_s] - c(t - s) = 0.$$

■

A stochastic process $X := (X_t, t \in \mathbb{T})$ is said to be a *uniformly integrable martingale* with respect to a filtration $\mathcal{F}$ if X is both a martingale and is uniformly integrable (UI). See Definition 3.32 for the definition and Proposition 3.34 for sufficient and necessary conditions for uniform integrability. The main example of a UI martingale is given next.

■ **Example 5.11 (Standard UI Martingale)** Let Z be an integrable random variable. Then, the stochastic process $X := (X_t, t \in \mathbb{T})$ defined by

(5.12) $$X_t := \mathbb{E}_t Z$$

is an UI martingale. We can check this as follows:

1. $X_t \in \mathcal{F}_t$ by the definition of conditional expectation. So X is adapted to the filtration $\mathcal{F}$.

2. Using Jensen's inequality (see Lemma 2.45) and repeated conditioning, we have

$$\mathbb{E}|X_t| = \mathbb{E}|\mathbb{E}_t Z| \leq \mathbb{E}\mathbb{E}_t|Z| = \mathbb{E}|Z| < \infty,$$

which shows that X_t is integrable for every t.

3. Again using repeated conditioning, we have for $s < t$:

$$\mathbb{E}_s X_t = \mathbb{E}_s \mathbb{E}_t Z = \mathbb{E}_s Z = X_s,$$

so that X is a martingale.

4. Proving uniform integrability is a bit more involved and requires Proposition 3.34, Point 3; that is, (X_t) is UI if and only if there is an increasing convex function f such that $f(x)/x \to \infty$ and

$$\sup_t \mathbb{E}f(|X_t|) < \infty.$$

Since the random variable Z is UI (since Z is integrable), there exists an f with the above property such that

$$\mathbb{E}f(|Z|) < \infty.$$

Since $X_t = \mathbb{E}_t Z$, we have by Jensen's inequality (Lemma 2.45):

$$|X_t| = |\mathbb{E}_t Z| \le \mathbb{E}_t |Z|,$$

and since f is increasing and convex,

$$f(|X_t|) \le f(\mathbb{E}_t |Z|) \underbrace{\le}_{\text{Jensen}} \mathbb{E}_t f(|Z|).$$

Taking the supremum of the expectations on both sides (unconditioning) gives:

$$\sup_t \mathbb{E}f(|X_t|) < \infty,$$

so that (X_t) is uniformly integrable by Proposition 3.34, Point 3.

∎

5.3 Optional Stopping

The main result of this section is Doob's "optional" stopping theorem, which, under certain conditions, extends the martingale property (5.6) to stopping times $S < T$, to give

(5.13) $\mathbb{E}_S[X_T - X_S] = 0,$

and consequently (assuming $0 \in \mathbb{T}$):

(5.14) $\mathbb{E}X_T = \mathbb{E}X_0.$

This is certainly not true for *all* stopping times, as the following example illustrates:

∎ **Example 5.15 (Symmetric Random Walk Continued)** Let X be a symmetric random walk on the integers, as in Example 5.2 and, as in that example, let T be the first time that X hits level 1; this is a stopping time. Then,

$$1 = \mathbb{E}X_T \ne \mathbb{E}X_0 = 0.$$

Note that T is not bounded, because $\mathbb{P}(T > a) > 0$ for all $a > 0$. Exercise 10 shows that, interestingly, $\mathbb{P}(T < \infty) = 1$ but $\mathbb{E}T = \infty$.   ∎

The question is under what conditions (5.13) and (5.14) *do* hold. For this it will be convenient to first study certain integral transformations of martingales. Indeed, such stochastic integration techniques underpin many elegant proofs in probability.

5.3.1 Stochastic Integration

Stochastic analysis often involves the evaluation of a *stochastic integral*; that is, an integral of the form

(5.16)
$$\int_0^t F_s \, dX_s,$$

where the *integrator* $X := (X_t)$ and *integrand* $F := (F_t)$ are real-valued stochastic processes on some probability space $(\Omega, \mathcal{H}, \mathbb{P})$. Let $f_t := F_t(\omega)$ be the realization of F_t for a specific $\omega \in \Omega$ — and similarly let $x_t := X_t(\omega)$, but with the additional assumption that $x := (x_t)$ is right-continuous and left-limited.

If for each $\omega \in \Omega$ the real-valued function x is increasing, and if the function $f := (f_t)$ is measurable, then the integral (5.16) can be defined *pathwise*; that is, for each ω the integral

(5.17)
$$\int_0^t f_s \, dx_s$$

is well-defined. Namely, since there corresponds to the increasing function x a unique measure μ, we can define (5.17) as the Lebesgue integral $\int_0^t \mu(ds) \, f_s$. An example of such a stochastic integral is (2.81) in Section 2.8.2, where the integrator is the Poisson random measure.

The arguments above can be extended to the case when the function x can be written as the difference of two increasing functions, say $x = y - z$. Then, (5.16) can again be defined pathwise as

$$\int_0^t f_s \, dy_s - \int_0^t f_s \, dz_s.$$

The corresponding pathwise integral is called the *Lebesgue–Stieltjes* integral.

It is not too difficult to show that such functions x are of *bounded variation* on $[0, t]$, meaning that their *total variation* on $[0, t]$ is finite. The latter is defined as

(5.18)
$$V_x(t) := \sup_{(\Pi_n)} \sum_{k=0}^{n-1} |x_{s_{k+1}} - x_{s_k}|,$$

where the supremum is taken over any *segmentation*[1] Π_n of the interval $[0, t]$,

(5.19) $\Pi_n := \{s_0, \ldots, s_n : 0 = s_0 < s_1 < \cdots < s_n = t\}, \quad n \in \mathbb{N},$

such that the *mesh* of Π_n, defined as $\|\Pi_n\| := \max_{0 \le k \le n-1}(s_{k+1} - s_k)$, goes to 0 as $n \to \infty$. Any right-continuous function of bounded variation can be written as a difference of two increasing functions; see Exercise 11 for a proof.

The upshot of this is that for a process X that does not "wiggle" too much, the stochastic integral can be defined pathwise as a Lebesgue–Stieltjes integral.

---

[1]Also called *partition*. However, note that in our usage a partition is a collection of sets, whereas a segmentation is a collection of points.

■ **Example 5.20 (Integral With Respect to a Poisson Process)** As an example of a stochastic integral that can be evaluated pathwise like (5.17), consider the integral

$$\int_0^t N_{s-} \, dN_s,$$

where $(N_t, t \geq 0)$ is a Poisson counting process with rate 1 and N_{s-} denotes the left limit $\lim_{r\uparrow s} N_r$. For each $\omega \in \Omega$, the integral is readily evaluated as a Lebesgue integral. In particular, denoting the arrival times during $[0,t]$ by $T_1(\omega), T_2(\omega), \ldots, T_{N_t(\omega)}(\omega)$, let μ be the corresponding counting measure. At these N_t arrival times, the integrand (N_{s-}) takes the values $0, 1, \ldots, N_t - 1$, respectively. Hence, the integral is $0 + 1 + \cdots + N_t - 1 = N_t(N_t - 1)/2$. In the same way, $\int_0^t N_s \, dN_s = (N_t + 1)N_t/2$, so the two integrals are not the same. In particular, we see that the elementary "change of variable" rule $\int_0^t g(s) \, dg(s) = g^2(t)/2$ does not hold here. Next, consider the integral

$$Z_t := \int_0^t N_{s-} \, dM_s = \int_0^t N_{s-} \, dN_s - \int_0^t N_{s-} \, ds, \quad t \geq 0,$$

where $M_s := N_s - s$. We have already evaluated the first integral on the right-hand side. Likewise, the second integral can be evaluated pathwise as a Lebesgue integral. Due to the insensitivity of the integral (Proposition 1.54), we may replace $s-$ with s in the second integral. In terms of the $\{T_n\}$, the integral can be expressed as the area below the graph of $(N_s, 0 \leq s \leq t)$:

$$\sum_{k=0}^{N_t}(k - 1)(T_k - T_{k-1}) + (t - T_{N_t})N_t,$$

where the sum is assumed to be empty if $N_t \leq 1$. A typical path of the integral-transform process (Z_t) is given in Figure 5.21.

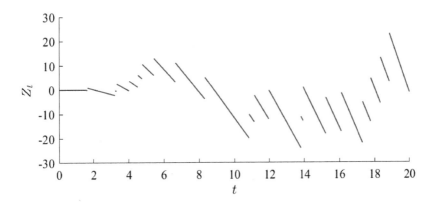

Figure 5.21: The integral-transform process (Z_t) is a martingale.

We saw in Example 5.10 that $(M_t, t \geq 0)$ is a martingale with respect to the natural filtration $(\mathcal{F}_t)$ of (N_t). It turns out that the process $(Z_t, t \geq 0)$ is also a martingale with respect to $(\mathcal{F}_t)$. To prove the martingale property (that is, $\mathbb{E}_r(Z_t - Z_r) = 0$ for every $0 \leq r \leq t$, where $\mathbb{E}_r$ denotes the conditional expectation given $\mathcal{F}_r$), we need to show that

$$\mathbb{E}_r \int_r^t N_{s-} \, dN_s = \mathbb{E}_r \int_r^t [N_r + (N_s - N_r)] \, ds.$$

Since $Z_t = (N_t^2 - N_t)/2$, the left-hand side yields $\mathbb{E}_r[(N_t^2 - N_t) - (N_r^2 - N_r)]/2 = N_r(t-r) + (t-r)^2/2$, where we used the independence and stationarity of the Poisson process increments. Similarly, for the right-hand side we have $N_r(t-r) + \int_r^t \mathbb{E}_r(N_s - N_r) \, ds = N_r(t-r) + \int_r^t(s-r) \, ds = N_r(t-r) + (t-r)^2/2$, so the two expected integrals are equal given $\mathcal{F}_r$, and hence the martingale property holds. This example illustrates an important recurring theme in stochastic integration: if the integrator process is a martingale, and the integrand is adapted and left-continuous, then the resulting integral forms again a martingale. ∎

When X has infinite variation, such as the Wiener process, then the pathwise definition of stochastic integration is no longer applicable. In this case, the most used stochastic integral is the *Itô integral*, which we will discuss in more detail in Chapter 7; a brief mention suffices at this point. For the Wiener process $(W_t, t \geq 0)$, the Itô integral on the interval $[0, t]$ can be defined as the limit

$$\int_0^t F_s \, dW_s := \lim_{n \to \infty} \sum_{k=0}^{n-1} F_{s_k}(W_{s_{k+1}} - W_{s_k}),$$

where (Π_n) is again of the form (5.19) with $\|\Pi_n\| \to 0$. For this integral to be well-defined, some conditions must be placed on the integrand. In particular, if F is adapted and left-continuous on $(0, t]$, then it can be shown to be a *predictable* process in the following sense; see also (4.58).

Definition 5.22: Predictable Process

Let $\mathcal{F} := (\mathcal{F}_t, t \geq 0)$ be a filtration. A real-valued stochastic process F is said to be $\mathcal{F}$-*predictable* (or simply *predictable*) if it is measurable with respect to the σ-algebra

$$\mathcal{F}^p := \sigma(H \times (a, b] : a, b \in \mathbb{R}_+, H \in \mathcal{F}_a) \vee \sigma(H \times \{0\} : H \in \mathcal{F}_0).$$

Loosely speaking, this means that we are able to predict the value of F_t from all the information available before t.

For discrete processes $X := (X_n, n \in \mathbb{N})$ and $F := (F_n, n \in \mathbb{N})$, the treatment of stochastic integration becomes much easier. We simply define

$$(5.23) \qquad \int_0^n F_k \, dX_k := F_0 X_0 + F_1 (X_1 - X_0) + \cdots + F_n (X_n - X_{n-1}), \qquad n = 1, 2, \ldots.$$

For $n = 0$, the integral is defined as $F_0 X_0$. The justification is as follows. If X is an increasing process, then for any $\omega \in \Omega$, there corresponds to $X(\omega)$ a discrete measure with mass $m(k) := X_k(\omega) - X_{k-1}(\omega)$, at $k = 1, \ldots, n$, and $m(0) := X_0(\omega)$. The corresponding Lebesgue integral with respect to $F(\omega)$ is then given by (5.23), as a function of ω. If X is arbitrary, we can write X as the difference of two increasing processes and then take the integral as the difference between the positive and negative part. This is equivalent to associating with $X(\omega)$ a *signed* measure (i.e., the difference of two measures) with the same masses $m(k), k = 0, \ldots, n$, given above. This thus defines the Lebesgue–Stieltjes integral in the discrete case; i.e., the Stieltjes integral interpreted as a Lebesgue integral.

Note that (5.23) defines a new stochastic process

$$(5.24) \qquad\qquad\qquad Z_n := \int_0^n F_k \, dX_k, \qquad n \in \mathbb{N}.$$

Think of F_n as the number of shares owned at period n — say the time interval $(n-1, n]$ in years — and X_n as the price at the end of period n. Then, $(X_n - X_{n-1}) F_n$ is the profit made during period n. Moreover, Z_n is the total capital at the end of period n, starting with an initial capital of $X_0 F_0$. If (X_n) is a martingale, then, by definition, conditional on the past until time $n - 1$, the expected increase in share price is 0; that is, $\mathbb{E}_{n-1}(X_n - X_{n-1}) = 0$. We shall see that if F is a predictable process, then the expected profit during period n, given the information up to period $n - 1$ is also 0. Here is the precise definition of predictability for the discrete case.

Definition 5.25: Discrete Predictable Process

Given a filtration $\mathcal{F} := (\mathcal{F}_n, n \in \mathbb{N})$, a process $F := (F_n, n \in \mathbb{N})$ is said to be *predictable* if $F_0 \in \mathcal{F}_0$ and $F_n \in \mathcal{F}_{n-1}, n = 1, 2, \ldots$.

Continuing the financial analogy, if the shares process F is predictable with respect to the natural filtration of the share price — that is, if the number of shares owned at period n is completely determined by the evolution of the share price in the preceding periods, then $\mathbb{E}_{n-1}(X_n - X_{n-1}) F_n = F_n \mathbb{E}_{n-1}(X_n - X_{n-1}) = 0$, as reported. The total capital process (Z_n) defined in (5.24) is in fact a martingale, as the following theorem shows:

Theorem 5.26: Integral Transform of a Martingale

Let F be a bounded predictable process and X a martingale adapted to a filtration $\mathcal{F}$. Then, the integral transform $Z_n := \int_0^n F_k \, dX_k, n \in \mathbb{N}$ is an $\mathcal{F}$-martingale as well.

Proof. Obviously, the process $Z := (Z_n)$ is adapted to $\mathcal{F}$, as both F and X are. The boundedness assumption on F ensures that Z is integrable. Finally, Z satisfies the martingale property:

$$\mathbb{E}_n(Z_{n+1} - Z_n) = \mathbb{E}_n[(X_{n+1} - X_n)F_{n+1}] = F_{n+1}\, \mathbb{E}_n(X_{n+1} - X_n) = 0$$

for all $n \in \mathbb{N}$. $\square$

■ **Example 5.27 (Isometry Property)** Let $X := (X_n, n \in \mathbb{N})$ be an L^2-martingale and let $F := (F_n, n \in \mathbb{N})$ be a bounded predictable process, both with respect to some filtration $(\mathcal{F}_n, n \in \mathbb{N})$. Consider the integral process

$$Z_n := \int_0^n F_k \, dX_k = Z_0 + \sum_{k=1}^n F_k(X_k - X_{k-1}), \quad n \in \mathbb{N},$$

with $Z_0 := F_0 X_0$. By Theorem 5.26, the expectation of Z_n is $\mathbb{E}Z_0 = \mathbb{E}F_0 X_0$ for all n. What is the variance of Z_n? Expanding Z_n^2, we have

$$Z_n^2 - Z_0^2 = \sum_{k=1}^n F_k^2(X_k - X_{k-1})^2 + 2Z_0 \sum_{k=1}^n F_k(X_k - X_{k-1}) + 2\sum_{j=1}^{n-1}\sum_{k=j+1}^n F_j(X_k - X_{k-1}).$$

For $j < k$,

$$\mathbb{E}[F_j(X_k - X_{k-1})] = \mathbb{E}\,\mathbb{E}_j F_j(X_k - X_{k-1}) = F_j \mathbb{E}_j(X_k - X_{k-1}) = 0,$$

where we used repeated conditioning in the first equality, the fact that X is adapted in the second equality, and the martingale property of X in the final equality. A similar argument shows $\mathbb{E}[Z_0(X_k - X_{k-1})] = 0$ for all $k \geq 1$. Therefore,

(5.28) $$\mathbb{E}\left[Z_n^2 - Z_0^2\right] = \sum_{k=1}^n \mathbb{E}\left[F_k^2(X_k - X_{k-1})^2\right].$$

This is the *isometry* property of the discrete integral (Z_n). We will encounter its continuous-time equivalent for Itô integrals in Theorem 7.6.

For the special case $F_k = 1, k \in \mathbb{N}$, we have $Z_n = X_n, n \in \mathbb{N}$, and

(5.29) $$\mathbb{E}\left[X_n^2 - X_0^2\right] = \sum_{k=1}^n \mathbb{E}(X_k - X_{k-1})^2.$$

If, in addition, $\mathbb{E}X_0 = 0$, then (5.29) simply states that the variance of X_n is the sum of the variances of its increments:

$$\mathbb{V}\text{ar}\, X_n = \mathbb{V}\text{ar}\, X_0 + \sum_{k=1}^{n} \mathbb{V}\text{ar}(X_k - X_{k-1}),$$

even though the increments are not necessarily independent.   ∎

It is sometimes useful to consider a stochastic process that is stopped at some random time T.

Definition 5.30: Stopped Stochastic Process

Let $X := (X_t, t \in \mathbb{T})$ be a stochastic process and T a random time with values in $\overline{\mathbb{T}}$. By the process X *stopped at time* T we mean the process $(X_t^{\dagger}, t \in \mathbb{T})$ defined by

$$X_t^{\dagger}(\omega) := X_{T(\omega) \wedge t}(\omega) = \begin{cases} X_t(\omega) & \text{if } t \leq T(\omega), \\ X_{T(\omega)}(\omega) & \text{if } t > T(\omega). \end{cases}$$

Stopping a martingale at a random stopping time, gives again a martingale.

Theorem 5.31: Stopped Martingale

Let $M := (M_n, n \in \mathbb{N})$ be a martingale and T a stopping time with values in $\overline{\mathbb{N}}$. Then, the process M stopped at time T is again a martingale.

Proof. We can write the stopped process $M^{\dagger}$ as a stochastic integral $M^{\dagger} := \int F\,dM$ with respect to the predictable process $F_n := \mathbb{1}_{\{T \geq n\}}, n \in \mathbb{N}$. That (F_n) is predictable follows from the fact that each F_n is a Bernoulli random variable with $\{F_n = 0\} = \{T \leq n - 1\} \in \mathcal{F}_{n-1}$. Now apply Theorem 5.26 to complete the proof.   □

■ **Example 5.32 (Symmetric Random Walk Continued)** We continue Example 5.15, and consider the stopped martingale $(X_{T \wedge n}, n \in \mathbb{N})$. Since this is a martingale, we have

$$\mathbb{E}X_{T \wedge n} = \mathbb{E}X_{T \wedge 0} = \mathbb{E}X_0 = 0.$$

We mentioned that T can be shown to be almost surely finite. Hence, almost surely $X_{T \wedge n}$ converges to X_T. However,

$$0 = \lim_n \mathbb{E}X_{T \wedge n} \neq \mathbb{E} \lim_n X_{T \wedge n} = \mathbb{E}X_T = 1.$$

So, we may not swap the limit and expectation. Although $X_{T \wedge n}$ converges to X_T, it does so in a non-monotone way; otherwise, we would have been able to invoke the Monotone Convergence Theorem 2.34. ∎

■ **Example 5.33 (Asymmetric Random Walk)** Let $X := (X_n, n \in \mathbb{N})$ be an asymmetric random walk on the integers, defined by $X_0 := a$ for some $a \in \mathbb{N}$, and

$$X_{n+1} := X_n + 2B_n - 1, \quad n \in \mathbb{N},$$

where $B_0, B_1, \ldots \sim_{\text{iid}} \text{Ber}(p)$. Defining $q := 1 - p$, let

$$M_n := (q/p)^{X_n}, \quad n \in \mathbb{N}.$$

This is a martingale, since each M_n is integrable and

$$\mathbb{E}_n M_{n+1} = M_n \, \mathbb{E}_n (q/p)^{2B_n - 1} = M_n [(p/q)q + (q/p)p] = M_n.$$

Suppose a is strictly positive and let T be the first time at which either 0 or b is reached, for some $b \in \mathbb{N}$ with $b > a$; that is,

$$T := \min\{n : X_n = b \text{ or } 0\}.$$

We can think of X_n as the fortune of a gambler after n bets in a game of chance that will either increase or decrease his/her earnings by 1 dollar, with probability p and $q := 1 - p$, respectively. The gambler plays until time T; that is, when she/he goes bankrupt or reaches the level b. The *gambler's ruin problem* is to calculate the probability that level b is reached before bankruptcy occurs.

Consider the stopped martingale $\widehat{M} := (M_{n \wedge T}, n \in \mathbb{N})$. By the martingale property of $\widehat{M}$, we have

$$\mathbb{E}\,\widehat{M}_n = \mathbb{E}\,\widehat{M}_0 = \mathbb{E}\,M_0 = (q/p)^a, \quad n \in \mathbb{N}.$$

Moreover, $\lim \widehat{M}_n = M_T$, almost surely, provided that $\mathbb{P}(T < \infty) = 1$. Since $\widehat{M}$ is bounded, the Bounded Convergence Theorem 2.36 lets us conclude that

$$\mathbb{E}\,M_T = (q/p)^a.$$

In particular,

$$\left(\frac{q}{p}\right)^a = \mathbb{E}\,M_T = \mathbb{P}(X_T = b)\left(\frac{q}{p}\right)^b + (1 - \mathbb{P}(X_T = b))\left(\frac{q}{p}\right)^0.$$

Rearranging gives

(5.34)
$$\mathbb{P}(X_T = b) = \frac{1 - \left(\frac{q}{p}\right)^a}{1 - \left(\frac{q}{p}\right)^b}$$

for $p \neq q$. For the symmetric case $p = q$, applying the same arguments to the martingale $(X_n, n \in \mathbb{N})$, we obtain $\mathbb{P}(X_T = b) = a/b$ by rearranging:

$$a = \mathbb{E}X_T = \mathbb{P}(X_T = b)\, b + \mathbb{P}(X_T = a)\, 0.$$

The only "loose end" in this derivation is that we have not shown that T is almost surely finite. We can prove it as follows. Consider the first nb Bernoulli random variables that are used in the construction of X, divided into n batches of size b. Define the events $A_n := \{T > nb\}$, $n = 1, 2, \ldots$. If the event A_n occurs, then none of the n batches contain only successes or only failures — otherwise, T would be less than or equal to nb. Thus,

$$\mathbb{P}(A_n) \leq \left(1 - p^b - q^b\right)^n.$$

Consequently, we have by the continuity from above property of $\mathbb{P}$ (see (2.5)), that

$$\mathbb{P}(T = \infty) = \mathbb{P}\left(\cap_{n=1}^{\infty} A_n\right) = \lim_{n \to \infty} \mathbb{P}(A_n) \leq \lim_{n \to \infty} \left(1 - p^b - q^b\right)^n = 0,$$

since $A_1 \supseteq A_2 \supseteq \cdots$.   ∎

For certain martingales in continuous time — for example, Poisson martingales as in Example 5.10 — we can construct martingale transforms similar to (5.23). Specifically, let $N := (N_t, t \in \mathbb{R}_+)$ be an increasing right-continuous process adapted to a filtration $\mathcal{F}$. Suppose that $v_t := \mathbb{E}N_t$ is finite, and that

$$\overline{N}_t := N_t - v_t, \quad t \geq 0,$$

is an $\mathcal{F}$-martingale. The following is closely related to Proposition 4.59. We use the predictability Definition 5.22.

Proposition 5.35: Stochastic Integral for Predictable Integrands

Let N be defined as above. Then, for any positive $\mathcal{F}$-predictable process (F_t),

(5.36) $$\mathbb{E}\int_{\mathbb{R}_+} F_t\, dN_t = \mathbb{E}\int_{\mathbb{R}_+} F_t\, dv_t.$$

Proof. As in the proof of Proposition 4.59, it suffices to prove (5.36) only for indicators of sets $H \times A$, where either $H \in \mathcal{F}_a$ and $A := (a, b]$, or $H \in \mathcal{F}_0$ and $A := \{0\}$. In the first case, we have

$$\mathbb{E}\int_{(a,b]} \mathbb{1}_H\, dN_t = \mathbb{E}\mathbb{1}_H(N_b - N_a) = \mathbb{E}\mathbb{1}_H\, \mathbb{E}_a(N_b - N_a)$$

$$= \mathbb{E}\mathbb{1}_H(v_b - v_a) = \mathbb{E}\int_{(a,b]} \mathbb{1}_H\, dv_t.$$

In the second case, $\mathbb{E}\mathbb{1}_H N_0 = \mathbb{E}\mathbb{1}_H \mathbb{E}_0 N_0 = \mathbb{E}\mathbb{1}_H \nu_0$. □

The following is a continuous version of Theorem 5.26. Note that we already saw an illustration of the theorem in Example 5.20.

> **Theorem 5.37: Integral Transform of a Continuous Martingale**
>
> Let $\overline{N}$ be an $\mathcal{F}$-martingale, as defined above. For any bounded $\mathcal{F}$-predictable process F, the integral transform
> $$Z_t := \int_0^t F_s \, d\overline{N}_s, \quad t \geq 0,$$
> is a martingale.

Proof. Adaptedness and integrability are straightforward. For the martingale property, write

$$\mathbb{E}_s[Z_t - Z_s] = \mathbb{E}_s \int_{(s,t]} F_u \, d\overline{N}_u = \mathbb{E}_s \int_{\mathbb{R}_+} \mathbb{1}_{(s,t]}(u) F_u \, dN_u - \mathbb{E}_s \int_{\mathbb{R}_+} \mathbb{1}_{(s,t]}(u) F_u \, d\nu_u.$$

To show that $\mathbb{E}_s[Z_t - Z_s] = 0$, it thus suffices to show that for any positive $V \in \mathcal{F}_s$ we have

$$\mathbb{E} V \int_{\mathbb{R}_+} \mathbb{1}_{(s,t]}(u) F_u \, dN_u = \mathbb{E} V \int_{\mathbb{R}_+} \mathbb{1}_{(s,t]}(u) F_u \, d\nu_u,$$

which is equivalent to showing that

$$\mathbb{E} \int_{\mathbb{R}_+} G_u F_u \, dN_u = \mathbb{E} \int_{\mathbb{R}_+} G_u F_u \, d\nu_u,$$

where $G_u := V\mathbb{1}_{(s,t]}(u)$. The process G is positive and predictable and so are GF^+ and GF^-. The result now follows from Proposition 5.35. □

■ **Example 5.38 (Arrival Times of a Poisson Process)** Let $N := (N_t, t \geq 0)$ be a Poisson process with rate c. Denote the arrival times by $T_1, T_2, \ldots$ and set $T_0 := 0$. Let $\mathcal{F} := (\mathcal{F}_t, t \geq 0)$ be the natural filtration of N. We want to prove that the interarrival times are iid and $\mathsf{Exp}(c)$ distributed; thus, that for each $n \in \mathbb{N}$, $T_{n+1} - T_n$ is independent of $\mathcal{F}_{T_n}$ and is $\mathsf{Exp}(c)$ distributed. Consider thereto the martingale $\overline{N}_t = N_t - ct, t \geq 0$ and stopping times $S := T_n$ and $T := T_{n+1}$. For an arbitrary positive $V \in \mathcal{F}_S$ and $r \geq 0$, define

$$F_t := V \mathbb{1}_{(S,T]}(t) \, re^{-rt}, \quad t \geq 0.$$

The process (F_t) is positive and $\mathcal{F}$-predictable. Thus, by Proposition 5.35,

$$\mathbb{E} \int F_t \, dN_t = \mathbb{E} \int F_t \, c \, dt,$$

which is equivalent to

$$\mathbb{E}_S \int_{(S,T]} r \mathrm{e}^{-rt} \, dN_t = \mathbb{E}_S \int_{(S,T]} r \mathrm{e}^{-rt} c \, dt.$$

Evaluating both integrals (for the left-hand side, use the fact that (N_t) only has one jump of size 1 at time T in the interval $(S, T]$) gives

$$r \, \mathbb{E}_S \, \mathrm{e}^{-rT} = c \, \mathbb{E}_S (\mathrm{e}^{-rS} - \mathrm{e}^{-rT}).$$

Multiplying both sides by e^{rS} and passing the latter inside the conditional expectation, we obtain after rearranging:

$$\mathbb{E}_S \, \mathrm{e}^{-r(T-S)} = \frac{c}{c+r}, \qquad r > 0,$$

showing that $T - S = T_{n+1} - T_n$ is independent of $\mathcal{F}_{T_n}$ and is $\mathsf{Exp}(c)$ distributed. ■

5.3.2   Doob's Stopping Theorem

The following treatment of Doob's stopping theorem is restricted to the discrete case and relies on the integral transformation technique of the previous section. The stopping times involved are required to be bounded. For uniformly integrable martingales this boundedness requirement is no longer necessary; see Theorem 5.59.

Theorem 5.39: Doob's Stopping Theorem

A stochastic process $M := (M_n, n \in \mathbb{N})$ is a martingale if and only if for every pair of *bounded* stopping times S and T with $S \leq T$ the random variables M_S and M_T are integrable, and

(5.40) $$\mathbb{E}_S(M_T - M_S) = 0.$$

Proof. Let M be a martingale. Take two bounded stopping times S and T, with $S \leq T \leq n$ for some fixed n. We want to show that (5.40) holds. Let V be a bounded random variable in $\mathcal{F}_S$. Consider the predictable process $(F_k, k \in \mathbb{N})$ defined by

$$F_k := \begin{cases} V & \text{if } S < k \leq T, \\ 0 & \text{otherwise.} \end{cases}$$

Its integral with respect to M is the process $X := (X_n) = \left(\int_0^n F_k \, dM_k\right)$, with

$$X_n - \underbrace{X_0}_{M_0 F_0} = (M_{S+1} - M_S)V + \cdots + (M_T - M_{T-1})V = (M_T - M_S)V.$$

Since F is predictable and bounded, X is a martingale, by Theorem 5.26. By taking $V := 1$ and $S := 0$, we see that M_T is integrable, and $V := 1$ and $T := n$ shows M_S is integrable. Finally, since $V \in \mathcal{F}_S$, we have

$$\mathbb{E} \, V \, \mathbb{E}_S(M_T - M_S) \underbrace{=}_{\substack{\text{projection} \\ \text{property}}} \mathbb{E} \, V(M_T - M_S) = \mathbb{E}(X_n - X_0) = 0.$$

Since V is an arbitrary bounded random variable in $\mathcal{F}_S$, (5.40) must hold.

Conversely, suppose that (5.40) holds for every pair of bounded stopping times S and T with $S \le T$ and that M_S and M_T are integrable. We want to show that $M := (M_n, n \in \mathbb{N})$ is a martingale; i.e., an adapted process, where each M_n is integrable and satisfying the martingale property. Adaptedness is by assumption and integrability follows from the integrability assumption of M_T, by taking $T := n$. To check the martingale property, take a pair (m, n) with $m < n$ and an event $H \in \mathcal{F}_m$. We want to show

$$\mathbb{E} \mathbb{1}_H \mathbb{E}_m(M_n - M_m) = 0.$$

Take $S := m$, and $T := n \mathbb{1}_H + m \mathbb{1}_{\Omega \setminus H}$. Obviously, S is a stopping time. To see that T is also a stopping time, apply the alarm test: For every ω we are able to construct an alarm that sounds at $T(\omega)$, because $T(\omega) \ge m$ and at time m we will be able to determine whether $T(\omega) = n$ (when H occurs) or $T(\omega) = m$ (when H does not occur). So, we have a pair of stopping times (S, T) such that $S \le T \le n$ and, by construction, $M_T - M_S = \mathbb{1}_H(M_n - M_m)$. Applying (5.40), we have

$$\mathbb{E} \mathbb{1}_H \mathbb{E}_m(M_n - M_m) = \mathbb{E} \mathbb{E}_m \mathbb{1}_H(M_n - M_m) = 0$$

for all $H \in \mathcal{F}_m$ and $m < n$, which can only be true if $\mathbb{E}_m(M_n - M_m) = 0$, i.e., if M has the martingale property. $\square$

■ **Remark 5.41 (Submartingale Inequality)** When M is instead a submartingale, we can repeat the above proof almost verbatim to conclude that

$$\mathbb{E}_S(M_T - M_S) \ge 0$$

for bounded stopping times $S \le T$. ■

Akin to Markov's inequality (3.9), the next lemma bounds the tail probability of the running maximum of a positive submartingale in terms of the expected value of its final member.

Lemma 5.42: Maximum Inequality for Positive Submartingales

Let $(Y_k, k \in \mathbb{N})$ be a positive submartingale and define $Y_n^* := \max_{k \le n} Y_k$. Then, for any $b > 0$,

(5.43) $$b \, \mathbb{P}(Y_n^* \ge b) \le \mathbb{E}[Y_n \mathbb{1}_{\{Y^* \ge b\}}] \le \mathbb{E} Y_n.$$

Proof. Define $T := \inf\{n \ge 0 : Y_n \ge b\}$. It holds that $\{Y_n^* \ge b\} = \{T \le n\}$ and hence

$$b \, \mathbb{1}_{\{T \le n\}} \le Y_{T \wedge n} \, \mathbb{1}_{\{T \le n\}} \le \mathbb{1}_{\{T \le n\}} \, \mathbb{E}_{T \wedge n} Y_n = \mathbb{E}_{T \wedge n}[Y_n \mathbb{1}_{\{T \le n\}}],$$

where the second inequality is due to Doob's stopping theorem for submartingales; see Remark 5.41. Taking expectations, yields (5.43). □

Recall Kolmogorov's inequality in Theorem 3.11: If $X_1, X_2, \ldots$ are independent with zero mean, and $S_n := \sum_{i=1}^{n} X_i$, then

$$\mathbb{P}\left(\max_{k \le n} |S_k| > b \right) \le \frac{\mathbb{V}\mathrm{ar}\, S_n}{b^2}.$$

The following extends this result to the maximum of a martingale:

Theorem 5.44: Doob–Kolmogorov Inequality

Let $M := (M_k)$ be a martingale in L^p for some $p \in [1, \infty)$. Then, for $b > 0$,

(5.45) $$\mathbb{P}\left(\max_{k \le n} |M_k| > b \right) \le \frac{\mathbb{E}|M_n|^p}{b^p}.$$

Proof. If M is a martingale in L^p, then $Y := |M|^p$ is a positive submartingale. Let $a := b^p$. By Lemma 5.42, $a \, \mathbb{P}(\max_{k \le n} Y_k > a) \le \mathbb{E} Y_n$. Translated back, using M and b, this gives (5.45). □

To show that Kolmogorov's inequality in Theorem 3.11 follows directly from Theorem 5.44, note that $(S_n, n \in \mathbb{N})$ is a martingale. In particular, if $\mathbb{E} S_n^2 = \infty$, then the result holds trivially. And if $\mathbb{E} S_n^2 < \infty$ then S is a square-integrable martingale, so (5.45) holds with $p = 2$.

5.4 (Sub)Martingale Convergence

The second major use of martingales is in proofs of convergence for stochastic processes. A first step in the analysis is to characterize the number of times a stochastic process "upcrosses" an interval.

5.4.1 Upcrossings

Let $X := (X_k, k \in \mathbb{N})$ be an adapted process with respect to some filtration and consider the integral transformation $Z_n := \int_0^n F_k dX_k, n \in \mathbb{N}$, where (F_k) is a predictable process. Continuing the financial analogy in Section 5.3.1, think of X_k as the share price at time k and F_k as the number of shares owned at period k. It is completely determined by the share price history before period k. That is, the share buying strategy is determined by the information available before the kth period. For example, the strategy could have the following rules: (1) we are allowed at most one share, (2) when the share price drops to level a or lower, we buy one share (if our portfolio is empty), and (3) when the share price reaches b or higher, for some fixed $b > a$, we sell the share, if we own one. The total profit at period n is $Z_n - Z_0$. If the share price at n (if any) is higher than what it was bought for, then the total profit $Z_n - Z_0$ is at least $(b - a)U_n$, where U_n is the number of times the process X has *upcrossed* the interval (a, b) during the periods $0, 1, \ldots, n$.

As a special case, suppose that $X \geq 0$ is a *positive submartingale* and that $a = 0$. Thus, if at any time k when the price drops to 0, we buy a share (if we have no share), and we sell a share (if we own one) when the price hits b or any price higher. Again, we are allowed at most one share. We thus have

$$(5.46) \qquad\qquad Z_n - Z_0 \geq bU_n,$$

where U_n is the number of upcrossings of $(0, b)$; i.e., the total number of shares sold up to period n. Now observe that for each $k = 0, 1, 2, \ldots$, we have:

$$\mathbb{E}_k[Z_{k+1} - Z_k] = \mathbb{E}_k[(X_{k+1} - X_k)F_{k+1}] = \underbrace{F_{k+1}}_{\leq 1} \underbrace{\mathbb{E}_k[X_{k+1} - X_k]}_{\geq 0} \leq \mathbb{E}_k[X_{k+1} - X_k].$$

Summing the above terms from $k = 0$ to $n - 1$ and taking expectations gives

$$\mathbb{E}[Z_n - Z_0] \leq \mathbb{E}[X_n - X_0].$$

Combining this with (5.46) gives:

$$(5.47) \qquad\qquad b\,\mathbb{E}U_n \leq \mathbb{E}[X_n - X_0].$$

For a general interval (a, b) and a submartingale X, the above result is generalized as follows. Recall that for a random variable V, $V^+ := V \vee 0 = \max\{V, 0\}$.

Theorem 5.48: Expected Upcrossings of a Submartingale

Suppose that X is a submartingale. Then, the expected number of upcrossings $U_n(a, b)$ of (a, b) satisfies:

$$(5.49) \qquad (b - a)\,\mathbb{E}U_n(a, b) \leq \mathbb{E}[(X_n - a)^+ - (X_0 - a)^+].$$

Proof. The number of upcrossings of (a, b) for X is the same as the number of upcrossings of $(0, b - a)$ for the process $(X - a)^+$. The latter is a positive submartingale, for which, by (5.47), we have (5.49). □

The upcrossing theorem will be useful for proving convergence results for (sub)martingales. Recall that submartingale has a tendency to increase. If this increase is constrained in some way, it is reasonable that the process would converge. The following theorem restrains the growth of the submartingale by means of a finiteness condition on the supremum of the $\mathbb{E}X_n^+$:

Theorem 5.50: (Sub)Martingale Convergence

Let $X := (X_n, n \in \mathbb{N})$ be a submartingale. Suppose that $\sup_n \mathbb{E}X_n^+ < \infty$. Then, X converges almost surely to an integrable random variable X_∞.

Proof. Suppose that the sequence $(X_n(\omega))$ does *not* converge for some ω. Then, there exist $a, b \in \mathbb{Q}$ with $\liminf X_n(\omega) < a < b < \limsup X_n(\omega)$ such that the total number of upcrossings of (a, b) by $(X_n(\omega))$ is ∞. The corresponding random variable is defined as $U(a, b) := \lim U_n(a, b)$. Consider the event

$$\bigcup_{\substack{a, b \in \mathbb{Q} \\ a < b}} \{U(a, b) = \infty\}.$$

We need to show that the probability of this event is in fact 0. Or, equivalently, that $\mathbb{P}(U(a, b) < \infty) = 1$ for every $a, b \in \mathbb{Q}$ with $a < b$. Thus, it suffices to show that $\mathbb{E}U(a, b) < \infty$ for any such a pair (a, b). For this, we apply Theorem 5.48 and take the supremum over n on both sides of (5.49) to get

$$(b - a) \sup_n \mathbb{E}U_n(a, b) \le \sup_n \mathbb{E}(X_n - a)^+.$$

Since $(U_n(a, b))$ is increasing, we have $\sup_n \mathbb{E}U_n(a, b) = \mathbb{E}U(a, b)$ by the Monotone Convergence Theorem 2.34. Also, $(X_n - a)^+ \le X_n^+ + |a|$, so $\sup_n \mathbb{E}(X_n - a)^+ \le \sup_n \mathbb{E}X_n^+ + |a| < \infty$ by the theorem assumption. Consequently, $\mathbb{E}U(a, b) < \infty$, so that $X_\infty := \lim X_n$ exists almost surely. It remains to show that X_∞ is integrable. This follows from Fatou's Lemma 2.35:

(5.51) $\qquad \mathbb{E}|X_\infty| = \mathbb{E}\liminf |X_n| \le \liminf \mathbb{E}|X_n| \le 2 \sup_n \mathbb{E}X_n^+ - \mathbb{E}X_0 < \infty.$

The second-last inequality follows from the facts that

$$\mathbb{E}|X_n| = \mathbb{E}X_n^+ + \mathbb{E}X_n^- = 2\,\mathbb{E}X_n^+ - \mathbb{E}X_n$$

and that $\mathbb{E}X_n \ge \mathbb{E}X_0$, since X is a submartingale. □

A uniformly integrable submartingale satisfies the condition of Theorem 5.50. This is because $\mathbb{E}X_n^+ \leq \mathbb{E}|X_n|$, and hence $\sup_n \mathbb{E}X_n^+ \leq \sup_n \mathbb{E}|X_n| < \infty$, by the property of uniform integrability (see Proposition 3.34, Point 5). The following affirms the central role of uniform integrability in connection with (sub)martingales. We already saw in Theorem 3.38 that a sequence of real-valued random variables converges in L^1 if and only if it converges in probability and is uniformly integrable.

Theorem 5.52: (Sub)Martingale Convergence and Uniform Integrability

Let $X := (X_n, n \in \mathbb{N})$ be a submartingale. Then, X converges almost surely and in L^1 if and only if it is uniformly integrable. Moreover, if X converges, setting $X_\infty := \lim X_n$ extends X to a submartingale $\overline{X} := (X_n, n \in \overline{\mathbb{N}})$.

Proof. If X converges almost surely (and hence in probability) and in L^1, it must be uniformly integrable by Theorem 3.38.

Conversely, suppose that X is a UI submartingale. Then, as mentioned above, it satisfies the condition of Theorem 5.50 and hence it converges almost surely to an integrable random variable X_∞. Again by Theorem 3.38, X also converges to X_∞ in L^1 sense. To show that the extended process $\overline{X}$ is a submartingale over $\overline{\mathbb{N}}$, we need to show that:

1. $X_\infty \in \mathcal{F}_\infty$,

2. $\mathbb{E}|X_\infty| < \infty$,

3. $\mathbb{E}_m(X_\infty - X_m) \geq 0$ for all $m \in \mathbb{N}$.

The first point follows from the fact that X_∞ is the limit of the X_n, all of whom belong to $\mathcal{F}_\infty$. The second point was already shown in (5.51). For the final point, take $H \in \mathcal{F}_m$ for some arbitrary m. By conditioning on $\mathcal{F}_m$ we have for every $n \geq m$:

$$\mathbb{E}\mathbb{1}_H(X_n - X_m) = \mathbb{E}\mathbb{1}_H\mathbb{E}_m(X_n - X_m) \geq 0,$$

where the last inequality follows from the submartingale property of X. Since $X_n - X_m$ converges in L^1 to $X_\infty - X_m$ as $n \to \infty$ and $\mathbb{1}_H$ is bounded, we have

$$(5.53) \quad \underbrace{\mathbb{E}\mathbb{1}_H\mathbb{E}_m(X_\infty - X_m)}_{\text{cond.exp.}} = \mathbb{E}\mathbb{1}_H(X_\infty - X_m) = \lim_n \mathbb{E}\mathbb{1}_H(X_n - X_m) \geq 0.$$

If the random variable $\mathbb{E}_m(X_\infty - X_m)$ were strictly negative on some set $A \in \mathcal{F}_m$ with $\mathbb{P}(A) > 0$, then taking $H = A$ would give $\mathbb{E}\mathbb{1}_H\mathbb{E}_m(X_\infty - X_m) < 0$, which contradicts (5.53). Hence, $\mathbb{E}_m(X_\infty - X_m) \geq 0$ almost surely. □

We have seen in Example 5.11 that if Z is an integrable random variable, then $M_n := \mathbb{E}_n Z, n = 0, 1, 2, \dots$ defines a uniformly integrable martingale. We can think of Z as some kind of ultimate truth that is revealed at the end of time. For any finite time n, M_n conveys what we know about Z based on the history up to time n. Any UI martingale is actually of this form, as detailed in the following theorem:

Theorem 5.54: Uniformly Integrable Martingale

A process $M := (M_n, n \in \mathbb{N})$ is a UI martingale if and only if there exists an integrable random variable Z such that

(5.55) $$M_n = \mathbb{E}_n Z, \quad n \in \mathbb{N}.$$

Moreover, then M converges almost surely and in L^1 to an integrable random variable

(5.56) $$M_\infty := \mathbb{E}_\infty Z,$$

and $(M_n, n \in \overline{\mathbb{N}})$ is again a UI martingale. If $Z \in \mathcal{F}_\infty$, then $M_\infty = Z$.

Proof. To prove sufficiency, suppose that M satisfies (5.55). Then, it is a UI martingale, as shown in Example 5.11. Hence, by Theorem 5.52, M converges almost surely and in L_1 to a random variable M_∞ and $\overline{M} := (M_n, n \in \overline{\mathbb{N}})$ is again a martingale. It is in fact UI, because M is UI and M_∞ is integrable. We still need to show that $M_\infty = \mathbb{E}_\infty Z$, but we leave this to the end.

For the necessity part, suppose M is an arbitrary UI martingale. Again, by Theorem 5.52, M converges almost surely and in L_1 to a random variable M_∞ and $\overline{M} := (M_n, n \in \overline{\mathbb{N}})$ is a UI martingale. In particular, by the martingale property,

$$\mathbb{E}_n M_\infty = M_n, \quad n = 0, 1, 2, \ldots,$$

so (5.55) holds with $Z = M_\infty$.

It remains to show that (5.56) holds if $M_n = \mathbb{E}_n Z$, $n \in \mathbb{N}$. Take an arbitrary $H \in \mathcal{F}_\infty$. We want to show

(5.57) $$\mathbb{E} 1_H M_\infty = \mathbb{E} 1_H Z,$$

using a monotone class argument. Let $\mathcal{D}$ be the collection of all $H \in \mathcal{F}_\infty$ for which (5.57) holds. This is a d-system (check yourself). Moreover, it contains $\mathcal{F}_n$ for each n, because for any $H \in \mathcal{F}_n$, it holds that

$$\mathbb{E} 1_H M_\infty = \mathbb{E} 1_H \underbrace{\mathbb{E}_n M_\infty}_{M_n} = \mathbb{E} 1_H \underbrace{\mathbb{E}_n Z}_{M_n} = \mathbb{E} 1_H Z.$$

Thus, $\mathcal{D}$ is a d-system that contains the p-system $\cup_n \mathcal{F}_n$. It follows by the Monotone Class Theorem 1.12 that $\mathcal{D}$ contains the σ-algebra generated by $\cup_n \mathcal{F}_n$, i.e., $\mathcal{F}_\infty$. So

(5.57) holds for all $H \in \mathcal{F}_\infty$. Consequently, for all $H \in \mathcal{F}_\infty$,

$$\mathbb{E}1_H \underbrace{\mathbb{E}_\infty Z}_{\text{cond.exp.}} = \mathbb{E}1_H Z = \mathbb{E}1_H M_\infty,$$

which can only be true if $M_\infty = \mathbb{E}_\infty Z$. Finally, if $Z \in \mathcal{F}_\infty$, then $\mathbb{E}_\infty Z = Z$, in which case $M_\infty = Z$. □

If the time set $\mathbb{T} = \mathbb{N}$ and the martingale $(M_n \in \mathbb{N})$ is UI, then we can think of M_n as our best estimate at time n of the final truth M_∞ that will be revealed at the end of time. As time proceeds, more and more information becomes available about this final truth.

What if the martingale has been going since time immemorial and we observe it today at time $n = 0$, using all the information of the past? To explore this, consider the following *martingale in reversed time*. This is simply a martingale $M := (M_n, n \in \mathbb{T})$ with time set $\mathbb{T} := \{\dots, -2, -1, 0\}$. As before, we have a filtration $\mathcal{F} := (\mathcal{F}_n)$ that is increasing in n; that is, as time increases more information becomes available.

Theorem 5.58: Convergence for Martingales in Reversed Time

For $\mathbb{T} = \{\dots, -2, -1, 0\}$, let $M := (M_n, n \in \mathbb{T})$ be a martingale with respect to the filtration $(\mathcal{F}_n, n \in \mathbb{T})$. Then, M is UI and, moreover, as $n \downarrow -\infty$ it converges almost surely and in L^1 to the integrable random variable $M_{-\infty} := \mathbb{E}_{-\infty} M_0$, where $\mathcal{F}_{-\infty} := \cap_{n \in \mathbb{T}} \mathcal{F}_n$.

Proof. From the martingale property, we have

$$M_n = \mathbb{E}_n M_0, \quad n \in \mathbb{T}.$$

In Example 5.11 we saw that martingales of such form are uniformly integrable. So M_0 is the "ground truth" that is revealed more and more as we go forward in time, starting from the "beginning of time". To show that the value of the martingale at the beginning of time, say $M_{-\infty}$, has meaning, we repeat the proof of Theorem 5.54, using the upcrossing theorem. In particular, by Theorem 5.48 the expected number of upcrossings of an interval (a, b) by the martingale $(M_n, M_{n+1}, \dots, M_0)$ is bounded by

$$\frac{1}{b-a}\mathbb{E}\left[(M_0 - a)^+ - (M_n - a)^+\right] \le \frac{1}{b-a}\mathbb{E}(M_0 - a)^+ < \infty.$$

This holds for any $n \in \mathbb{T}$. Thus, the number of upcrossings of (a, b) by M over $\mathbb{T}$ is almost surely finite for any interval (a, b), with $a < b$. Just as in the proof of

Theorem 5.50, this implies that M converges almost surely to a random variable $M_{-\infty}$ as $n \downarrow -\infty$. By uniform integrability of M, the convergence is in L^1 as well. Finally, as $M_n \in \mathcal{F}_k, k = n, n + 1, \ldots, 0$, the random variable $M_{-\infty}$ belongs to all $\mathcal{F}_n, n \in \mathbb{T}$, and hence it belongss to the intersection of these σ-algebras.   □

We now have gathered enough results to extend Doob's stopping theorem for UI martingales to arbitrary (not only bounded) stopping times.

Theorem 5.59: Stopping for Uniformly Integrable Martingales

If $M := (M_n, n \in \overline{\mathbb{N}})$ is a uniformly integrable martingale, then for *every* pair of stopping times S and T with $S \le T$ it holds that

$$\mathbb{E}_S M_T = M_S.$$

In particular,

$$\mathbb{E} M_T = \mathbb{E} M_0.$$

Proof. Since M is a UI martingale, we have

$$M_n = \mathbb{E}_n Z, \quad n \in \overline{\mathbb{N}}$$

for some integrable random variable Z. From Doob's Stopping Theorem 5.39 applied to the bounded stopping times $T \wedge n$ and n, we have

$$M_{T \wedge n} = \mathbb{E}_{T \wedge n} M_n = \mathbb{E}_{T \wedge n} \mathbb{E}_n Z = \mathbb{E}_n \mathbb{E}_T Z, \quad n \in \mathbb{N},$$

where we have used the repeated conditioning property of conditional expectations. Since M is UI on $\overline{\mathbb{N}}$, M_T is integrable, even if T can take the value ∞. By Theorem 5.54, the random variable $\mathbb{E}_n \mathbb{E}_T Z$ converges to $\mathbb{E}_T Z$ almost surely. But $M_{T \wedge n} = \mathbb{E}_n \mathbb{E}_T Z$ also converges almost surely to M_T. Hence, $M_T = \mathbb{E}_T Z$. For any stopping time $S \le T$ we have by repeated conditioning:

$$\mathbb{E}_S M_T = \mathbb{E}_S \mathbb{E}_T Z = \mathbb{E}_S Z = M_S,$$

as had to be shown. Finally, by taking expectations (i.e., unconditioning) with $S = 0$, we have

$$\mathbb{E} M_T = \mathbb{E} M_0.$$

□

■ **Example 5.60 (Asymmetric Random Walk Continued)** Let $(X_n, n \in \mathbb{N})$ be the asymmetric random walk on the integers from Example 5.33, to which corresponds the martingale

$$M_n = (q/p)^{X_n}, \quad n \in \mathbb{N}.$$

Let T be the first time that (X_n), starting at some strictly positive integer a, reaches either 0 or $b > a$. The stopped martingale $(M_{n \wedge T}, n \in \overline{\mathbb{N}})$ is uniformly integrable, as all its values lie between 1 and $(q/p)^b$. So, we can immediately conclude from Theorem 5.59 that $\mathbb{E} M_T = \mathbb{E} M_0$. In particular, there is no need to ascertain the almost sure finiteness of T, as we did in Example 5.33. ∎

5.5 Applications

In this section, we discuss a number of beautiful applications of martingale theory.

5.5.1 Kolmogorov's 0–1 Law

Let $X_1, X_2, \ldots$ be a sequence of independent random variables, with natural filtration $\mathcal{F} := (\mathcal{F}_n)$, where $\mathcal{F}_n := \sigma(X_1, \ldots, X_n)$. Recall the notation $\mathbb{E}_n Z = \mathbb{E}[Z \mid \mathcal{F}_n] = \mathbb{E}[Z \mid X_1, \ldots, X_n]$. Define $\mathcal{T}_n := \sigma(X_{n+1}, X_{n+2}, \ldots)$ and $\mathcal{T} := \cap_n \mathcal{T}_n$. The σ-algebra $\mathcal{T}$ is called the *tail σ-algebra* of $X_1, X_2, \ldots$. It consists of events that are not affected by a finite number of the $\{X_i\}$; for example, the event

$$\left\{ \limsup \frac{1}{n}(X_1 + \cdots + X_n) > x \right\}$$

belongs to $\mathcal{T}$.

> **Theorem 5.61: Kolmogorov's 0–1 Law**
>
> If $H \in \mathcal{T}$, then $\mathbb{P}(H)$ is either 0 or 1. Consequently, a numerical random variable Z that is $\mathcal{T}$-measurable must be almost surely constant.

Proof. For every event H,

$$\mathbb{E}_n \mathbb{1}_H \overset{\text{a.s.}}{\to} \mathbb{E}_\infty \mathbb{1}_H,$$

by Theorem 5.54. Suppose $H \in \mathcal{T}$. Since $\mathcal{T}$ is independent of $\mathcal{F}_n$, we have

$$\mathbb{E}_n \mathbb{1}_H = \mathbb{E} \mathbb{1}_H = \mathbb{P}(H).$$

On the other hand, since $\mathcal{T}_n \subseteq \mathcal{F}_\infty$ for all n, we have $\mathcal{T} \subseteq \mathcal{F}_\infty$, which implies that

$$\mathbb{E}_\infty \mathbb{1}_H = \mathbb{1}_H.$$

Combining the three displayed results gives

$$\mathbb{P}(H) = \mathbb{1}_H, \quad \text{almost surely.}$$

This means that $\mathbb{P}(H)$ is either 0 or 1. Finally, suppose that Z is a numerical random variable in $\mathcal{T}$ (i.e., Z is $\mathcal{T}/\overline{\mathcal{B}}$-measurable). Then, $\{Z = z\} \in \mathcal{T}$ has either probability 0 or 1. Hence, there must be a $c \in \overline{\mathbb{R}}$ such that $\mathbb{P}(Z = c) = 1$. □

5.5.2 Strong Law of Large Numbers

We give a proof of the Law of Large Numbers (Theorem 3.44), under the condition that $X_1, X_2, \ldots$ is an iid sequence of random variables with finite expectation c. We thus want to prove that the sample mean process $\overline{X}_n, n = 1, 2, \ldots$, with

$$\overline{X}_n := \frac{X_1 + \cdots + X_n}{n},$$

converges almost surely to c as $n \to \infty$. We consider thereto the filtration $\mathcal{F} := (\mathcal{F}_{-n}, n = 1, 2, \ldots)$, where $\mathcal{F}_{-n} := \sigma(\overline{X}_n, \overline{X}_{n+1}, \ldots)$. Thus, $\mathcal{F}_{-n}$ contains present and future information of the sample means at time n. Because the $(X_1, \ldots, X_n)$ is independent of $X_{n+1}, X_{n+2}, \ldots$, we have for each $k \in \{1, \ldots, n\}$:

$$\mathbb{E}_{-n}X_k := \mathbb{E}[X_k \mid \mathcal{F}_{-n}] = g_k(\overline{X}_n)$$

for certain measurable numerical functions $g_k, k = 1, \ldots, n$. In fact, since the distribution of $(X_1, \ldots, X_n)$ is invariant under permutations, all $\{g_k\}$ must be the same. Moreover, as $\mathbb{E}_{-n}(X_1 + \cdots + X_n) = n\overline{X}_n$, we must have

$$\mathbb{E}_{-n}X_k = \overline{X}_n, \quad k = 1, \ldots, n$$

and, in particular,

$$\mathbb{E}_{-n}X_1 = \overline{X}_n, \quad n = 1, 2, \ldots.$$

This shows that $(\overline{X}_n, n = 1, 2, \ldots)$ is a uniformly integrable martingale. By Theorem 5.58 it converges almost surely and in L^1 to an integrable random variable $\overline{X}_\infty$ as $n \to \infty$. To show that $\overline{X}_\infty = c$ almost surely, note that

$$\overline{X}_\infty = \lim_n \frac{1}{n}(X_{k+1} + \cdots + X_{k+n}),$$

which shows that $\overline{X}_\infty$ belongs to the tail σ-algebra $\sigma(X_{k+1}, X_{k+2}, \ldots)$ for every $k = 0, 1, \ldots$ and hence it belongs to the intersection of these. By Kolmogorov's 0–1 law, $\overline{X}_\infty$ must be almost surely constant. Finally, by L^1 convergence, this constant must be c, as $\mathbb{E}\overline{X}_\infty = \lim \mathbb{E}\overline{X}_n = c$.

5.5.3 Radon–Nikodym Theorem

Recall that the Radon–Nikodym Theorem 1.59 is closely connected to the concept of conditional expectation; see Section 4.2. We provide a proof of the theorem using martingale techniques, under the condition of separability of the related σ-algebra.

Definition 5.62: Separable σ-Algebra

A σ-algebra on Ω is said to be *separable* if it is generated by a sequence (H_n) of subsets of Ω.

■ **Example 5.63 (The Borel σ-Algebra is Separable)** Let $\Omega := (0, 1]$ and consider the following sequence of partitions that become increasingly finer:

$$\mathcal{P}_0 := (0, 1]$$
$$\mathcal{P}_1 := (0, 1/2], (1/2, 1]$$
$$\vdots$$
$$\mathcal{P}_n := (0, 1/2^n], (1/2^n, 2/2^n], \ldots, (1 - 2^{-n}, 1]$$
$$\vdots$$

Let (H_n) be the sequence of intervals $(0, 1], (0, 1/2], (1/2, 1], \ldots$, in the order in which these sets appear above. Then, every interval $(a, b] \subseteq (0, 1]$ can be written as a countable union of elements in (H_n), and hence $\sigma(H_1, H_2, \ldots) = \mathcal{B}_{(0,1]}$; that is, the Borel σ-algebra on $(0, 1]$ is separable.   ■

The above example proffers the idea that the appropriate way to look at a separable σ-algebra $\mathcal{G}$ is through a sequence of ever-finer partitions $\mathcal{P}_0, \mathcal{P}_1, \ldots$. Given the sequence (H_n), it is always possible to construct such a sequence of partitions. Namely, for every n, the σ-algebra $\mathcal{F}_n$ defined as $\sigma(H_1, \ldots, H_n)$ has only a finite number of elements, and so we can find a finite partition $\mathcal{P}_n$ such that every element of $\mathcal{F}_n$ can be written as a finite union of elements in $\mathcal{P}_n$. In particular, $\sigma(\mathcal{P}_n) = \mathcal{F}_n$. Now, $(\mathcal{F}_n)$ is a filtration and $\mathcal{F}_\infty = \lim \mathcal{F}_n = \vee_n \mathcal{F}_n = \mathcal{G}$.

The following is a slightly restricted version — applied to a probability space $(\Omega, \mathcal{H}, \mathbb{P})$ — of the Radon–Nikodym Theorem 1.59, which we can prove rigorously using martingale theory:

Theorem 5.64: Radon–Nikodym

Let $(\Omega, \mathcal{H}, \mathbb{P})$ be a probability space, where $\mathcal{H}$ is separable, and let Q be a finite measure on $(\Omega, \mathcal{H})$ with $Q \ll \mathbb{P}$. Then, there exists a positive random variable $Z \in \mathcal{H}$, written $Z = dQ/d\mathbb{P}$, such that $Q = Z \mathbb{P}$; that is,

$$(5.65) \qquad Q(H) = \int_H Z \, d\mathbb{P} = \mathbb{E}\mathbb{1}_H Z, \quad H \in \mathcal{H}.$$

Proof. The proof is mostly by construction, apart from a few technical points. Since $\mathcal{H}$ is separable, we can find a sequence of ever-finer finite partitions $(\mathcal{P}_n)$ and a filtration $\mathcal{F} := (\mathcal{F}_n)$ with $\mathcal{F}_n := \sigma(\mathcal{P}_n)$, as discussed after Example 5.63. We first construct a stochastic process $X := (X_n)$ as follows:

$$X_n := \sum_{H \in \mathcal{P}_n} \frac{Q(H)}{\mathbb{P}(H)} \mathbb{1}_H, \quad n \in \mathbb{N},$$

with the convention that $0/0 := 0$. Since $\mathcal{P}_n$ is a partition, there exists for each specific $\omega \in \Omega$ a unique set $H_\omega \in \mathcal{P}_n$ that contains ω. The above definition thus implies that $X_n(\omega) = Q(H_\omega)/\mathbb{P}(H_\omega)$. Obviously, $X_n \geq 0$ for all $n \in \mathbb{N}$; and $X_n \in \mathcal{F}_n$, since X_n only takes finitely many values and the inverse image of each of these values is a set in $\mathcal{P}_n$. Since any $H \in \mathcal{F}_n$ can be written as a finite union $H = \cup_k H_k$ of disjoint sets in $\mathcal{P}_n$, we have

(5.66)
$$\mathbb{E}\mathbb{1}_H X_n = \sum_k \frac{Q(H_k)}{\mathbb{P}(H_k)}\mathbb{P}(H_k) = \sum_k Q(H_k) = Q(H).$$

We want to show that X is a positive martingale with respect to the filtration $\mathcal{F}$ and that it converges almost surely to a random variable $Z \in \mathcal{H}$. Positivity and adaptedness have already been shown, and taking $H = \Omega$ in (5.66) shows integrability, since $\mathbb{E}X_n = Q(\Omega) < \infty$. The martingale property follows from

$$\mathbb{E}\mathbb{1}_H X_n = Q(H) = \mathbb{E}\mathbb{1}_H X_{n+1}, \quad H \in \mathcal{F}_n,$$

where we have used the fact that also $H \in \mathcal{F}_{n+1}$. Since X is a positive martingale, $-X$ is a negative submartingale with $\sup_n \mathbb{E}(-X_n^+) \leq 0 < \infty$, and we can apply Theorem 5.52 immediately to conclude that X converges almost surely to an integrable random variable $Z \in \mathcal{H}$.

In fact, X is a uniformly integrable martingale, and thus the convergence to Z is in L^1 sense as well. To prove uniform integrability, we need to show that for every $\varepsilon > 0$ there is a b such that

$$\sup_n \mathbb{E}X_n\mathbb{1}_{\{X_n>b\}} \leq \varepsilon.$$

By (5.66) we have
$$\mathbb{E}X_n\mathbb{1}_{\{X_n>b\}} = Q(X_n > b).$$
So it suffices to show that for all $\varepsilon > 0$ there exists a b such that $Q(X_n > b) \leq \varepsilon$ for all n. Since Q is absolutely continuous with respect to $\mathbb{P}$ (i.e., $\mathbb{P}(H) = 0$ implies $Q(H) = 0$), a small value of $\mathbb{P}(H)$ ought to imply a small value of $Q(H)$. This suggests that for every $\varepsilon > 0$ we seek a $\delta > 0$ such that

(5.67)
$$\mathbb{P}(H) \leq \delta \Rightarrow Q(H) \leq \varepsilon \quad \text{for all } H \in \mathcal{H}.$$

If for every $\varepsilon > 0$ (5.67) holds for some $\delta > 0$, then take $b := Q(\Omega)/\delta$ and $H := \{X_n > b\}$, so that by Markov's inequality (3.9) and (5.66):

$$\mathbb{P}(X_n > b) \leq \frac{1}{b}\mathbb{E}X_n = \frac{1}{b}Q(\Omega) = \delta$$

and hence $Q(H) \leq \varepsilon$. To prove (5.67), take an $\varepsilon > 0$ and suppose that (5.67) does *not* hold. Thus, there exists $H_n \in \mathcal{H}$ such that

$$\mathbb{P}(H_n) \leq 2^{-n} \quad \text{while} \quad Q(H_n) > \varepsilon.$$

Define H such that $\mathbb{1}_H = \limsup \mathbb{1}_{H_n}$. By the Borel–Cantelli Lemma 3.14, we have $\limsup \mathbb{1}_{H_n} = 0$, almost surely, which means that $\mathbb{P}(H) = 0$. But also, by Fatou's Lemma 2.35,

$$\int \limsup \mathbb{1}_{H_n} dQ = -\int \liminf(-\mathbb{1}_{H_n}) dQ \geq -\liminf \int (-\mathbb{1}_{H_n}) dQ$$

$$= \limsup \int \mathbb{1}_{H_n} dQ = \limsup Q(H_n) \geq \varepsilon.$$

This is in contradiction to the absolute continuity of Q with respect to $\mathbb{P}$, so there *does* exist a δ such that (5.67) holds.

Thus, we have established that X is uniformly integrable and that it converges to Z in both almost sure and L^1 sense. It remains to show (5.65). Define for every event H:

$$\widehat{Q}(H) := \int_H Z \, d\mathbb{P} = \mathbb{E}\mathbb{1}_H Z = \lim \mathbb{E}\mathbb{1}_H X_n,$$

where the last equality follows from L^1 convergence of X. But for $H \in \mathcal{F}_n$ we have $\widehat{Q}(H) = Q(H)$, by (5.66). Since Q and $\widehat{Q}$ coincide on the p-system $\cup_n \mathcal{F}_n$ that generates $\mathcal{H}$, they must coincide on $\mathcal{H}$.   □

5.6 Martingales in Continuous Time

We conclude this chapter with a preparatory discussion of martingales in continuous time. We already encountered an important example in the form of the process $(N_t - ct, t \geq 0)$, where (N_t) is a Poisson counting process with rate c; see Example 5.10. In a similar way, we can obtain martingales from other Lévy processes by subtracting their mean.

Chapters 6 and 7 will showcase many continuous-time martingales related to the Wiener process. The Wiener process itself provides the main example of a continuous-time martingale.

■ **Example 5.68 (Wiener Process is a Martingale)** Let $(W_t, t \geq 0)$ be the Wiener process, as previously examined in Examples 2.77 and 2.96. It is adapted to its natural filtration, integrable (as $\mathbb{E}|W_t| < \infty$ for all t) and satisfies the martingale property

$$\mathbb{E}_s W_t = \mathbb{E}_s[W_s + W_t - W_s] = W_s + \underbrace{\mathbb{E}_s[W_t - W_s]}_{= 0} = W_s$$

for all $s < t$.   ■

The objective of this section is to extend the results for discrete-time martingales to the continuous-time case. To that end, recall the main findings for a martingale $M := (M_n, n \in \mathbb{T})$ with time set $\mathbb{T} = \mathbb{N}$ or $\overline{\mathbb{N}}$:

1. *Doob's Stopping Theorem*: For every pair of bounded stopping times $S \leq T$,

   (5.69)
   $$\mathbb{E}_S M_T = M_S,$$

   provided M_S and M_T are integrable; see Theorem 5.39.

2. *Doob's Stopping Theorem for UI Martingales*: For a UI martingale M on $\mathbb{T} = \overline{\mathbb{N}}$, (5.69) holds for every pair of stopping times $S \leq T$; see Theorem 5.59.

3. *Martingale Convergence for UI martingales*: A UI martingale M on $\mathbb{N}$ is of the form
   $$M_n = \mathbb{E}_n Z, \quad n \in \mathbb{N}.$$

   It converges almost surely and in L^1 to an integrable random variable M_∞ and can be extended to a UI martingale on $\overline{\mathbb{N}}$; see Theorem 5.54.

5.6.1 Local Martingales and Doob Martingales

When analysing a stochastic process $M := (M_t, t \in \mathbb{T})$ with $\mathbb{T} = \mathbb{R}_+$ or $\overline{\mathbb{R}}_+$ for some filtration $\mathcal{F}$ on a probability space $(\Omega, \mathcal{H}, \mathbb{P})$, it is often convenient (and sometimes necessary) to impose various regularity conditions on the process itself, the underlying probability space, and/or the filtration that is used. Typical regularity conditions are:

1. The probability space $(\Omega, \mathcal{H}, \mathbb{P})$ be *complete*, meaning that $\mathcal{H}$ contains every *negligible* set; that is, if $A \subset H$ with $\mathbb{P}(H) = 0$, then $A \in \mathcal{H}$.

2. The filtration $\mathcal{F}$ be *augmented*, meaning that $(\Omega, \mathcal{H}, \mathbb{P})$ is complete and all the negligible events in $\mathcal{H}$ are also in $\mathcal{F}_0$ (and hence in all $\mathcal{F}_t$).

3. The paths of M be right-continuous and have left-limits.

Even when these regularity assumptions are in place, it is sometimes necessary to extend the notion of the natural filtration of a process ever so slightly, as illustrated by the following example:

■ **Example 5.70 (Not a Stopping Time)** Let X_t be the position of a particle that starts at position 0 and moves with velocity 1. At a random time T it stops. A typical trajectory is shown in Figure 5.71. As we cannot say at any time t whether the particle has actually stopped, T is not a stopping time for the natural filtration of (X_t). However, if we were able to look an infinitesimal amount of time ahead at every t, then we would be able to discern whether the particle has stopped at time t.

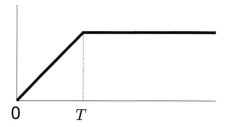

Figure 5.71: T is not a stopping time.

◾

The previous example suggests that we use filtrations for which the information at time t is exactly the same as the information at time t if we can also "peek ahead" an infinitesimal amount of time. This leads to the following definition:

Definition 5.72: Right-continuous Filtration

A filtration $\mathcal{F} := (\mathcal{F}_t)$ is said to be *right-continuous* if for all t,

$$\bigcap_{\varepsilon > 0} \mathcal{F}_{t+\varepsilon} = \mathcal{F}_t.$$

It is easy to construct a right-continuous filtration, $\mathcal{F}^+$ from a natural filtration $\mathcal{F}$, by defining

$$(5.73) \qquad\qquad \mathcal{F}_t^+ := \bigcap_{u > t} \mathcal{F}_u.$$

In Example 5.70, T is a stopping time of $\mathcal{F}^+$, but not of the natural filtration $\mathcal{F}$ of the process (X_t).

From Theorem 5.83 we have seen that a martingale $(M_t, t \geq 0)$ that is stopped at a random stopping time T, i.e., $(M_{t \wedge T}, t \geq 0)$, is again a martingale. The converse is generally not true: a stopped process that is a martingale may itself not be a martingale.

Definition 5.74: Local Martingale

A process $(Z_t, t \geq 0)$ is called a *local martingale* if there exists a sequence of stopping times $(T_n, n \in \mathbb{N})$, called a *localizing sequence*, such that $T_n \to \infty$ almost surely and each stopped process $(Z_{t \wedge T_n}, t \geq 0)$ is a martingale.

Similarly, a process $(Z_t, t \geq 0)$ is said to be of *locally bounded variation* if each $(Z_{t \wedge T_n}, t \geq 0)$ is of bounded variation.

■ **Example 5.75 (Local Martingale)** Suppose that $X \sim t_1$ has a Cauchy distribution and is independent of the Wiener process $(W_t, t \geq 0)$. Then, the process $Z := (X W_t, t \geq 0)$ is not a martingale, because $\mathbb{E}|X W_t| = \infty$. Nevertheless, Z is a local martingale with respect to the filtration $(\mathcal{F}_t, t \geq 0)$ with $\mathcal{F}_t := \sigma\{W_s, s \leq t\} \vee \sigma X$. A particular localizing sequence is $T_n := \inf\{t \geq 0 : |X W_t| \geq n\}$, $n \in \mathbb{N}$. ■

Theorem 5.76: Properties of Local Martingales

Let $Z := (Z_t, t \geq 0)$ be a local martingale with localizing sequence $(T_n, n \in \mathbb{N})$. Then, the following hold:

1. If $Z \geq 0$ and $\mathbb{E}Z_0 < \infty$, then Z is a supermartingale.

2. If for each $t \geq 0$ the sequence $(Z_{t \wedge T_n}, n \in \mathbb{N})$ is uniformly integrable, then Z is a martingale.

Proof. Define $Z_t^{(n)} := Z_{t \wedge T_n}, t \geq 0, n \in \mathbb{N}$. By the martingale property of $(Z_t^{(n)}, t \geq 0)$, we have $\mathbb{E}_s Z_t^{(n)} = Z_s^{(n)}$ for $s \leq t$. Since $T_n \overset{a.s.}{\to} \infty$, we also have $Z_t^{(n)} \overset{a.s.}{\to} Z_t$ as $n \to \infty$. Hence, applying Fatou's Lemma 2.35 with regard to $\mathbb{E}_s$, we find

$$\mathbb{E}_s Z_t = \mathbb{E}_s \liminf_n Z_t^{(n)} \leq \liminf_n \mathbb{E}_s Z_t^{(n)} = \liminf_n Z_s^{(n)} = Z_s,$$

so the supermartingale property holds for Z. In addition, the process is adapted to its natural filtration, and integrability follows from $\mathbb{E}|Z_t| \leq \mathbb{E}\,\mathbb{E}_0|Z_t| = \mathbb{E}\,Z_0 < \infty$, so Z is a supermartingale. This proves property 1.

For property 2, almost sure convergence $Z_t^{(n)} \overset{a.s.}{\to} Z_t$ (and thus $Z_t^{(n)} \overset{\mathbb{P}}{\to} Z_t$) combined with the uniform integrability of $(Z_t^{(n)}, n \in \mathbb{N})$ imply that $Z_t^{(n)} \overset{L^1}{\to} Z_t$, by Theorem 3.38. In other words, $\mathbb{E}_s|Z_{t \wedge T_n} - Z_t| \to 0$, which shows that the limit of $\mathbb{E}_s Z_t^{(n)}$ exists and is equal to $\mathbb{E}_s Z_t$. However, taking limits on both sides of the martingale equality $\mathbb{E}_s Z_t^{(n)} = Z_s^{(n)}$ also shows that $\lim_n \mathbb{E}_s Z_t^{(n)} = Z_s$. Thus, $\mathbb{E}_s Z_t = Z_s$, so that Z has the martingale property. Adaptedness is again automatic, and integrability follows from Fatou and the uniform integrability of $(Z_t^{(n)}, n \in \mathbb{N})$:

$$\mathbb{E}|Z_t| = \mathbb{E} \liminf_n |Z_t^{(n)}| \leq \liminf_n \mathbb{E}|Z_t^{(n)}| \leq \sup_n \mathbb{E}|Z_t^{(n)}| < \infty.$$

□

■ **Example 5.77 (Continuous Local Martingale)** If $Z := (Z_t, t \geq 0)$ is a continuous local martingale with $Z_0 = 0$, then we can always choose the localizing sequence $(T_n, n \in \mathbb{N})$ to be:

$$T_n := \inf\{t \geq 0 : |Z_t| \geq n\}, \quad n \in \mathbb{N}.$$

To see this, let (τ_m) be the localizing sequence for Z. Since $(Z_{t\wedge\tau_m}, t \geq 0)$ is a continuous martingale, the stopped process $(Z_{t\wedge\tau_m\wedge T_n}, t \geq 0)$ is also a martingale by Theorem 5.83. The bound $\mathbb{E}|Z_{t\wedge\tau_m\wedge T_n}| \leq n$ implies that $(Z_{t\wedge\tau_m\wedge T_n}, n \in \mathbb{N})$ is UI for each fixed n. Therefore, the second part of Theorem 5.76 implies that $(Z_{t\wedge T_n}, t \geq 0)$ is a martingale for each n. ∎

Below, $\mathbb{T} = \mathbb{R}_+$ or $\overline{\mathbb{R}}_+$. For simplicity, we assume that $M := (M_t, t \in \mathbb{T})$ is a martingale with respect to an augmented and right-continuous filtration, and has right-continuous and left-limited paths.

Definition 5.78: Doob Martingale

Let ζ be a stopping time. A process $M := (M_t, t \in \mathbb{T})$ is called a *Doob martingale* on $[0, \zeta]$ if for all stopping times $0 \leq S \leq T \leq \zeta$ it holds that

$$(5.79) \qquad \mathbb{E}_S M_T = M_S,$$

where M_S and M_T are integrable.

The following provides an equivalent description of a Doob martingale:

Theorem 5.80: Characterization of a Doob Martingale

Let ζ be a stopping time. A process $M := (M_t, t \in \mathbb{T})$ is a Doob martingale on $[0, \zeta]$ if and only if for every stopping time $T \leq \zeta$,

$$(5.81) \qquad \mathbb{E}M_T = \mathbb{E}M_0.$$

Proof. Suppose M is a Doob martingale on $[0, \zeta]$. From (5.79), with $(S, T) := (T, \zeta)$, we have

$$(5.82) \qquad \mathbb{E}_T M_\zeta = M_T.$$

This holds true for any $T \leq \zeta$. Taking expectations in (5.82) with $(T, \zeta) := (0, T)$ then gives (5.81).

To show necessity, suppose that (5.81) is true for any stopping time $T \leq \zeta$. We want to show that (5.82) holds. Take an event $H \in \mathcal{F}_T$ and define

$$U := T\mathbb{1}_H + \zeta(1 - \mathbb{1}_H).$$

Then, U is a stopping time with $T \leq U \leq \zeta$, and

$$M_\zeta - M_U = (M_\zeta - M_T)\mathbb{1}_H.$$

It follows, using the assumed property (5.81), that

$$0 = \mathbb{E}M_\zeta - \mathbb{E}M_U = \mathbb{E}[M_\zeta - M_U] = \mathbb{E}(M_\zeta - M_T)\mathbb{1}_H$$

for any $H \in \mathcal{F}_T$; that is, (5.82) holds. By applying the latter property to stopping times S and T with $S \leq T \leq \zeta$, we find

$$\mathbb{E}_S M_T = \mathbb{E}_S \mathbb{E}_T M_\zeta = \mathbb{E}_S M_\zeta = M_S.$$

That is, M is a Doob martingale on $[0, \zeta]$.   □

The following is the continuous-time version of Doob's Stopping Theorem 5.39:

Theorem 5.83: Doob's Stopping Theorem for Continuous Martingales

A martingale $M := (M_t, t \in \mathbb{R}_+)$ is a Doob martingale on $[0, b]$ for every $b \in \mathbb{R}_+$ and satisfies

(5.84) $$\mathbb{E}_S M_T = M_S$$

for every pair (S, T) of bounded stopping times with $S \leq T$.

Proof. Let M be a martingale and take $b \in \mathbb{R}_+$. To show that it is a Doob martingale on $[0, b]$, we need to show (using Theorem 5.80) that $\mathbb{E}M_T = \mathbb{E}M_0$ for every stopping time $T \leq b$. We do this by applying Doob's Stopping Theorem 5.39 to a discrete-time martingale (M_{T_n}), where (T_n) is a sequence of stopping times decreasing to T, as defined in Exercise 26. For each n, define $\mathbb{T}_n$ as the countable time set $\{k/2^n, k \in \mathbb{N}, n \in \mathbb{N}\} \cup \{b+1\}$ and note that T_n is bounded by $b+1$ and takes values in a finite subset of $\mathbb{T}_n$. Now, consider the discrete-time martingale $(M_t, t \in \mathbb{T}_n)$. By Theorem 5.39, with $(S, T) := (T_n, b+1)$, we have

(5.85) $$M_{T_n} = \mathbb{E}_{T_n} M_{b+1}.$$

Defining $X_{-n} := M_{T_n}, n \in \mathbb{N}$, (5.85) shows that $(X_{-n}, n \in \mathbb{N})$ is a reversed-time (and hence uniformly integrable) martingale with respect to the filtration $(\mathcal{F}_{T_n})$ — noting that $\ldots \leq T_2 \leq T_1 \leq T_0$. It follows from Theorem 5.58 that M_{T_n} converges almost surely and in L^1 to an integrable random variable. But since (T_n) decreases to T and M is (assumed to be) right-continuous, this limiting random variable must be M_T. Finally, by L^1 convergence, we have, in view of (5.85),

$$\mathbb{E}M_T = \lim \mathbb{E}M_{T_n} = \mathbb{E}M_{b+1} = \mathbb{E}M_0.$$

Thus, M is a Doob martingale on $[0, b]$ for any $b \in \mathbb{R}_+$ and, by definition, (5.84) holds for any pair (S, T) of bounded stopping times with $S \leq T$.   □

As in the discrete case (see Remark 5.41), when M is instead a submartingale, the above theorem and proof can be modified slightly to yield the inequality:

$$\mathbb{E}_S(M_T - M_S) \geq 0.$$

5.6.2 Martingale Inequalities

It is possible to formulate inequalities for the maxima of continuous submartingales, similar to the Doob–Kolmogorov inequality in Theorem 5.44. The following inequality is of particular use:

Proposition 5.86: Doob's Maximal Inequality

Let $X := (X_t, t \geq 0)$ be a submartingale that is positive and continuous. Then, for $p \geq 1$ and $b > 0$,

$$(5.87) \qquad \mathbb{P}\Big(\max_{0 \leq s \leq t} X_s \geq b \Big) \leq \frac{\mathbb{E} X_t^p}{b^p}.$$

Proof. First, note that by the Extreme Value Theorem the continuous process X attains its maximum on the closed bounded set $[0, t]$, so that here $\sup_{0 \leq s \leq t} X_s = \max_{0 \leq s \leq t} X_s$. Second, since X is a submartingale, X^p is also a submartingale for any $p \geq 1$, provided that $\mathbb{E} X_t^p < \infty$; see Theorem 5.7. Third, let $D_{n,t} := \{tk2^{-n},\ k = 0, 1, \ldots, 2^n\}$ and $M_{n,t} := \max_{s \in D_{n,t}} X_s$. Applying Lemma 5.42 to the positive discrete-time submartingale $(X_s^p,\ s \in D_{n,t})$, we obtain the inequality:

$$\mathbb{P}(M_{n,t} \geq b) = \mathbb{P}\Big(\max_{s \in D_{n,t}} X_s^p \geq b^p \Big) \leq \mathbb{E} X_t^p / b^p.$$

Next, if $A_n := \{M_{n,t} \geq b\}$, then (A_n) is an increasing sequence of events, so that

$$\lim_n \mathbb{P}(A_n) = \mathbb{P}(\cup_{k=1}^\infty A_k) = \mathbb{P}(\lim_n M_{n,t} \geq b) \leq \mathbb{E} X_t^p / b^p.$$

Finally, the continuity of X implies that $\lim_n M_{n,t} = \max_{0 \leq s \leq t} X_s$, giving (5.87). $\square$

Sometimes we want to bound the L^p norm of the maximum of a continuous martingale. The following inequality then becomes useful. Note that it can be easily modified to the discrete-time case.

Proposition 5.88: Doob's Norm Inequality

Let $M := (M_t, t \geq 0)$ be a continuous martingale in L^p for some $p > 1$ and let $Z_t := \max_{s \leq t} |M_s|$. Then,

$$(5.89) \qquad \mathbb{E} Z_t^p \leq q^p\, \mathbb{E} |M_t|^p, \quad \text{with } q := p/(p-1).$$

Proof. To simplify the notation, let $Z := Z_t$. Because $|M|$ is a positive submartingale, we can apply the inequality (5.43) (which obviously holds for the continuous-time case as well) to conclude that for all $x \geq 0$,

$$\mathbb{E} x \mathbb{1}_{\{Z \geq x\}} \leq \mathbb{E} |M_t| \mathbb{1}_{\{Z \geq x\}}.$$

Using this inequality, we have

$$
\mathbb{E} Z^p = \int_0^\infty dx \; p x^{p-2} \mathbb{E} x \mathbb{1}_{\{Z \geq x\}}
$$

$$
\leq \mathbb{E}|M_t| \int_0^\infty dx \; p x^{p-2} \mathbb{1}_{\{Z \geq x\}} = \mathbb{E}|M_t| q Z^{p-1}.
$$

Finally, with $X := |M_t|$ and $Y := Z^{p-1}$, we have by Hölder's inequality $\|XY\|_1 \leq \|X\|_p \|Y\|_q$, with $1/p + 1/q = 1$:

$$
\mathbb{E}|M_t| q Z^{p-1} \leq q (\mathbb{E}|M_t|^p)^{1/p} \, (\mathbb{E} Z^p)^{1/q},
$$

where we have used that $(p-1)q = p$. Since the left-hand side is greater than or equal to $a := \mathbb{E} Z^p$, we have $a \leq q (\mathbb{E}|M_t|^p)^{1/p} a^{1/q}$; that is, $a \leq q^p \, \mathbb{E}|M_t|^p$. □

As an application of Doob's norm inequality, the following lemma shows that the only interesting continuous-time martingales with continuous sample paths are the ones whose paths have infinite total variation:

Lemma 5.90: Continuous Martingales with Finite Variation

Let $X := (X_t, t \geq 0)$ be a martingale with continuous paths and total variation $V_t < \infty$ on each interval $[0, t]$. Then, X is almost surely constant.

Proof. Without loss of generality, we assume that $\mathbb{E}X_0 = 0$, and so $\mathbb{E}X_t = 0$ for all $t \geq 0$, by the martingale property. Take an $m \in \mathbb{N}$, and let $T := \inf\{t \geq 0 : V_t \geq m\}$. Note that the function $t \mapsto V_t$ is continuous. Take a sequence (Π_n) of segmentations of $[0, t]$, whose mesh goes to 0 as $n \to \infty$; as in (5.19). Applying (5.89) to the martingale $(X_{t \wedge T})$, with $p = 2$ (and hence $q = 2$), gives $\mathbb{E} \max_{s \leq t} X_{s \wedge T}^2 \leq 4 \mathbb{E} X_{t \wedge T}^2$, so that

$$
\frac{1}{4} \mathbb{E} \max_{s \leq t} X_{s \wedge T}^2 \leq \mathbb{E} \sum_{k=0}^{n-1} \left[X_{s_{k+1} \wedge T}^2 - X_{s_k \wedge T}^2 \right] = \mathbb{E} \sum_{k=0}^{n-1} \left(X_{s_{k+1} \wedge T} - X_{s_k \wedge T} \right)^2
$$

$$
\leq \underbrace{\mathbb{E}\left[V_T \max_k \left| X_{s_{k+1} \wedge T} - X_{s_k \wedge T} \right| \right]}_{=:Z_n} \leq m \, \mathbb{E} Z_n,
$$

where we have used (5.29) in the equality above. Since the paths of X are continuous, they are uniformly continuous on $[0, t]$, and hence $Z_n \to 0$ almost surely as $n \to \infty$. Since $Z_n \leq V_{t \wedge T} \leq m$, the Bounded Convergence Theorem 2.36 implies that $\mathbb{E}Z_n \to 0$ as $n \to \infty$, and hence that $\mathbb{E} \max_{s \leq t} X_{s \wedge T}^2 = 0$. It follows that $X_s = 0$, $s \in [0, t \wedge T]$ for an arbitrary t. Since $T \to \infty$ as $m \to \infty$, almost surely $X_s = 0$ for all s. □

5.6.3 Martingale Extensions

In various applications of continuous-time martingales, we are given a martingale M on $\mathbb{R}$ and a stopping time T that is allowed to take the value ∞; for example, T could be the time that M enters some set, which may never happen. The question is then whether the Doob's stopping theorem still holds for T. The following shows that indeed it does, as long as M is a martingale on $\overline{\mathbb{R}}_+$:

Proposition 5.91: Doob Martingale on $\overline{\mathbb{R}}_+$

A process M is a Doob martingale on $\overline{\mathbb{R}}_+$ if and only if it is a martingale on $\overline{\mathbb{R}}_+$. If so, then

$$\mathbb{E}_S M_T = M_S$$

for arbitrary stopping times with $S \le T$.

Proof. Necessity is obvious. To prove sufficiency, suppose that M is a martingale on $\overline{\mathbb{R}}_+$. We shall show that

$$(5.92) \qquad M_T = \mathbb{E}_T M_\infty$$

for every stopping time T. The characterization Theorem 5.80 then shows that M is a Doob martingale on $\overline{\mathbb{R}}_+$. Also, for any $S \le T$, taking expectations with respect to $\mathbb{E}_S$ in (5.92) gives $\mathbb{E}_S M_T = M_S$, which then completes the proof.

To show (5.92), first observe that for any $n \in \mathbb{N}$:

$$(5.93) \qquad M_{T \wedge n} = \mathbb{E}_{T \wedge n} M_n.$$

This follows from Theorem 5.83, with $(S, T) := (T \wedge n, n)$. Next, because M is a martingale on $\overline{\mathbb{R}}_+$, the process $(M_n, n \in \overline{\mathbb{N}})$ is a martingale. In particular, by the martingale property,

$$(5.94) \qquad M_n = \mathbb{E}_n M_\infty.$$

Combining (5.93) and (5.94) and using the rules for repeated conditioning, we conclude that

$$M_{T \wedge n} = \mathbb{E}_{T \wedge n} \mathbb{E}_n M_\infty = \mathbb{E}_n \mathbb{E}_T M_\infty.$$

Taking $n \to \infty$, $M_{T \wedge n}$ converges almost surely to M_T, whereas $\mathbb{E}_n \mathbb{E}_T M_\infty$ converges to $\mathbb{E}_\infty \mathbb{E}_T M_\infty$; that is, the conditional expectation of the random variable $\mathbb{E}_T M_\infty$, given $\widetilde{\mathcal{F}}_\infty := \vee_{n \in \mathbb{N}} \mathcal{F}_n$. The latter is the same as $\mathcal{F}_\infty := \vee_{t \in \mathbb{R}} \mathcal{F}_t$, so that $\mathbb{E}_\infty \mathbb{E}_T M_\infty = \mathbb{E}_T M_\infty$, which establishes (5.92). $\square$

The following is the continuous-time version of Theorem 5.54:

Theorem 5.95: Extension of UI Martingales

A martingale M on $\mathbb{R}_+$ can be extended to a Doob martingale $\overline{M}$ on $\overline{\mathbb{R}}_+$ if and only if it is uniformly integrable; that is, if and only if

$$M_t = \mathbb{E}_t Z, \quad t \in \mathbb{R}_+$$

for some integrable random variable Z. Moreover, then, it converges almost surely and in L^1 to an integrable random variable M_∞ and can be extended to a UI martingale on $\overline{\mathbb{R}}_+$.

Proof. Suppose that the martingale M on $\mathbb{R}_+$ can be extended to a Doob martingale $\overline{M}$ on $\overline{\mathbb{R}}_+$. In particular, there exists a random variable M_∞ in $\mathcal{F}_\infty$ such that

$$M_t = \mathbb{E}_t M_\infty$$

for every $t \in \mathbb{R}_+$. Thus, by Example 5.11, M is a uniformly integrable martingale.

Conversely, suppose that M is UI on $\mathbb{R}_+$, then also $(M_n, n \in \mathbb{N})$ is UI on $\mathbb{N}$ and by Theorem 5.54 this sequence converges almost surely and in L^1 to an integrable random variable $M_\infty \in \mathcal{F}_\infty$, and $M_n = \mathbb{E}_n M_\infty$ for all n. To show that the same holds in continuous time, pick any $t \in \mathbb{R}_+$ and $n > t$. Then, by the martingale property of M, and using the properties of repeated conditioning, we have for $t \in \mathbb{R}_+$:

$$M_t = \mathbb{E}_t M_n = \mathbb{E}_t \mathbb{E}_n M_\infty = \mathbb{E}_t M_\infty,$$

which shows that $(M_t, t \in \overline{\mathbb{R}}_+)$ is a martingale and, in view of Proposition 5.91, a Doob martingale on $\overline{\mathbb{R}}_+$. $\qquad\square$

Proposition 5.96: Sufficient Condition for Being Doob

Let M be a martingale on $\mathbb{R}_+$ and ζ a stopping time. If, almost surely,

$$\sup_{t \in [0, \zeta] \cap \mathbb{R}_+} |M_t| \leq Z,$$

where $\mathbb{E}Z < \infty$ (i.e., M is dominated by an integrable random variable Z), then $M_\zeta = \lim M_{\zeta \wedge t}$ exists and is integrable. Moreover, M is a Doob martingale on $[0, \zeta]$.

Proof. Let $\widehat{M} := (M_{t \wedge \zeta}, t \in \mathbb{R}_+)$ be the process M stopped at time ζ. Then, $\widehat{M}$ is a martingale on $\mathbb{R}_+$ — we leave the proof as Exercise 27.

By assumption, $\widehat{M} \leq Z$, almost surely, so that $\widehat{M}$ is uniformly integrable and hence it converges almost surely and in L^1 to $\lim \widehat{M}_t = \lim M_{\zeta \wedge t} = M_\zeta$, which must be integrable. By Theorem 5.95, $\widehat{M}$ can be extended to a Doob martingale on $\overline{\mathbb{R}}_+$ by defining $\widehat{M}_\infty := M_\zeta$. Then, for every stopping time $T \leq \zeta$, we have $M_T = \widehat{M}_T$ and

$$\mathbb{E}M_T = \mathbb{E}\widehat{M}_T = \mathbb{E}\widehat{M}_0 = \mathbb{E}M_0.$$

Theorem 5.80 then implies that M is a Doob martingale on $[0, \zeta]$.   □

Exercises

1. Show that $\mathcal{F}_T$ in Definition 5.3 is a σ-algebra on Ω that is contained in $\mathcal{F}_\infty$. Show also that the stopping time T itself is $\mathcal{F}_T$-measurable.

2. Let T be a stopping time of a filtration $\mathcal{F} := (\mathcal{F}_n, n \in \mathbb{N})$. We say that the stochastic process $X := (X_n, n \in \mathbb{N})$ *belongs to* $\mathcal{F}$ if $X_n \in \mathcal{F}_n$ for all n; we write $X \in \mathcal{F}$. Show that

$$\mathcal{F}_T = \{X_T : X \in \mathcal{F}\}.$$

3.* Let X be the asymmetric random walk starting at a and T the hitting time of 0 or b, as defined in Example 5.33. Assume again that $p \neq q$ and that $a, b \in \mathbb{N}$, with $0 < a < b$. Consider the process $Z_n := X_n - n(p - q), n \in \mathbb{N}$.

  (a) Prove that (Z_n) is a martingale.

  (b) Show that $\mathbb{E}Z_T = \mathbb{E}Z_0$, and use this together with (5.34) to find an explicit expression for $\mathbb{E}T$ in terms of p, q, a, and b.

4.* Let X be a symmetric random walk, as in Example 5.2, and define $T := \inf\{n : |X_n| = a\}$ for some positive integer a.

  (a) Show that $(X_n^2 - n, n \in \mathbb{N})$ is a martingale.

  (b) Prove that $\mathbb{E}T = a^2$.

5.* The stochastic process $X := (X_n, n \in \mathbb{N})$ is such that $(X_{n+1} \mid X_0, \ldots, X_n) \sim \mathcal{U}(X_n, 1)$, $n \in \mathbb{N}$, with $X_0 := 0$. Show that $Y := (Y_n)$, with $Y_n := 2^n(1 - X_n)$, is a martingale with respect to the natural filtration of X.

6. Let $\xi_1, \xi_2, \ldots$ be a sequence of independent $\mathsf{Exp}(1)$ random variables. Define $S_n := \sum_{k=1}^n \xi_k$ (with $S_0 := 0$) and $T := \inf\{n : \xi_n > 1\}$. Show that T is a stopping time with respect to the natural filtration of the $\{\xi_n\}$ and use martingale arguments to compute $\mathbb{E}S_T$.

7. Let $\xi_1, \xi_2, \ldots$ be a sequence of independent random variables with $\mathbb{P}(\xi_k = 1) = \mathbb{P}(\xi_k = -1) = 1/2$ for all k. Define $S_n := \sum_{k=1}^{n} \xi_k$ (with $S_0 := 0$) and $T := \inf\{n : S_n \in \{-1, 9\}\}$.

(a) Use martingale arguments to compute the distribution of S_T.

(b) Use martingale arguments to compute the expected value of T. Hint: $(S_n^2 - n)$ is also a martingale.

8. Let $M := (M_n, n \in \mathbb{N})$ be a martingale for which $\mathbb{E}M_n^2 < \infty$ for every $n \in \mathbb{N}$. Show that $\sup_n \mathbb{E}M_n^2 < \infty$ if and only if $\sum_{k=1}^{\infty} \mathbb{E}(M_k - M_{k-1})^2 < \infty$. Hint: write $M_n = M_0 + \sum_{k=1}^{n}(M_k - M_{k-1})$.

9. Building upon Example 5.10, let X be a real-valued compound Poisson process with Lévy measure v, satisfying $v(\mathbb{R}) = c < \infty$. Show that

$$M_t := X_t - t \int v(dx)\, x, \quad t \geq 0,$$

is a martingale.

10.* In Example 5.15, denote the probability generating function of T by G; in particular, $G(z) = \mathbb{E}\, z^T$ for $|z| \leq 1$.

(a) By conditioning on X_1, show that

$$G(z) = \frac{1 - \sqrt{1 - z^2}}{z}.$$

(b) Expanding $G(z)$ via Newton's formula (see Exercise 4.8) deduce that $T < \infty$ with probability 1, but that $\mathbb{E}T = \infty$.

11. This is to show that a right-continuous function f is of bounded variation on an interval $[0, t]$ if and only if $f = g - h$, where g and h are real-valued positive functions that are increasing on $[0, t]$. Let $V_f(t)$ be the total variation of f on $[0, t]$, as defined in (5.18).

(a) Show that if f is increasing, then $V_f(t) = f(t) - f(0)$.

(b) Prove that if f and g are of bounded variation, then $f - g$ and $f + g$ are also of bounded variation.

(c) Suppose that f is of bounded variation on $[0, t]$. Consider the functions g and h, defined for $r \in [0, t]$ by

$$g(r) := V_f(r) \quad \text{and} \quad h(r) := V_f(r) - f(r).$$

Show that g and h are increasing and that $f = g - h$.

12. Let $(Z_k, k = 1, 2, \ldots)$ be a sequence of iid random variables with $\mathbb{P}(Z_k = 1) = p > \frac{1}{2}$ and $\mathbb{P}(Z_k = -1) = 1 - p$. Let $\mathcal{F}_n := \sigma(Z_1, \ldots, Z_n)$ and $\mathcal{F} := (\mathcal{F}_n)$. Define $X_0 := 1$ and

$$X_{n+1} := X_n + C_{n+1} Z_{n+1}, \quad n \in \mathbb{N},$$

where $C := (C_n)$ is an $\mathcal{F}$-predictable process such that $C_n \in [0, (1 - \varepsilon) X_{n-1}]$ for some $\varepsilon > 0$. Show that the process $M := (\ln X_n - n\alpha, n \in \mathbb{N})$, where

$$\alpha := p \ln p + (1 - p) \ln(1 - p) + \ln 2$$

is an $\mathcal{F}$-supermartingale. For which predictable process C is M an $\mathcal{F}$-martingale?

13.* Let $M := (M_n)$ be a martingale with $M_0 = 0$ and such that, for some sequence of constants c_n, $|M_n - M_{n-1}| \leq c_n$ for all n. Then, for $x > 0$:

$$\mathbb{P}\left(\max_{k \leq n} M_k \geq x\right) \leq \exp\left(\frac{-x^2}{2 \sum_{k=1}^{n} c_k^2}\right).$$

This is the *Azuma–Hoeffding inequality*. We can prove it using the following steps:

(a) Show that the process (Y_n) with $Y_n := e^{\theta M_n}$ is a positive submartingale for any $\theta \in \mathbb{R}$.

(b) Apply (a) and (5.43) to conclude that

$$\mathbb{P}\left(\max_{k \leq n} M_k \geq x\right) \leq e^{-\theta x} \mathbb{E} e^{\theta M_n}.$$

(c) Write $M_n = M_0 + \sum_{k=1}^{n}(M_k - M_{k-1})$ and use the condition $|M_n - M_{n-1}| \leq c_n$ and the exponential bound in Exercise 2.17 to show that

$$\mathbb{E} e^{\theta M_n} \leq \mathbb{E} e^{\theta M_{n-1}} e^{\frac{1}{2}\theta^2 c_n^2}.$$

(d) Complete the proof.

14. Let $X_1, X_2, \ldots$ be a sequence of independent binomial random variables such that $X_k \sim \text{Bin}(k, 1/k)$. Define $M_n := \prod_{k=1}^{n} X_k$ for $n = 1, 2, \ldots$.

(a) Show that $M := (M_n, n \geq 1)$ is martingale with respect to the natural filtration $\mathcal{F} := (\mathcal{F}_n, n \geq 1)$ of the process $(X_n, n \geq 1)$.

(b) Show that M_n converges almost surely and find its limit.

(c) Show that $(M_n, n \geq 1)$ is not an UI martingale.

15.* In Example 5.60, because $(M_{n \wedge T}, n \in \overline{\mathbb{N}})$ is uniformly integrable, it converges almost surely and in L^1 to some integrable random variable. What is the probability distribution of this random variable?

16. Assume C is a family of uniformly integrable random variables on some probability space $(\Omega, \mathcal{H}, \mathbb{P})$. Let $\mathcal{D}$ be the family of random variables where $Y \in \mathcal{D}$ if there exists an $X \in C$ and sub σ-algebra $\mathcal{G} \subset \mathcal{H}$ such that $Y = \mathbb{E}_{\mathcal{G}} X$. Show that the family $\mathcal{D}$ is uniformly integrable.

17. Suppose Q and P are finite measures, with $Q \ll P$ and $P \ll Q$. Such measures P and Q are said to be *equivalent*. Show that the Radon–Nikodym derivative $Z = dQ/dP$ satisfies $P(Z \le 0) = 0$.

18. Consider the measure space $((0,1], \mathcal{B}_{(0,1]}, \mathrm{Leb}_{(0,1]})$. Let $\mathcal{F}_n$ be the sub-σ-algebra of $\mathcal{H}$ that is generated by the partition $\mathcal{P}_n$ given in Example 5.63, and let $\mathcal{F} := (\mathcal{F}_n)$ be the corresponding filtration. Let F be a bounded increasing function on $(0, 1]$. Define

$$f_n(x) := 2^n \left(F\left(\frac{\lceil 2^n x \rceil}{2^n} \right) - F\left(\frac{\lfloor 2^n x \rfloor}{2^n} \right) \right),$$

where $\lceil a \rceil$ rounds a up to the nearest integer and $\lfloor a \rfloor$ rounds a down to the nearest integer.

(a) Show that (f_n) is a martingale with respect to $\mathcal{F}$ and show that it converges almost surely.

(b) Show that (f_n) converges in L^1 when $F(x) := x^{1/2}$. Hint: It may be useful to recall $\sqrt{b} - \sqrt{a} = (b-a)/(\sqrt{b}+\sqrt{a})$.

19. Let $M := (M_n)$ be a martingale with $\mathbb{E} M_n^2 < \infty$ for each n. Prove that if $\sup_n \mathbb{E} M_n^2 < \infty$, then $M_n \to M_\infty$ almost surely and in L^2. Hint: use Exercise 8.

20. Show that if (ξ_n) is a martingale and $\xi_n \xrightarrow{L^1} a$ for some $a \in \mathbb{R}$, then $\xi_n = a$ almost surely for each n.

21. Let $\{X_{n,k}, n = 1, 2, \ldots, k = 1, 2, \ldots\}$ be a collection of iid random variables taking values in $\mathbb{N}$. Assume that $\mathbb{E} X_{n,k} = \mu < \infty$ and $\mathrm{Var}\, X_{n,k} = \sigma^2 < \infty$ for all n, k. Define the process (Z_n) such that $Z_0 := 1$ and

$$Z_{n+1} := \sum_{k=1}^{Z_n} X_{n+1,k}, \quad n \in \mathbb{N}.$$

If $Z_n = 0$, then $Z_m = 0$ for all $m \ge n$. The process (Z_n) is called a *branching process*, as illustrated in Figure 5.97.

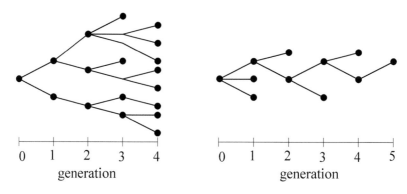

Figure 5.97: Left: population grows exponentially. Right: population dies out.

It can be used to model population dynamics: Z_n is the total number of individuals in the nth generation. At each generation, each individual of that generation creates offspring according to a common offspring distribution, with expectation μ and variance σ^2, independently of the other individuals in the present and past generations. The number of offspring of the kth individual in the nth generation is $X_{n+1,k}$. Depending on the offspring distribution, the population can grow exponentially or can become extinct.

(a) Show that the process (M_n) defined by $M_n := Z_n/\mu^n$ is a martingale with respect to the natural filtration of (Z_n).

(b) Show that $\mathbb{E}_n Z_{n+1}^2 = \mu^2 Z_n^2 + \sigma^2 Z_n$.

(c) Prove that if $\mu > 1$, then $\sup \mathbb{E} M_n^2 < \infty$.

(d) Prove that if $\mu > 1$, then $M_n \xrightarrow{L^1} M_\infty$ and $\mathbb{V}\mathrm{ar}\, M_\infty = \sigma^2(\mu(\mu - 1))^{-1}$.

22. Let $(X_n, n \in \mathbb{N})$ be a sequence of positive integrable random variables on some probability space $(\Omega, \mathcal{H}, \mathbb{P})$ with filtration $\mathcal{F}_n := \sigma(X_0, \ldots, X_n), n \in \mathbb{N}$. Suppose that $\mathbb{E}_n X_{n+1} \le X_n + y_n$, where $(y_n, n \in \mathbb{N})$ is a sequence of positive constants such that $\sum_{n=0}^{\infty} y_n < \infty$. Show that $X_n \xrightarrow{\text{a.s.}} X_\infty$, where X_∞ is some integrable random variable.

23.* Let $(X_n, n = 1, 2, \ldots)$ be a sequence of independent random variables such that $\mathbb{P}(X_n = -1) = \mathbb{P}(X_n = 1) = \frac{1}{2}$. Show that the sequence of random variables $M_n = \sum_{j=1}^{n} X_j j^{-1}, n = 1, 2, \ldots$ converges almost surely and in L^1 to an integrable random variable M_∞.

24. Let $\xi_1, \xi_2, \ldots$ be a sequence of independent random variables with

$$\mathbb{P}\left(\xi_n = \frac{n+1}{n}\right) = 1 - \mathbb{P}\left(\xi_n = \frac{n}{n+1}\right) = \frac{n}{2n+1}.$$

Define $a_n := \mathbb{E}\,\xi_n^{1/2}$, $N_0 := 1$, and

$$N_n := \prod_{k=1}^{n} \frac{\xi_k^{1/2}}{a_k}.$$

(a) Show that (N_n) is a martingale with respect to the natural filtration, satisfying $\mathbb{E}\,N_n^2 < \infty$ for all n.

(b) Use N_n and the discrete version of Doob's norm inequality (5.89) to show that the martingale (M_n) defined by

$$M_0 := 1, \quad M_n := \prod_{k=1}^{n} \xi_k, \quad n = 1, 2, \ldots,$$

converges almost surely and in L^1 to a random variable M_∞ with $\mathbb{E}\,M_\infty = 1$.

25. Check that $\mathcal{F}^+$ defined by (5.73) is indeed a right-continuous filtration.

26. Let T be a stopping time with respect to some filtration $(\mathcal{F}_t)$. Define for each $n \in \mathbb{N}$:
$$d_n(t) := \frac{k+1}{2^n} \quad \text{if} \quad \frac{k}{2^n} \le t < \frac{k+1}{2^n} \quad \text{for some } k \in \mathbb{N}.$$
Define $T_n := d_n(T)$, $n \in \mathbb{N}$. Show that $T_n \in \mathcal{F}_T$ and that (T_n) is a sequence of stopping times decreasing to T.

27. Let M be a martingale on $\mathbb{R}_+$ and ζ a stopping time. Let $\widehat{M} := (M_{t \wedge \zeta}, t \in \mathbb{R}_+)$. Complete the proof of Proposition 5.96 by showing that $\widehat{M}$ is a martingale on $\mathbb{R}_+$.

CHAPTER 6

WIENER AND BROWNIAN MOTION PROCESSES

In this chapter, we further explore the Wiener and Brownian motion processes. We prove the existence of the Wiener process via Lévy's construction, and discuss many of its features, including its Gaussian, martingale, Markov, and path properties. We discuss the close relation between Brownian motions and the Laplace operator. We also show that the maximum and hitting time processes have intimate connections with Lévy processes and Poisson random measures.

6.1 Wiener Process

We already encountered the Wiener process in Chapters 2 and 4. In Example 2.77, we defined the Wiener process as a Gaussian process as follows:

Definition 6.1: Wiener Process

The *Wiener process* $(W_t, t \geq 0)$ is a zero-mean Gaussian process with continuous sample paths and covariance function

$$\gamma_{s,t} = \min\{s, t\} =: s \wedge t, \quad s, t \geq 0.$$

In Example 2.96 we recognized that the Wiener process belongs to the family of *Lévy processes* — Markov processes that have independent and identically distributed increments, and right-continuous and left-limited sample paths that start from 0; see Sections 2.8.4 and 4.5. The compound Poisson process is an example of a *pure-jump* Lévy process. In contrast, a Wiener process is a Lévy process with

---

continuous sample paths, as shown in the following result, which can also serve as an equivalent definition of the Wiener process:

Theorem 6.2: Wiener Process as a Lévy Process

$W := (W_t, t \geq 0)$ is a Wiener process if and only if the three properties hold:

1. *(Independent increments)*: For any $t_1 < t_2 \leq t_3 < t_4$, the increments $W_{t_4} - W_{t_3}$ and $W_{t_2} - W_{t_1}$ are independent.

2. *(Gaussian stationarity)*: For all $t, u \geq 0$, $W_{t+u} - W_t \sim \mathcal{N}(0, u)$.

3. *(Continuity of paths)*: W has continuous paths, with $W_0 = 0$.

Proof. The continuity of paths is already part of Definition 6.1 and therefore does not need to be discussed further here.

Assuming that W is a Wiener process, we now demonstrate the Gaussian stationarity and independent increments properties in a slightly different way from Example 2.96. Since the Wiener process is a Gaussian process, any increment $W_{t+u} - W_t$ has a Gaussian distribution. Its expectation is 0 and its variance follows from:

$$\mathbb{Cov}(W_{t+u} - W_t, W_{t+u} - W_t) = \mathbb{Cov}(W_{t+u}, W_{t+u}) + \mathbb{Cov}(W_t, W_t) - 2\,\mathbb{Cov}(W_{t+u}, W_t)$$
$$= t + u + t - 2t = u.$$

This proves the Gaussian stationarity. To show the independence of the increments, take $t_1 < t_2 \leq t_3 < t_4$. The distribution of $\mathbf{W} := [W_{t_1}, \ldots, W_{t_4}]^\top$ is multivariate normal with mean vector $\mathbf{0}$ and covariance matrix

$$\Sigma := \begin{bmatrix} t_1 & t_1 & t_1 & t_1 \\ t_1 & t_2 & t_2 & t_2 \\ t_1 & t_2 & t_3 & t_3 \\ t_1 & t_2 & t_3 & t_4 \end{bmatrix}.$$

We can easily verify that $\Sigma = \mathbf{L}\mathbf{D}\mathbf{L}^\top$, with

$$\mathbf{L} := \begin{bmatrix} 1 & 0 & 0 & 0 \\ 1 & 1 & 0 & 0 \\ 1 & 1 & 1 & 0 \\ 1 & 1 & 1 & 1 \end{bmatrix} \quad \text{and} \quad \mathbf{D} := \begin{bmatrix} t_1 & 0 & 0 & 0 \\ 0 & t_2 - t_1 & 0 & 0 \\ 0 & 0 & t_3 - t_2 & 0 \\ 0 & 0 & 0 & t_4 - t_3 \end{bmatrix}.$$

This means, from the theory of Gaussian random vectors, that $\mathbf{W}$ has the same distribution as $\mathbf{L}\mathbf{D}^{1/2}\mathbf{Z}$, where $\mathbf{Z} := [Z_1, Z_2, Z_3, Z_4]^\top$ is a vector of independent standard Gaussians. Consequently, the increments $W_{t_2} - W_{t_1}$ and $W_{t_4} - W_{t_3}$ have the same distribution as $\sqrt{t_2 - t_1}\, Z_2$ and $\sqrt{t_4 - t_3}\, Z_4$, which are independent of each other.

Next, assuming that the three properties hold, we prove that W is a zero-mean Gaussian process with covariance $\mathbb{C}\text{ov}(W_s, W_t) = s \wedge t$. This is equivalent to proving that for an arbitrary n and $t_1 < t_2 < \cdots < t_n$, the vector $\boldsymbol{W} = [W_{t_1}, \ldots, W_{t_n}]^\top$ has the characteristic function: $\psi_{\boldsymbol{W}}(\boldsymbol{r}) := \exp(-\boldsymbol{r}^\top \boldsymbol{\Sigma} \boldsymbol{r}/2)$, where $\boldsymbol{\Sigma}$ is the $n \times n$ covariance matrix with (i, j)-th entry $t_i \wedge t_j$. To this end, let $\mathbf{L} \in \mathbb{R}^{n \times n}$ be a lower triangular matrix with all entries equal to 1 and $\mathbf{D} \in \mathbb{R}^{n \times n}$ be a diagonal matrix with entries $t_1, t_2 - t_1, \ldots, t_n - t_{n-1}$ down the main diagonal. Then, $\boldsymbol{\Sigma} = \mathbf{L}\mathbf{D}\mathbf{L}^\top$ and the Gaussian stationarity and the independence of increments imply that

$$\boldsymbol{X} := \mathbf{L}^{-1}\boldsymbol{W} = [W_{t_1}, W_{t_2} - W_{t_1}, \ldots, W_{t_n} - W_{t_{n-1}}]^\top \sim \mathcal{N}(\mathbf{0}, \mathbf{D}).$$

The proof is then complete by the following calculation:

$$\psi_{\boldsymbol{W}}(\boldsymbol{r}) = \psi_{\boldsymbol{X}}(\mathbf{L}^\top \boldsymbol{r}) = \exp(-(\mathbf{L}^\top \boldsymbol{r})^\top \mathbf{D}(\mathbf{L}^\top \boldsymbol{r})/2) = \exp(-\boldsymbol{r}^\top \boldsymbol{\Sigma} \boldsymbol{r}/2).$$

$\square$

The following algorithm uses the same ideas as in the previous proof to simulate the Wiener process at specific time points $t_1, \ldots, t_n$:

■ **Algorithm 6.3 (Simulating a Wiener Process)**

1. Let $0 = t_0 < t_1 < \cdots < t_n$ be the times at which the process needs to be simulated.

2. Simulate $Z_1, \ldots, Z_n \overset{\text{iid}}{\sim} \mathcal{N}(0, 1)$.

3. Return $W_0 := 0$ and $W_{t_k} := \sum_{i=1}^{k} Z_k \sqrt{t_k - t_{k-1}}$, $k = 1, \ldots, n$.

The Wiener process plays a central role in probability and forms the basis of many other stochastic processes. The Wiener process can also be viewed as a continuous version of a random walk process. Two typical sample paths are depicted in Figure 6.4.

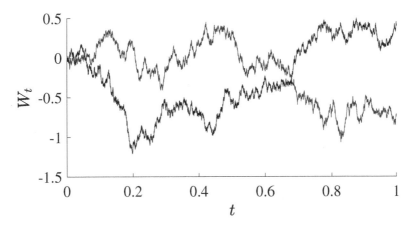

Figure 6.4: Two realizations of the Wiener process on the interval [0,1].

■ **Example 6.5 (Brownian Bridge)** Suppose the values of a Wiener process at times t_1 and $t_2 > t_1$ are $W_{t_1} := a$ and $W_{t_2} := b$. We can then easily simulate W_t at some $t \in [t_1, t_2]$, conditional on $W_{t_1} = a$ and $W_{t_2} = b$, because the conditional distribution of W_t is Gaussian with mean $a + (b - a)(t - t_1)/(t_2 - t_1)$ and variance $(t - t_1)(t_2 - t)/(t_2 - t_1)$.

To see this, consider the vector $[W_{t_1}, W_{t_2}, W_t]^\top$. It is jointly Gaussian, with mean vector $\mathbf{0}$ and covariance matrix

$$\Sigma := \begin{bmatrix} t_1 & t_1 & t_1 \\ t_1 & t_2 & t \\ t_1 & t & t \end{bmatrix}.$$

Its Cholesky factor is

$$\mathbf{B} := \begin{bmatrix} \sqrt{t_1} & 0 & 0 \\ \sqrt{t_1} & \sqrt{t_2 - t_1} & 0 \\ \sqrt{t_1} & \frac{t - t_1}{\sqrt{t_2 - t_1}} & \sqrt{\frac{(t - t_1)(t_2 - t)}{t_2 - t_1}} \end{bmatrix},$$

so that W has the same distribution as $\mathbf{B}[Z_1, Z_2, Z_3]^\top$, where $Z_1, Z_2, Z_3 \sim_{\text{iid}} \mathcal{N}(0, 1)$. In other words, we can write

$$W_{t_1} = \sqrt{t_1}\, Z_1$$

$$W_{t_2} = \sqrt{t_1}\, Z_1 + \sqrt{t_2 - t_1}\, Z_2$$

$$W_t = \sqrt{t_1}\, Z_2 + \frac{t - t_1}{\sqrt{t_2 - t_1}} Z_2 + \sqrt{\frac{(t - t_1)(t_2 - t)}{t_2 - t_1}}\, Z_3$$

$$= W_{t_1} + (W_{t_2} - W_{t_1}) \frac{t - t_1}{t_2 - t_1} + \sqrt{\frac{(t - t_1)(t_2 - t)}{t_2 - t_1}}\, Z_3,$$

which implies that, given $W_{t_1} = a$ and $W_{t_2} = b$, W_t has mean $a + (b-a)(t-t_1)/(t_2-t_1)$ and variance $(t - t_1)(t_2 - t)/(t_2 - t_1)$. ■

In Example 2.77, we defined a Brownian motion as an affine transformation of a Wiener process.

Definition 6.6: Brownian Motion

A stochastic process $(B_t, t \geq 0)$ satisfying

$$B_t = B_0 + \mu t + \sigma W_t, \quad t \geq 0,$$

where (W_t) is a Wiener process independent of B_0, is called a *Brownian motion* with *drift* μ and *diffusion coefficient* σ.

A *standard Brownian motion* is one where $\mu = 0$ and $\sigma = 1$. The only difference with a Wiener process is thus its (random) starting position.

Many properties of the Brownian motion process follow directly from those of the Wiener process. The generation of a Brownian motion at times $t_1, \ldots, t_n$ follows directly from its definition.

■ **Algorithm 6.7 (Simulating a Brownian Motion)**

1. Simulate the starting position B_0.

2. Simulate outcomes $W_{t_1}, \ldots, W_{t_n}$ of a Wiener process at times $t_1, \ldots, t_n$.

3. Return $B_{t_i} := B_0 + \mu\, t_i + \sigma\, W_{t_i}$, $i = 1, \ldots, n$ as the outcomes of the Brownian motion at times $t_1, \ldots, t_n$.

Multidimensional Brownian motions are likewise obtained from an affine transformation $\boldsymbol{B}_t = \boldsymbol{B}_0 + \boldsymbol{\mu}\, t + \sigma \boldsymbol{W}_t$ of a multidimensional Wiener process, where σ is a matrix.

Definition 6.8: Multidimensional Wiener Process

Let $(W_{t,i}, t \geq 0)$, $i = 1, \ldots, d$ be independent Wiener processes and let $\boldsymbol{W}_t := [W_{t,1}, \ldots, W_{t,d}]^\top$. The process $(\boldsymbol{W}_t, t \geq 0)$ is called a *d-dimensional Wiener process*.

If $(\boldsymbol{W}_t)$ is a d-dimensional Wiener process, then $\boldsymbol{W}_t$ has a $\mathcal{N}(\boldsymbol{0}, t\mathbf{I}_d)$ multivariate normal distribution, where $\mathbf{I}_d$ denotes the d-dimensional identity matrix.

■ **Example 6.9 (Three-dimensional Wiener Process)** The following MATLAB program generates a realization of the three-dimensional Wiener process at times $0, 1/N, 2/N, \ldots, 1$, for $N = 10^4$. Figure 6.10 shows a typical realization.

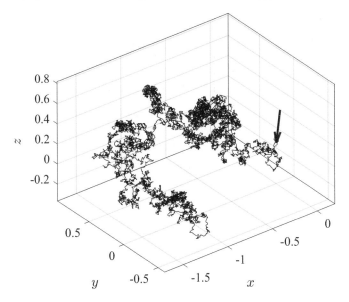

Figure 6.10: Three-dimensional Wiener process. The arrow points to the origin.

```
N=10^4; T=1; dt=T/N; %step size
X=cumsum([0,0,0;randn(N,3)*sqrt(dt)],1);
plot3(X(:,1),X(:,2),X(:,3))
```

■

We next prove a number of properties of the Wiener process $W := (W_t, t \geq 0)$; there are many more to follow!

Theorem 6.11: Time-reversal and Time-shifting

If $(W_s, s \in [0, t])$ is a Wiener process on $[0, t]$, then so is the time-reversed process $(\widehat{W}_s, s \in [0, t])$ defined by $\widehat{W}_s := W_{t-s} - W_t$. Similarly, if $(W_s, s \geq 0)$ is a Wiener process, then for any $t \geq 0$ the process $(\widetilde{W}_s, s \geq 0)$ defined by $\widetilde{W}_s := W_{t+s} - W_t$ is also a Wiener process.

Proof. Obviously, $(\widehat{W}_s, s \in [0, t])$ is a Gaussian process with zero mean and continuous sample paths, as these properties are inherited from W. It remains to check that $\mathbb{E}\widehat{W}_s\widehat{W}_u = s$ for all $0 < s < u < t$. This follows from

$$\begin{aligned}
\mathbb{E}\widehat{W}_s\widehat{W}_u &= \mathbb{E}(W_{t-s} - W_t)(W_{t-u} - W_t) \\
&= \mathbb{E}W_{t-s}W_{t-u} + \mathbb{E}W_t^2 - \mathbb{E}W_tW_{t-u} - \mathbb{E}W_{t-s}W_t \\
&= (t-u) + t - (t-u) - (t-s) \\
&= s.
\end{aligned}$$

A similar argument shows that $(\widetilde{W}_s, s \geq 0)$ is a Wiener process.   □

Theorem 6.12: Scaling

If (W_t) is a Wiener process, then so is (X_t), with $X_t := W_{at}/\sqrt{a}$, $t \geq 0$ for any $a > 0$.

Proof. That the scaled process is continuous, Gaussian, and has zero-mean is obvious. For $s < t$, the covariance function satisfies

$$\mathbb{E}X_sX_t = \mathbb{E}\frac{W_{as}}{\sqrt{a}}\frac{W_{at}}{\sqrt{a}} = \frac{as}{a} = s,$$

and so is the same as for the Wiener process.   □

Lemma 6.13: Probability Bound on the Supremum

For a Wiener process (W_t), it holds for every $\varepsilon > 0$ that

$$\mathbb{P}\left(\sup_{0 \leq t \leq 1} |W_t| > \varepsilon \right) \leq 2\,e^{-\varepsilon^2/2}.$$

Proof. Let $q_0, q_1, \ldots$ be an enumeration of the rational numbers in $[0, 1]$ and let D_n be the first n of these numbers, with $\{1\}$ appended if not already in D_n. For some fixed parameter $r > 0$, let $Y_t := e^{rW_t}$, so that the process $(Y_t, t \in D_n)$ is a positive discrete-time submartingale. Then, an application of Lemma 5.42 immediately yields the upper bound:

$$\mathbb{P}\left(\sup_{t \in D_n} Y_t > e^{r\varepsilon} \right) \leq \frac{\mathbb{E}Y_1}{e^{r\varepsilon}} = e^{\frac{1}{2}r^2 - r\varepsilon},$$

which is minimized at $r = \varepsilon$, giving the minimum $e^{-\varepsilon^2/2}$. Since $t \mapsto e^{rW_t}$ is continuous, $\sup_{t \in D_n} Y_t \uparrow \sup_{t \in [0,1]} Y_t$. Thus, using the continuity from below property (2.3):

$$\mathbb{P}\left(\sup_{t \in [0,1]} Y_t > e^{r\varepsilon} \right) = \lim_n \mathbb{P}\left(\sup_{t \in D_n} Y_t > e^{r\varepsilon} \right).$$

Hence, for $r := \varepsilon$ we obtain:

$$\mathbb{P}\left(\sup_{t \in [0,1]} W_t > \varepsilon \right) = \mathbb{P}\left(\sup_{t \in [0,1]} Y_t > e^{r\varepsilon} \right) \leq e^{-\varepsilon^2/2},$$

and the proof is complete by noting that $\mathbb{P}(\sup_t |W_t| > \varepsilon) \leq 2\,\mathbb{P}(\sup_t W_t > \varepsilon)$. □

Theorem 6.14: Reciprocal Time

If (W_t) is a Wiener process, then so is (X_t), with $X_t := t\,W_{1/t}$, $t > 0$, $X_0 := 0$.

Proof. Evidently, $(X_t, t > 0)$ is a zero-mean Gaussian process with continuous paths and with $\mathbb{E}X_s X_t = s$ for all $0 < s < t$. The only thing to check is that $(X_t, t \geq 0)$ — that is, including $t = 0$ — has the same properties; in particular, that the process is almost surely continuous at 0. In other words, we need to show that as $t \to \infty$:

(6.15)
$$\frac{W_t}{t} \xrightarrow{\text{a.s.}} 0.$$

Take any $t > 0$ and let $n := \lfloor t \rfloor$. Then, we can write

$$W_t = \sum_{k=1}^{n} Z_k + (W_t - W_n),$$

where the increments $Z_k := W_k - W_{k-1} \sim_{iid} \mathcal{N}(0, 1)$ are independent of $W_t - W_n$. This leads to the bound

$$\left|\frac{W_t}{t}\right| \leq \frac{1}{n}\left|\sum_{k=1}^{n} Z_k + (W_t - W_n)\right| \leq \left|\frac{1}{n}\sum_{k=1}^{n} Z_k\right| + \frac{1}{n}\sup_{0 \leq s \leq 1}|\widetilde{W}_s|,$$

where $\widetilde{W}_s := W_{n+s} - W_n$. By the strong Law of Large numbers (Theorem 3.44), $\sum_{k=1}^{n} Z_k/n$ converges almost surely to 0. Since $(\widetilde{W}_s, s \geq 0)$ is a Wiener process by Theorem 6.11, we have by Lemma 6.13 that for any $\varepsilon > 0$:

$$\mathbb{P}\left(\frac{1}{n}\sup_{0 \leq s \leq 1}|\widetilde{W}_s| > \varepsilon\right) \leq 2e^{-n^2\varepsilon^2/2}.$$

Since $\sum_n e^{-n^2\varepsilon^2/2} < \infty$, it follows that $\frac{1}{n}\sup_{0 \leq s \leq 1}|\widetilde{W}_s| \xrightarrow{cpl.} 0$. Hence, Theorem 3.40 implies that $\frac{1}{n}\sup_{0 \leq s \leq 1}|\widetilde{W}_s| \xrightarrow{a.s.} 0$ and consequently that (6.15) holds. □

■ **Remark 6.16 (Starting Position)** Since the Wiener process starts at position 0 by definition, it is sometimes useful to consider instead a standard Brownian motion process starting from some arbitrary state x under a probability measure $\mathbb{P}^x$. The corresponding expectation operator is then denoted by $\mathbb{E}^x$. ■

6.2 Existence

In Example 4.32 and Theorem 4.34 we showed that there indeed exists a Wiener process with the above properties. In this section, we show existence in a more direct way, constructing the Wiener process on $[0, 1]$ from a linear combination of random variables $Z_0, Z_1, \ldots \sim_{iid} \mathcal{N}(0, 1)$. Suppose that

$$W_t^{(n)} := \sum_{k=0}^{n-1} c_k(t) Z_k$$

for some sequence (c_k) of continuous deterministic functions on $[0, 1]$. Then, each sample path of the process $W^{(n)} := (W_t^{(n)}, t \in [0, 1])$ is continuous by construction and the random variable $W_t^{(n)}$ is normally distributed, because it is a linear combination of Gaussian random variables. Further, $\mathbb{E}W_t^{(n)} = \sum_{k=0}^{n-1} c_k(t) \times 0 = 0$ and

$$\mathbb{C}\text{ov}(W_s^{(n)}, W_t^{(n)}) = \sum_{k=0}^{n-1}\sum_{j=0}^{n-1} c_j(s) c_k(t) \underbrace{\mathbb{C}\text{ov}(Z_j, Z_k)}_{= \mathbb{1}_{\{j=k\}}} = \sum_{k=0}^{n-1} c_k(s) c_k(t).$$

In other words, the approximation process $W^{(n)}$ is a zero-mean Gaussian process with continuous sample paths and covariance function $\gamma_{s,t}^{(n)} := \sum_{k=0}^{n-1} c_k(s)c_k(t)$.

This meets all the requirements in Definition 6.1, except that the covariance function does not necessarily equal $\gamma_{s,t} := s \wedge t$ for all $s, t \in [0, 1]$. This requirement can be met if we choose the functions $\{c_k\}$ such that $\gamma_{s,t}^{(n)} \to s \wedge t$ as $n \to \infty$, so that $W^{(n)}$ will have the correct covariance function as $n \to \infty$. However, since the continuity of $W^{(n)}$ does not necessarily imply the continuity of the limiting process, we will have to demonstrate that the limit $W := \lim_n W^{(n)}$ not only exists, but also retains the continuity of its sample paths. This is the essence of the sought after construction, and all that is needed is to make the argument rigorous.

■ **Example 6.17 (Construction using Trigonometric Functions)** The trigonometric expansion in (B.26) suggests that we can use the functions $c_0(t) := t$ and $c_k(t) := \sqrt{2} \sin(k\pi t)/(k\pi)$, $t \in [0, 1]$ for $k = 1, 2, \ldots$ to represent the covariance function as the limit of

$$\gamma_{s,t}^{(n)} := st + \sum_{k=1}^{n-1} \frac{\sqrt{2}\sin(k\pi s)}{k\pi} \frac{\sqrt{2}\sin(k\pi t)}{k\pi} \to s \wedge t = \gamma_{s,t}.$$

This suggests that the sine series representation :

$$W_t^{(n)} := Z_0 t + \sum_{k=1}^{n-1} Z_k \frac{\sqrt{2}\sin(k\pi t)}{k\pi}, \quad t \in [0, 1]$$

converges to a Wiener process on $[0, 1]$. Similarly, the representation of the covariance function $s \wedge t$ in (B.27) suggests the use of $c_k(t) := 0$ for even k and $c_k(t) := 2\sqrt{2}\sin(k\pi t/2)/(k\pi)$, $t \in [0, 1]$ for odd k. This choice of $c_0, c_1, \ldots$ yields the *Karhunen–Loève expansion* of the Wiener process on $[0, 1]$:

$$W_t = \lim_n W_t^{(n)} = \sum_{k=1,3,\ldots} Z_k \frac{2\sqrt{2}\sin(k\pi t/2)}{k\pi}, \quad t \in [0, 1].$$

■

While the trigonometric expansions in the above example can be shown to provide a valid construction of the Wiener process, the theoretical arguments are much simpler if we instead work with the so-called *Haar basis* expansion.

Recall from Appendix B that we can choose the inner product (B.20) and an orthonormal basis $\{u_i\}$ for the Hilbert space $L^2[0, 1]$, so that any f in this space can be approximated arbitrarily well via $f^{(n)}(x) := \sum_{i=0}^{n-1} \langle f, u_i \rangle u_i(x)$ in the sense that $\|f - f^{(n)}\|_2 \to 0$ for $n \to \infty$, where $\|\cdot\|_2$ is the L^2 norm. Since the constant 1 and the Haar functions,

$$h_{k,j}(x) := 2^{k/2}\left(\mathbb{1}_{[j/2^k,\,(j+\frac{1}{2})/2^k]}(x) - \mathbb{1}_{[(j+\frac{1}{2})/2^k,\,(j+1)/2^k]}(x)\right),$$

where $j = 0, 1, \ldots, 2^k - 1$ and $k = 0, 1, 2, \ldots$, are a basis for $L^2[0, 1]$, we can approximate the indicator function $\mathbb{1}_{[0,t]}$ via

$$\mathbb{1}_{[0,t]}^{(n)}(x) := \langle \mathbb{1}_{[0,t]}, 1 \rangle + \sum_{k=0}^{n-1} \sum_{j=0}^{2^k-1} c_{k,j}(t)\, h_{k,j}(x), \quad x \in [0, 1],$$

where

(6.18) $\qquad c_{k,j}(t) := \langle \mathbb{1}_{[0,t]}, h_{k,j} \rangle = 2^{-k/2-1}[1 - |2^{k+1}t - 2j - 1|]^+$

are the tent-shaped Schauder functions depicted on Figure B.30. Observe that here the notation uses a double index, rather than the single index as in the trigonometric expansions in Example 6.17. The expansion of $\mathbb{1}_{[0,t]}^{(n)}$ shows that the covariance function $s \wedge t$ can be written as the limit

$$\lim_n \langle \mathbb{1}_{[0,s]}^{(n)}, \mathbb{1}_{[0,t]}^{(n)} \rangle = \langle \mathbb{1}_{[0,s]}, \mathbb{1}_{[0,t]} \rangle = s \wedge t, \quad s, t \in [0, 1],$$

and hence $s \wedge t$ has the representation: $s \wedge t = st + \sum_{k,j} c_{k,j}(s)\, c_{k,j}(t)$ for $s, t \in [0, 1]$. The foregoing discussion then suggests that the limit as $n \to \infty$ of

(6.19) $\qquad W_t^{(n)} := t\, Z_0 + \sum_{k=0}^{n-1} \sum_{j=0}^{2^k-1} c_{k,j}(t)\, Z_{j+2^k}, \qquad t \in [0, 1]$

will yield the desired Wiener process.

Let D be the countable set of *dyadic numbers* as in (4.36):

$$D := \cup_{k=0}^{\infty} D_k \quad \text{with} \quad D_k := \left\{ \frac{j}{2^k}, j = 0, 1, \ldots, 2^k \right\}.$$

The set D is *dense* in $[0, 1]$; that is, every point in $[0, 1]$ can be approximated arbitrarily well by a point in D.

The Haar basis construction $W^{(n)}$ matches the properties of the Wiener process at all $t \in D_n$. In particular, $(W_t^{(n)}, t \in D_n)$ is a zero-mean Gaussian process; and, since $W_t^{(n+k)} = W_t^{(n)}, t \in D_n$ for $k = 1, 2, \ldots$, the process $(W_t^{(n)}, t \in D_n)$ has covariance function $\mathbb{E}\, W_s^{(n)} W_t^{(n)} = s \wedge t$ for all $s, t \in D_n$; see Exercise 3.

The Haar basis expansion (6.19) can be used to construct the Wiener process on $t \in D$ by simply taking $n \to \infty$. This construction is usually referred to as Lévy's construction of the Wiener process on the set D. One way to visualize Lévy's construction is to consecutively generate the process at points in $D_0 = \{0, 1\}$, then $D_1 \setminus D_0 = \{1/2\}$, followed by $D_2 \setminus \cup_{k=0}^{1} D_k = \{1/4, 3/4\}$, and so on. More precisely, from the Brownian bridge in Example 6.5, if $t_1 - t_0 = 2^{-k+1}$ and $d = (t_0 + t_1)/2 \in D$, then conditional on $W_{t_0} = a$ and $W_{t_1} = b$, we have $W_d \sim \mathcal{N}((a+b)/2, 2^{-k-1})$, so that Lévy's construction on D proceeds as:

$$W_0 := 0,$$

$$W_1 := Z_0,$$

$$W_{1/2} := \frac{W_0 + W_1}{2} + \frac{Z_1}{2^{(1+1)/2}},$$

$$W_{1/4} := \frac{W_0 + W_{1/2}}{2} + \frac{Z_2}{2^{(2+1)/2}},$$

$$W_{3/4} := \frac{W_{1/2} + W_1}{2} + \frac{Z_3}{2^{(2+1)/2}},$$

$$\vdots$$

After completing the construction on $D_n \backslash \cup_{k=0}^{n-1} D_k$, we can linearly interpolate the set of points $(W_d, d \in \cup_{k=0}^{n} D_k)$ to obtain the stochastic process $W^{(n)} := (W_t^{(n)}, t \in [0, 1])$ which, by construction, has continuous sample paths. In fact, this linear interpolation can be represented via (6.19). Figure 6.20 shows a particular realization of the paths of the processes $W^{(1)}$ (dotted) and $W^{(2)}$.

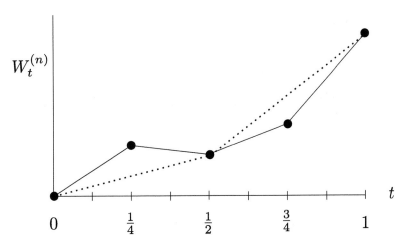

Figure 6.20: Lévy's construction for step $n = 1$ (dotted) and $n = 2$ (solid).

All that remains to validate the construction is to show that as $n \to \infty$ the sequence of processes $(W^{(n)})$ converges almost surely in an appropriate norm to a zero-mean Gaussian process on $[0, 1]$ with continuous sample paths and covariance function $s \wedge t$. Such a norm is the *supremum* or *uniform norm* $\|f\| := \sup_{t \in [0,1]} |f(t)|$ on $[0, 1]$, because if a continuous function $f^{(n)}$ converges in the uniform norm, then the limit $f(t) := \lim_n f^{(n)}(t)$, $t \in [0, 1]$ is also a continuous function; see Example B.13. The details are summarized in the following theorem:

Theorem 6.21: Existence of the Wiener Process

There exists a probability space $(\Omega, \mathcal{H}, \mathbb{P})$ and a zero-mean Gaussian process $W := (W_t, t \in [0, 1])$ with covariance function $\mathbb{E}\, W_s W_t = s \wedge t$ and uniformly continuous sample paths.

Proof. Let μ be the standard Gaussian distribution on $(\mathbb{R}, \mathcal{B})$ and take $(\Omega, \mathcal{H}, \mathbb{P}) := (\mathbb{R}, \mathcal{B}, \mu)^{\mathbb{N}}$ with $Z_0, Z_1, \ldots$ being the coordinate variables. Thus, on the probability space $(\Omega, \mathcal{H}, \mathbb{P})$ the $Z_0, Z_1, \ldots$ are iid $\mathcal{N}(0, 1)$ random variables that are used to define the processes $W^{(n)}, n = 0, 1, 2, \ldots$ via (6.19). Our goal is to show that $(W^{(n)})$ is almost surely a Cauchy sequence in the uniform norm: $\lim_{m,n \to \infty} \|W^{(m)} - W^{(n)}\| = 0$. This implies that $(W_t^{(n)})$ is a Cauchy sequence for every $t \in [0, 1]$, which in turn, due to the completeness of $\mathbb{R}$, implies the existence of its limit W_t for every $t \in [0, 1]$, and hence the existence of a process W to which $(W^{(n)})$ converges almost surely in the uniform norm.

First, let us show that for n large enough, $\|W^{(n+1)} - W^{(n)}\|$ is small. The precise relation between $W^{(n)}$ and $W^{(n+1)}$ for $n \geq 1$ is:

$$W_t^{(n+1)} = W_t^{(n)} + \sum_{j=0}^{2^n - 1} c_{n,j}(t)\, Z_{j+2^n}, \quad t \in [0, 1].$$

The function $c_{n,j}$ has its maximum at $t = (2j + 1)/2^{n+1}$, with maximum value $2^{-n/2-1} < 2^{-n/2}$. Thus,

$$\|W^{(n+1)} - W^{(n)}\| \leq 2^{-n/2} \underbrace{\max_{0 \leq j \leq 2^n - 1} |Z_{j+2^n}|}_{=:\, M_n}.$$

The random variable M_n is thus the maximum of the absolute values of 2^n iid $\mathcal{N}(0, 1)$ random variables. It is not difficult to show (see Exercise 6) that

$$\mathbb{P}(|Z| > z) \leq e^{-z^2/2}/z,$$

so that $\mathbb{P}(M_n > z) \leq 2^n \mathbb{P}(|Z| > z) \leq e^{-z^2/2 - \ln z + n \ln 2}$. Hence, for any $\varepsilon_n > 0$,

$$\mathbb{P}(\|W^{(n+1)} - W^{(n)}\| > \varepsilon_n) \leq \mathbb{P}(2^{-\frac{n}{2}} M_n > \varepsilon_n) \leq e^{-(2^n \varepsilon_n^2 + \ln \varepsilon_n^2 - n \ln 2)/2}.$$

If we choose $\varepsilon_n^2 := c\, n\, 2^{-n}$ for some constant $c > 2 \ln 2$, then $\sum_{k \geq n} \varepsilon_k = O(\sqrt{n}\, 2^{-\frac{n}{2}})$, because:

$$\int_n^\infty dx\, \sqrt{x}\, 2^{-\frac{x}{2}} = \int_{\sqrt{n}}^\infty dx\, 2x^2 e^{-\frac{\ln 2}{2} x^2} \leq \frac{2}{\ln 2} \sqrt{n}\, e^{-\frac{\ln 2}{2} n}.$$

Moreover,

$$\sum_n \mathbb{P}(\|W^{(n+1)} - W^{(n)}\| > \varepsilon_n) \leq \sum_n e^{-(c/2 - \ln 2)n} < \infty.$$

Consequently, by the Borel–Cantelli Lemma 3.14, there exists an almost sure event Ω_0 such that for every $\omega \in \Omega_0$ there is a n_ω such that:

$$\|W^{(n+1)}(\omega) - W^{(n)}(\omega)\| \leq \varepsilon_n \quad \text{for all } n \geq n_\omega.$$

As in part 3 of Proposition 3.2, it follows that for $\omega \in \Omega_0$ and $i, j \geq n \geq n_\omega$:

$$(6.22) \qquad \|W^{(i)}(\omega) - W^{(j)}(\omega)\| \leq \sum_{k=n}^{\infty} \varepsilon_k = O(\sqrt{n}\, 2^{-\frac{n}{2}}),$$

where the right-hand side vanishes as $n \to \infty$. Hence, for $\omega \in \Omega_0$, the sequence $(W^{(n)}(\omega))$ is Cauchy convergent in the uniform norm, and hence has a limit $W(\omega)$ in this norm. By setting $W(\omega) := 0$ for $\omega \notin \Omega_0$, we have defined the process $W := (W_t, t \in [0, 1])$ for all $\omega \in \Omega$. Since the process W is the limit (in the uniform norm) of a continuous process $W^{(n)}$, its sample paths are continuous. Moreover, since $[0, 1]$ is a closed and bounded set, the continuity is *uniform*; see Example B.13 in the Appendix. To conclude that W is a Wiener process, it remains to show that for any choice of d and $0 \leq t_1 < t_2 < \cdots < t_d \leq 1$, the d-dimensional vector $X := [W_{t_1}, \ldots, W_{t_d}]^\top$ is multivariate Gaussian with mean vector $\mathbf{0}$ and a covariance matrix with (i, j)th element $t_{i \wedge j}$. We leave the proof to Exercise 4. □

Theorem 6.21 tells us how to construct a Wiener processes only on $[0, 1]$. Given a sequence $(W_{t,k}, t \in [0, 1])$, $k = 0, 1, 2, \ldots$ of independent Wiener processes on $[0, 1]$, we can combine them to construct a Wiener process W on $[0, \infty)$ via:

$$(6.23) \qquad W_t := W_{t - \lfloor t \rfloor, \lfloor t \rfloor} + \sum_{k=0}^{\lfloor t \rfloor - 1} W_{1,k} \quad \text{for all } t \geq 0,$$

where $\sum_{k=0}^{-1} W_{1,k} := 0$. Clearly, the process $(W_t, t \geq 0)$ has continuous sample paths and it is not difficult to verify (see Exercise 5) that it is a zero-mean Gaussian process with covariance function $\mathbb{E}\, W_s W_t = s \wedge t$.

6.3 Strong Markov Property

At the beginning of Section 5.6 we mentioned that for continuous-time processes it is useful to *complete* the underlying probability space and to have an *augmented* and *right-continuous* filtration. Let us briefly discuss how this is relevant for the Wiener process W on some probability space $(\Omega, \mathcal{H}, \mathbb{P})$. Let $\mathcal{F} := (\mathcal{F}_t)$ be the natural filtration of W. We can complete the probability space by (1) extending $\mathcal{H}$ to $\overline{\mathcal{H}}$, where the latter includes all negligible sets of $\mathcal{H}$, and (2) extending $\mathbb{P}$ to $\overline{\mathbb{P}}$, where $\overline{\mathbb{P}}(H) = \mathbb{P}(H)$ for all $H \in \mathcal{H}$ and $\overline{\mathbb{P}}(H) = 0$ for all negligible sets in $H \in \mathcal{H}$. This gives the probability space $(\Omega, \overline{\mathcal{H}}, \overline{\mathbb{P}})$. Let $\mathcal{N}$ be the σ-algebra generated by the negligible sets in $\overline{\mathcal{H}}$; that is, all events in the extended probability space that have probability 0. Define the *augmentation*, $\overline{\mathcal{F}} := (\overline{\mathcal{F}}_t)$ of $\mathcal{F}$, by

(6.24) $$\overline{\mathcal{F}}_t := \mathcal{F}_t \vee \mathcal{N}.$$

The next theorem shows that this is a right-continuous filtration. That is, by peeking ahead an infinitesimal amount of time at time t, we cannot learn more than we already know at time t.

Theorem 6.25: Right-continuous Filtration

For the Wiener process, the augmentation $(\overline{\mathcal{F}}_t)$ of the natural filtration is right-continuous; that is,

$$\overline{\mathcal{F}}_t = \overline{\mathcal{F}}_t^+ := \bigcap_{\varepsilon > 0} \overline{\mathcal{F}}_{t+\varepsilon}.$$

Proof. If W is a Wiener process over $(\Omega, \mathcal{H}, \mathbb{P})$, it is also a Wiener process over $(\Omega, \overline{\mathcal{H}}, \overline{\mathbb{P}})$, since $\overline{\mathbb{P}}$ coincides with $\mathbb{P}$ on $\mathcal{H}$. In particular, it has independent increments and continuous sample paths under this probability model. Let (ε_n) be a strictly decreasing sequence to 0. Define

$$\mathcal{H}_n := \sigma\{W_t - W_s : \varepsilon_n \leq s < t \leq \varepsilon_{n-1}\};$$

that is, the history of the increment process during the time interval $[\varepsilon_n, \varepsilon_{n-1}]$. By the independence of the increments, the σ-algebras $\mathcal{H}_1, \mathcal{H}_2, \ldots$ are independent. By Kolmogorov's $0-1$ law, see Section 5.5.1, the tail σ-algebra $\cap_n \mathcal{H}_n$ is trivial, i.e., contains only events H with $\overline{\mathbb{P}}(H) = 0$ or $\overline{\mathbb{P}}(H) = 1$. In other words, $\cap_n \mathcal{H}_n \subseteq \mathcal{N}$. But also, $\mathcal{H}_{n+1} \vee \mathcal{H}_{n+2} \vee \cdots = \mathcal{F}_{\varepsilon_n}$, so

$$\bigcap_n \mathcal{H}_n = \bigcap_n \mathcal{F}_{\varepsilon_n} = \bigcap_{\varepsilon > 0} \mathcal{F}_\varepsilon =: \mathcal{F}_0^+,$$

showing that

$$\mathcal{F}_0^+ \subseteq \mathcal{N}.$$

For general t, let $\widehat{\mathcal{F}}$ be the filtration of the Wiener process $\widetilde{W}_u := W_{t+u} - W_u, u \geq 0$. In particular, as we have just shown, $\widehat{\mathcal{F}}_0^+ \subseteq \mathcal{N}$. It follows that

$$\bigcap_{\varepsilon > 0} \overline{\mathcal{F}}_{t+\varepsilon} = \bigcap_{\varepsilon > 0} (\mathcal{F}_t \vee \mathcal{N} \vee \widehat{\mathcal{F}}_\varepsilon) = \mathcal{F}_t \vee \mathcal{N} \vee \widehat{\mathcal{F}}_0^+ = \mathcal{F}_t \vee \mathcal{N} = \overline{\mathcal{F}}_t.$$

□

The statement $\mathcal{F}_0^+ \subseteq \mathcal{N}$ in the proof above is referred to as *Blumenthal's $0 - 1$ law*. It states that any information that can be gleaned from peeking ahead is trivial — it is either certain to happen or impossible. The following example gives a striking application:

■ **Example 6.26** (**Set of Zeros of a Wiener Process**) Let

$$T := \inf\{t > 0 : W_t = 0\}$$

and $H_n := \{W_t = 0 \text{ for some } 0 < t < 1/n\} \in \mathcal{F}_{1/n}$. Then, $\{T = 0\} = \cap_{n=1}^{\infty} H_n \in \mathcal{F}_0^+$ and hence it has probability 0 or 1. Suppose that $\mathbb{P}(T = 0) = 0$. Then, almost surely, there is an $\varepsilon > 0$ such that the Wiener process never hits 0 during the time interval $(0, \varepsilon)$. But this is a contradiction, as $\mathbb{P}(\{W_{\varepsilon/2} > 0\} \cap \{W_\varepsilon < 0\}) > 0$, and thus, by the continuity of the paths, W hits 0 with positive probability during $(0, \varepsilon)$. Consequently,

(6.27) $$\mathbb{P}(T = 0) = 1.$$

Let C be the (random) set of points where $W_t = 0$. By the path-continuity of W, if $W_t \neq 0$, then there will be points in an arbitrary neighborhood of t for which the process is not 0 either. Thus, the complement of C is an *open* set, and hence, per definition, C is a *closed*[1] set. Moreover, by (6.27), for any t with $W_t = 0$ and any $\varepsilon > 0$, there exists another $s < \varepsilon$ with $W_s = 0$; that is, *no points of C are isolated*. The fractal-like properties of C make it similar to the Cantor set in Example 1.1. It can be shown that it is uncountable, and can be mapped 1-to-1 to the real line.   ■

For the rest of the discussions in this chapter, we assume that the Wiener process $W := (W_t, t \geq 0)$ has the right-continuous filtration $\mathcal{F}^+ := (\mathcal{F}_t^+, t \geq 0)$, with $\mathcal{F}_t^+ := \cap_{\varepsilon > 0} \mathcal{F}_{t+\varepsilon}$. As Theorem 6.25 shows, $\mathcal{F}^+$ is contained in $\overline{\mathcal{F}}$, defined in (6.24).

Take any time $t > 0$ and define the process $\widetilde{W} := (W_{t+u} - W_t, u \geq 0)$. Then, $\widetilde{W}$ is again a Wiener process, as it is a zero-mean Gaussian process with $\mathbb{E}\widetilde{W}_u\widetilde{W}_s = u \wedge s$. Moreover, by the independence of increments, $\widetilde{W}$ is independent of $\mathcal{F}_t$. This shows that W is Markovian with respect to its natural filtration $\mathcal{F}$. More importantly, we have the following:

Theorem 6.28: Wiener Process is Markov w.r.t $\mathcal{F}^+$

Let W be a Wiener process. For every $t \geq 0$, the process $(W_{t+u} - W_t, u \geq 0)$ is independent of $\mathcal{F}_t^+$. Consequently, W is Markovian with respect to $\mathcal{F}^+$.

---

[1]Another way to see that C is closed, is that it is the image of the closed set $\{0\}$ under the continuous function $W_t, t \geq 0$.

Proof. Take a strictly decreasing sequence $t_n \downarrow t$. By the continuity of the paths, $W_{t+u} - W_t = \lim_n (W_{t_n+u} - W_{t_n})$. For any selection $u_1, \dots, u_m \geq 0$ and n, the vector $(W_{t_n+u_1} - W_{t_n}, \dots, W_{t_n+u_m} - W_{t_n})$ is independent of $\mathcal{F}_t^+$, and so is its limit as $n \to \infty$. In other words, the process $W_{t+u} - W_t, u \geq 0$ is independent of $\mathcal{F}_t^+$. $\qquad\square$

As a matter of fact, in the theorem above, we can replace t with any (almost surely) finite stopping time T. This is the *strong Markov property* of the Wiener process.

Theorem 6.29: Strong Markov Property

For any finite stopping time T with respect to $\mathcal{F}^+$, the process $(W_{T+u} - W_T, u \geq 0)$ is a Wiener process independent of $\mathcal{F}_T^+$.

Proof. Let T be a stopping time with respect to $\mathcal{F}^+$. As in the proof of Doob's stopping theorem for continuous martingales, Theorem 5.83, we first consider a sequence of stopping times (T_n) decreasing to T, by defining for each $n \in \mathbb{N}$:

$$T_n := \frac{k+1}{2^n} \quad \text{if} \quad \frac{k}{2^n} \leq T < \frac{k+1}{2^n} \quad \text{for some } k \in \mathbb{N}.$$

Note that T_n takes values in the countable set $\{k/2^n, k = 1, 2, \dots\}$ and that $T_n \in \mathcal{F}_T^+$ for all n. Fix n and consider the processes $X^{(n,k)} := (W_{k/2^n+u} - W_{k/2^n}, u \geq 0)$ for $k = 1, 2, \dots$ and $X^{(n)} := (W_{T_n+u} - W_{T_n}, u \geq 0)$. Each $X^{(n,k)}$ is a Wiener process and is independent of $\mathcal{F}_{k/2^n}^+$. In particular, the probability of each event $\{X^{(n,k)} \in A\}$ does not depend on k and is the same as the probability of the event $\{X^{(n)} \in A\}$.

We want to show that for any n, every event $E \in \mathcal{F}_{T_n}^+$ is independent of every event $\{X^{(n)} \in A\}$. Using the properties of $X^{(n,k)}$ and $X^{(n)}$ mentioned above, this independence follows from

$$\mathbb{P}(\{X^{(n)} \in A\} \cap E) = \sum_{k=1}^{\infty} \mathbb{P}(\{X^{(n,k)} \in A\} \cap E \cap \{T_n = k/2^n\})$$

$$= \sum_{k=1}^{\infty} \mathbb{P}(X^{(n,k)} \in A)\, \mathbb{P}(E \cap \{T_n = k/2^n\})$$

$$= \mathbb{P}(X^{(n)} \in A) \sum_{k=1}^{\infty} \mathbb{P}(E \cap \{T_n = k/2^n\})$$

$$= \mathbb{P}(X^{(n)} \in A)\, \mathbb{P}(E).$$

As $T_n \in \mathcal{F}_T^+$, this shows that for each n, $X^{(n)}$ is a Wiener process independent of $\mathcal{F}_T^+$. Hence, the increments

(6.30) $$W_{T+u+v} - W_{T+u} = \lim_n (W_{T_n+u+v} - W_{T_n+u}) = \lim_n (X_{u+v}^{(n)} - X_u^{(n)})$$

of the process $(W_{T+u} - W_T, u \geq 0)$ are independent and normally distributed with mean 0 and variance v. As this process has continuous paths, it must be a Wiener process. Moreover, all increments (6.30) are independent of $\mathcal{F}_T^+$ as they are the limit of random variables independent of $\mathcal{F}_T^+$.   □

A neat corollary of the strong Markov process property is the *reflection principle* for the Wiener process.

Theorem 6.31: Reflection Principle

Let T be a stopping time. Then, $(\widetilde{W}_t, t \geq 0)$, defined by

(6.32) $\widetilde{W}_t := W_t \, \mathbb{1}_{\{t \leq T\}} + (2W_T - W_t) \, \mathbb{1}_{\{t > T\}}, \quad t \geq 0,$

is a Wiener process.

Proof. The proof is illustrated in Figure 6.33, where T is the time that (W_t) hits the level $x = 0.4$ for the first time. If $T < \infty$, then by the strong Markov property, $(W_{T+t} - W_T, t \geq 0)$ is a Wiener process, and so is $-(W_{T+t} - W_T, t \geq 0)$. Moreover, both processes are independent of $\mathcal{F}_T^+$. By concatenating $(W_t, t \leq T)$ and $-(W_{T+t} - W_T, t \geq 0)$, we obtain a Wiener process, and this process is $(\widetilde{W}_t, t \geq 0)$.   □

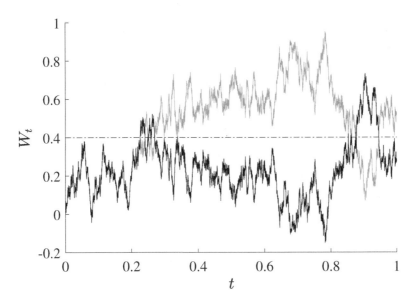

Figure 6.33: The reflected Wiener process is again a Wiener process.

The following is immediate by the independence and Gaussianity of increments of the Wiener process:

> ### Theorem 6.34: Transition Kernel of the Wiener Process
>
> The Wiener process W admits the transition function
>
> $$P_{t,u}(x, A) = \mathbb{P}(W_u \in A \mid W_t = x) = \int_A \mathrm{d}y \, p_{u-t}(y \mid x),$$
>
> where the *transition density* $p_t(y \mid x)$ is given by
>
> (6.35) $$p_t(y \mid x) := \frac{1}{\sqrt{2\pi t}} \, \mathrm{e}^{-\frac{(y-x)^2}{2t}}, \quad t \geq 0, \quad x, y \in \mathbb{R}.$$

Think of $\mathrm{d}y \, p_t(y \mid x)$ as the infinitesimally small probability that the process goes from x to a point in the interval $(y, y + \mathrm{d}y)$ during an interval of length t. For d-dimensional Wiener processes we have, similarly, the transition density

(6.36) $$p_t(y \mid x) = (2\pi t)^{-d/2} \, \mathrm{e}^{-\frac{\|y-x\|^2}{2t}}, \quad t \geq 0, \quad x, y \in \mathbb{R}^d.$$

Define the *Laplace operator* in the Cartesian coordinates $x \in \mathbb{R}^d$:

(6.37) $$\Delta := \sum_{i=1}^{d} \partial_{ii} = \sum_{i=1}^{d} \frac{\partial^2}{\partial x_i^2}.$$

We sometimes write Δ_x for Δ to emphasize that the partial derivatives are taken with respect to x. Then, direct substitution shows that the transition density (6.36) satisfies the *heat equation* on $\mathbb{R}^d$:

(6.38) $$\frac{\partial}{\partial t} p_t(y \mid x) = \frac{1}{2} \Delta_y p_t(y \mid x), \quad y \in \mathbb{R}^d, \, t \geq 0.$$

We can interpret $p_t(y \mid x)$ as the amount of heat at position y at time $t > 0$, if a heat impulse is released at position x at time $t = 0$ and is then allowed to freely diffuse through the medium.

6.4 Martingale Properties

Let $W := (W_t, t \geq 0)$ be a Wiener process adapted to the filtration $\mathcal{F}^+$, and recall that $\mathbb{E}_s$ denotes the conditional expectation with respect to $\mathcal{F}_s^+$.

We can associate many martingales with the Wiener process. For a start, the Wiener process itself is a martingale; see Example 5.68.

A second martingale associated with W is the process $(W_t^2 - t, t \geq 0)$. Adaptedness and integrability are again trivial. The martingale property follows from:

$$
\begin{aligned}
\mathbb{E}_s(W_t^2 - t) &= \mathbb{E}_s(W_s + W_t - W_s)^2 - t \\
&= \mathbb{E}_s[W_s^2 + 2W_s(W_t - W_s) + (W_t - W_s)^2 - t] \\
&= W_s^2 + 2W_s\mathbb{E}_s(W_t - W_s) + \mathbb{E}_s(W_t - W_s)^2 - t \\
&= W_s^2 + 0 + (t - s) - t = W_s^2 - s.
\end{aligned}
$$

Another useful martingale is the *exponential martingale* $(e^{rW_t - r^2 t/2}, t \geq 0)$ for any $r \in \mathbb{R}$. The exponential martingale characterizes the Wiener process.

Theorem 6.39: Exponential Martingale

A continuous process $X := (X_t, t \geq 0)$ is a Wiener process if and only if for each $r \in \mathbb{R}$ the process $S := (S_t)$ defined by

(6.40) $$S_t := e^{rX_t - \frac{1}{2}r^2 t}, \quad t \geq 0,$$

is a martingale.

Proof. Suppose $X = W$ is a Wiener process. We have for all $s < t$:

$$
\mathbb{E}_s \frac{S_t}{S_s} = \mathbb{E}_s e^{r(W_t - W_s) - \frac{1}{2}r^2(t-s)} = e^{-\frac{1}{2}r^2(t-s)}\mathbb{E}_s e^{r(W_t - W_s)} = e^{-\frac{1}{2}r^2(t-s)}e^{+\frac{1}{2}r^2(t-s)} = 1,
$$

since conditionally on W_u, $u \leq s$, the increment $W_t - W_s$ has a $\mathcal{N}(0, t-s)$ distribution and its MGF is $e^{\frac{1}{2}r^2}$, $r \in \mathbb{R}$. Consequently, since $\mathbb{E}S_t < \infty$ for all t, and

(6.41) $$\mathbb{E}_s S_t = S_s \, \mathbb{E}_s S_t / S_s = S_s,$$

it follows that S is a martingale.

Conversely, suppose that S is a martingale. Then, taking $t = s + u$ in (6.41), we have that

(6.42) $$\mathbb{E}_s e^{r(X_{s+u} - X_s)} = e^{\frac{1}{2}r^2 u}, \quad u \geq 0,$$

which shows that X has stationary and independent increments and $X_{s+u} - X_s \sim \mathcal{N}(0, u)$, so that it must be a Wiener process. $\qquad\square$

Next, we give a typical example of how Wiener martingales can be employed.

■ **Example 6.43 (Hitting Time)** For some $x \geq 0$, let

$$T_x := \inf\{t > 0 : W_t > x\}$$

be the *hitting time* of the open set (x, ∞). Recall that this is a stopping time of the right-continuous filtration $\mathcal{F}^+$, but not a stopping time of the natural filtration $\mathcal{F}$. Consider the exponential martingale (6.40):

$$S_t := e^{rW_t - \frac{1}{2}r^2 t}, \quad t \geq 0.$$

Note that (S_t) is bounded by e^{rx} on $[0, T_x] \cap \mathbb{R}_+$. Thus, by Proposition 5.96, W_{T_x} exists and is integrable, and (S_t) is a Doob martingale on $[0, T_x]$. Consequently, $\mathbb{E}_0 S_{T_x} = S_0$ and after taking expectations:

$$\mathbb{E} S_{T_x} = \mathbb{E} S_0 = 1.$$

Written out, this means

$$\mathbb{E}\left[e^{rx - \frac{1}{2}r^2 T_x} \mathbb{1}_{\{T_x < \infty\}} + 0 \, \mathbb{1}_{\{T_x = \infty\}} \right] = 1,$$

which shows that, with $s := r^2/2$,

$$(6.44) \qquad \mathbb{E}\, e^{-s T_x} = e^{-rx} = e^{-x\sqrt{2s}}, \quad s \geq 0,$$

yielding the Laplace transform of T_x. It is straightforward to check that this is the Laplace transform corresponding to the pdf

$$(6.45) \qquad f_{T_x}(t) = \frac{x\, e^{-x^2/(2t)}}{\sqrt{2\pi t^3}}, \quad t \geq 0$$

and that the corresponding expectation is ∞. Finally,

$$1 = \lim_{s \to 0} \mathbb{E}\, e^{-s T_x} = \lim_{s \to 0} \mathbb{E}\, e^{-s T_x} \mathbb{1}_{\{T_x < \infty\}} + \lim_{s \to 0} \mathbb{E}\, e^{-s T_x} \mathbb{1}_{\{T_x = \infty\}} = \mathbb{P}(T_x < \infty) + 0,$$

which shows that T_x is almost surely finite. Hence, starting from 0, the Wiener process hits any point $x > 0$ (and, by symmetry, any point $x < 0$) almost surely, but the expected hitting time is infinite. ∎

Finally, many martingales can be constructed from the Wiener process via the following theorem, involving the Laplace operator (6.37). In Chapter 7 we will recognize this as a consequence of Itô's formula. We formulate it in terms of a standard Brownian motion (B_t) rather than a Wiener process (W_t), to be able to start the process from different states. Recall that $\mathbb{E}^x$ is the expectation operator under which the process starts at x.

> ### Theorem 6.46: Martingales from Functions of a Brownian Motion
>
> Let $f : \mathbb{R}^d \to \mathbb{R}$ be twice continuously differentiable and let B be a d-dimensional standard Brownian motion. If for all $t > 0$ and $x \in \mathbb{R}^d$,
>
> $$\mathbb{E}^x |f(B_t)| < \infty \quad \text{and} \quad \mathbb{E}^x \int_0^t |\Delta f(B_s)|\, ds < \infty,$$
>
> then the process $(X_t, t \geq 0)$ defined by
>
> $$(6.47) \qquad X_t := f(B_t) - \frac{1}{2} \int_0^t \Delta f(B_s)\, ds$$
>
> is a martingale.

Proof. Take $0 \leq s < t$. The conditional expectation of X_t given $\mathcal{F}_s^+$ is, by the Markov property,

$$\mathbb{E}_s X_t = \mathbb{E}^{B_s} f(B_{t-s}) - \frac{1}{2} \int_0^s \Delta f(B_u)\, du - \int_0^{t-s} \mathbb{E}^{B_s} \frac{1}{2} \Delta f(B_u)\, du.$$

The integrand of the second integral can be written as

$$\mathbb{E}^{B_s} \frac{1}{2} \Delta f(B_u) = \frac{1}{2} \int_{\mathbb{R}^d} p_u(x \mid B_s) \Delta f(x)\, dx = \frac{1}{2} \int_{\mathbb{R}^d} \Delta p_u(x \mid B_s) f(x)\, dx,$$

where we have used the fact that $\int (p \Delta f - f \Delta p)\, dx = \int \nabla \cdot (p \nabla f - f \nabla p)\, dx = 0$ because $(p \nabla f - f \nabla p)$ vanishes at infinity. Next, since p satisfies (6.38), we can write

$$\mathbb{E}^{B_s} \frac{1}{2} \Delta f(B_u) = \int_{\mathbb{R}^d} \frac{\partial}{\partial u} p_u(x \mid B_s) f(x)\, dx.$$

Consequently, its integral from 0 to $t - s$ is

$$\int_0^{t-s} \left(\int_{\mathbb{R}^d} \frac{\partial}{\partial u} p_u(x \mid B_s) f(x)\, dx \right) du = \lim_{\varepsilon \downarrow 0} \int_{\mathbb{R}^d} \left(\int_\varepsilon^{t-s} \frac{\partial}{\partial u} p_u(x \mid B_s)\, du \right) f(x)\, dx$$

$$= \int_{\mathbb{R}^d} p_{t-s}(x \mid B_s) f(x)\, dx - \lim_{\varepsilon \downarrow 0} \int p_\varepsilon(x \mid B_s) f(x)\, dx = \mathbb{E}^{B_s} f(B_{t-s}) - f(B_s),$$

so that $\mathbb{E}_s X_t = X_s$, showing that (X_t) has the martingale property. Integrability of each X_t follows from the two conditions stated in the theorem, and adaptedness is obvious. $\square$

As an example of this theorem, taking $f(x) = x^2$ in the one-dimensional case gives the martingale $(B_t^2 - t)$.

6.5   Maximum and Hitting Time

For $x \in \mathbb{R}_+$, let T_x be the hitting time of the open set (x, ∞); see Example 6.43. In other words, we have

(6.48) $$T_x := \inf\{t > 0 : W_t > x\}, \quad x \geq 0.$$

Also, for $t \in \mathbb{R}_+$, let M_t be the *running maximum* of the Wiener process; that is,

(6.49) $$M_t := \max_{0 \leq s \leq t} W_s.$$

The two are related via

(6.50) $$\{M_t > x\} = \{T_x < t\}.$$

Indeed, we have

(6.51) $$T_x = \inf\{t > 0 : M_t > x\} \quad \text{and} \quad M_t = \inf\{x > 0 : T_x > t\},$$

so the processes $T := (T_x, x \geq 0)$ and $M := (M_t, t \geq 0)$ are functional inverses of each other. To obtain the path of one, just swap the axes of the other, as illustrated in Figure 6.52. Both processes are increasing and right-continuous. In fact, we will see that T is strictly increasing, whereas M has continuous sample paths that can remain constant in certain time intervals.

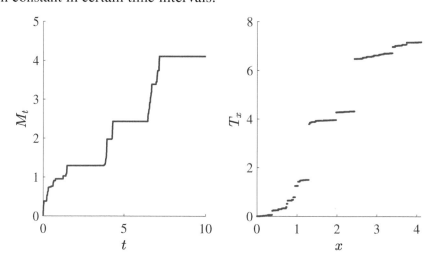

Figure 6.52: Processes M and T are functional inverses of each other.

By the continuity of W, we have $W_{T_x} = x$, and therefore, similar to (6.50), the event $\{T_x \leq t\}$ is equivalent to the event $\{M_t \geq x\}$. Consequently,

$$\mathbb{P}(M_t \geq x) = \mathbb{P}(T_x \leq t) = \mathbb{P}(T_x \leq t, W_t < x) + \mathbb{P}(T_x \leq t, W_t > x)$$

$$= 2\,\mathbb{P}(T_x \leq t, W_t > x) = 2\,\mathbb{P}(W_t > x)$$

(6.53) $$= 2 - 2\,\Phi\left(\frac{x}{\sqrt{t}}\right), \quad x \geq 0,$$

where Φ is the cdf of the standard normal distribution. The third equality above is due to the reflection principle, as from T_x onwards the Wiener process and its reflection around x have the same distribution. As $2\,\mathbb{P}(W_t > x) = \mathbb{P}(W_t > x) + \mathbb{P}(-W_t < -x) = \mathbb{P}(|W_t| > x)$, we have the following result:

Theorem 6.54: Distribution of M_t and $|W_t|$

For each *fixed* t, M_t and $|W_t|$ have the same distribution.

By differentiating (6.53) with respect to x, it follows that M_t has pdf

$$f_{M_t}(x) := \sqrt{\frac{2}{\pi t}}\ \exp\left(-\frac{1}{2}\frac{x^2}{t}\right), \quad x \ge 0.$$

That is, M_t has a $\mathcal{N}(0, t)$ distribution truncated to $[0, \infty)$. Similarly, by differentiating (6.53) with respect to t, we find that the pdf of T_x is

$$(6.55) \qquad f_{T_x}(t) := \frac{1}{\sqrt{2\pi t^3}}\, x\, \exp\left(-\frac{x^2}{2t}\right), \quad t \ge 0,$$

which is in agreement with (6.45). Thus, $T_x \sim \mathsf{InvGamma}(1/2, x^2/2)$. As, consequently, $1/T_x \sim \mathsf{Gamma}(1/2, x^2/2)$, this means that $1/T_x$ has the same distribution as Z^2/x^2 for $x > 0$, where $Z \sim \mathcal{N}(0, 1)$, so that T_x is distributed as x^2/Z^2. Either from Example 6.43, or Exercise 12, we know that $\mathbb{P}(T_x < \infty) = 1$ and $\mathbb{E}T_x = \infty$ for all $x > 0$.

Finally, M_t and W_t have joint cdf

$$(6.56) \qquad \mathbb{P}(M_t \le x, W_t \le y) = \Phi\left(\frac{2x - y}{\sqrt{t}}\right) - \Phi\left(\frac{-y}{\sqrt{t}}\right), \quad x \ge 0,\ x \ge y,$$

which can again be derived from the reflection principle, as

$$\mathbb{P}(M_t \ge x, W_t \le y) = \mathbb{P}(M_t \ge x, W_t \ge 2x - y)$$

$$= \mathbb{P}(W_t \ge 2x - y) = 1 - \Phi\left(\frac{2x - y}{\sqrt{t}}\right).$$

To further explore the behavior of the maximum process, consider Figure 6.57. The difference between the maximum process M and the Wiener process W — a positive process — is depicted in the lower panel.

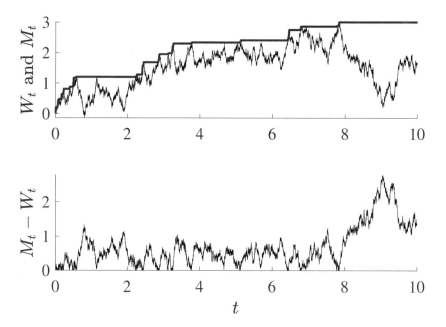

Figure 6.57: The difference between the maximum and Wiener process is a reflected Wiener process.

It turns out that this difference process has the same distribution as the process $|W|$ — called a *reflected Wiener process*.

Theorem 6.58: Difference Between M and W is Reflected Wiener Process

Let $Y_t := M_t - W_t$. The process $Y := (Y_t, t \geq 0)$ has the same distribution as a reflected Wiener process $|W| := (|W_t|, t \geq 0)$.

Proof. Obviously Y has continuous sample paths. Define

$$\widetilde{W}_t := W_{s+t} - W_s, \quad t \geq 0$$

and

$$\widetilde{M}_t := \max_{0 \leq u \leq t} \widetilde{W}_u, \quad t \geq 0.$$

We want to show that, conditional on $\mathcal{F}_s^+$, the random variable Y_{s+t} has the same distribution as $|Y_s + \widetilde{W}_t|$ for all t. This implies that Y is a Markov process with the same transition density as $|W|$, and so must have the same probability distribution as the latter. Fix $s, t \geq 0$ and write

$$Y_{s+t} = \max\{M_s, \ W_s + \widetilde{M}_t\} - W_s - \widetilde{W}_t$$
$$= \max\{Y_s, \ \widetilde{M}_t\} - \widetilde{W}_t.$$

Thus, the conditional distribution of Y_{s+t} given $\mathcal{F}_s^+$ is that of $\max\{Y_s, \widetilde{M}_t\} - \widetilde{W}_t$, where $\widetilde{M}_t$ and $\widetilde{W}_t$ are independent of $\mathcal{F}_s^+$. Since $\widetilde{M}_t$ and $\widetilde{W}_t$ are distributed as M_t and W_t, it therefore remains to be shown that for every y,

$$\mathbb{P}(\max\{y, M_t\} - W_t > a) = \mathbb{P}(|y + W_t| > a), \quad a > 0.$$

Write the first probability as the sum

$$\mathbb{P}(y - W_t > a) + \mathbb{P}(y - W_t \le a, M_t - W_t > a) =: p_1 + p_2.$$

We want to show:

(6.59) $\qquad p_1 = \mathbb{P}(y + W_t > a) \quad$ and $\quad p_2 = \mathbb{P}(y + W_t < -a),$

so that then $p_1 + p_2 = \mathbb{P}(|y + W_t| > a)$, as required. The first equality in (6.59) is easy, since W_t has the same distribution as $-W_t$. To prove the second equality in (6.59), consider the time-reversed Wiener process $R := (R_t) := (W_{t-u} - W_t, 0 \le u \le t)$. Denote its maximum in $[0, t]$ by $M_t^R := \max_{0 \le u \le t} R_u$. Then, $M_t^R = M_t - W_t$. Since $R_t = -W_t$, we have

$$p_2 = \mathbb{P}(y + R_t \le a, M_t^R > a).$$

Proceeding in the same way as in the proof of Theorem 6.54, we now apply the reflection principle to (R_t) at the first time that the process hits a, to find (with $\widetilde{R}$ the reflected Wiener process) that

$$p_2 = \mathbb{P}(\widetilde{R}_t \ge a + y),$$

which, by the Wiener properties of $\widetilde{R}$, is equal to $\mathbb{P}(-W_t \ge a + y) = \mathbb{P}(y + W_t \le -a)$, which needed to be shown. $\qquad\qquad\qquad\qquad\qquad\qquad\qquad\qquad\qquad\qquad\qquad\square$

We briefly touch on an interesting connection between the set of zeros of a Wiener process and the *local time* at 0. Think of a clock which only moves when the process is at 0 and stops every time when it is not. Consider the set of zeros for a reflected Wiener process in the lower panel of Figure 6.57. This is also the set of zeros for some Wiener process. Note that the path of the maximum process (M_t) in the upper panel increases every time (and only then) when the reflected Wiener process hits 0. In this way we can think of M_t as the time showing on the local clock when the standard clock shows time t. The functional inverse of M, i.e., T, then describes the time intervals when the local clock is not moving forward.

We now turn our attention to the process $T = (T_x, x \ge 0)$. The right panel of Figure 6.52 shows a typical path. This indicates that T is a pure-jump process, with a few large jumps and many very small jumps. We have already derived the pdf of T_x in (6.55). The process T turns out to be an increasing pure-jump Lévy process. In particular, T is time-homogeneous Markov processes with stationary and independent increments.

Theorem 6.60: T is a Lévy Subordinator

The process $(T_x, x \geq 0)$ is a strictly increasing pure-jump Lévy process with transition density

$$(6.61) \quad p_x(t \mid s) = \frac{x}{\sqrt{2\pi(t-s)^3}} \exp\left(-\frac{x^2}{2(t-s)}\right), \quad 0 < s < t, \quad x > 0.$$

Proof. Take any $y > x > 0$. To hit $(x + y, \infty)$ the process W has to hit (x, ∞) first, and then starting from x at time T_x, the process has to hit $(x + y, \infty)$, which takes an amount of time $\widetilde{T}_y$. Hence,

$$T_{x+y} = T_x + \widetilde{T}_y.$$

By the strong Markov property for W, $\widetilde{T}_y$ is independent of $\mathcal{F}_{T_x}^+$ and, by the stationarity of increments of W, it has the same distribution as T_y. Thus, T itself has stationary and independent increments. As the paths are right-continuous and left-limited and $T_0 = 0$, T is a Lévy process. In particular, T is a time-homogeneous Markov process. Its transition density to go from s to t in an interval of length x is exactly the pdf of T_x at $t - s$, so that (6.55) implies (6.61). $\square$

The distribution of T_1 is also called the *Lévy* distribution or Stable$(1/2, 1)$ distribution. We mentioned before that distribution of T_x is the same as the distribution of x^2/Z^2, where $Z \sim \mathcal{N}(0, 1)$. This implies that $(T_{cx}/c^2, x \geq 0)$ has the same distribution as $(T_x, x \geq 0)$ for all $c > 0$. A Lévy process $(X_t, t \geq 0)$ is said to be α-*stable* or *stable with index* α if $(c^{-1/\alpha} X_{ct}, t \geq 0)$ has the same distribution as $X_t, t \geq 0$ for all $c > 0$. The process $(T_x, x \geq 0)$ is thus stable with index $\alpha = 1/2$ and the Wiener process is stable with index $\alpha = 2$.

There is a fundamental association between pure-jump Lévy processes and Poisson random measures, as explained in Section 2.8.4. In particular, the increasing pure jump Lévy process $(T_x, x \geq 0)$ is of the form

$$T_x := \int_{[0,x] \times \mathbb{R}_+} N(\mathrm{d}u, \mathrm{d}y)\, y =: Nf,$$

where N is a Poisson random measure on $\mathbb{R}_+ \times \mathbb{R}_+$ and f is the function $f(u, y) := y \mathbb{1}_{[0,x]}(u)$ for $u, y > 0$; the corresponding integral is Nf. The mean measure of N is $\lambda := \mathrm{Leb} \otimes \nu$, where ν is the *Lévy measure* of N, which satisfies

$$\int_1^\infty \nu(\mathrm{d}y)\, y < \infty.$$

The Laplace transform of T_x can thus be written as

$$\mathbb{E}\, e^{-sT_x} = \mathbb{E}\, e^{-sNf} = e^{-\lambda(1-e^{-sf})} = e^{-\int_0^x \mathrm{d}u \int_0^\infty \nu(\mathrm{d}y)\,(1-e^{-sf(u,y)})} = e^{-x\int_0^\infty \nu(\mathrm{d}y)\,(1-e^{-sy})}.$$

In (6.44), we found that the Laplace transform of T_x is $e^{-x\sqrt{2s}}, s \geq 0$. This shows that the Lévy measure is

$$\nu(\mathrm{d}y) = \mathrm{d}y \, \frac{1}{\sqrt{2\pi y^3}}.$$

Figure 6.62 shows a typical realization of the atoms of N. There are a few atoms (u, y) where y is large, say of the order 1, but an infinity of atoms where y is small (close to 0). In fact, the atoms are accumulating near the horizontal axis. There is a one-to-one correspondence between the atoms in Figure 6.62 and the realization of $T := (T_x, x \geq 0)$ in Figure 6.52. Going from left to right (increasing x), every time u we hit an atom (u, y), we increase the process T by an amount y. Most of the time these increases are minuscule, but sometimes they are large.

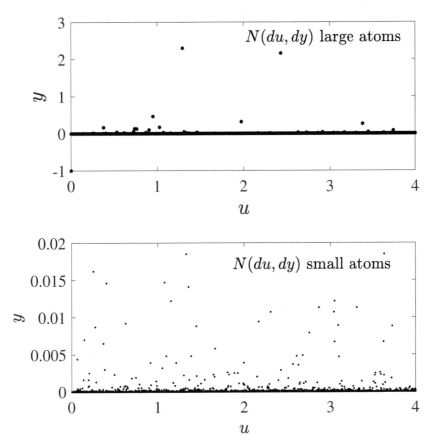

Figure 6.62: The hitting time process (T_x) can be constructed from the atoms of a Poisson random measure.

6.6 Brownian Motion and the Laplacian Operator

There is a fundamental connection between the d-dimensional standard Brownian motion process $\boldsymbol{B} := (\boldsymbol{B}_t, t \geq 0) = \boldsymbol{B}_0 + \boldsymbol{W}$, where $\boldsymbol{W} := (\boldsymbol{W}_t, t \geq 0)$ is a Wiener process, and the Laplacian operator Δ in (6.37). The key to this connection is Theorem 6.46, which states that (under mild conditions stated in the theorem) the process $(X_t, t \geq 0)$ defined by

$$(6.63) \qquad X_t := f(\boldsymbol{B}_t) - \int_0^t \frac{1}{2} \Delta f(\boldsymbol{B}_s) \, ds$$

is a martingale. Let $\mathbb{E}^x$ denote the expectation operator under which $\boldsymbol{B}$ starts at state $\boldsymbol{x}$. Also, for ease of notation, define $L := \frac{1}{2} \Delta$. Because of the martingale property of (X_t), we have $\mathbb{E}^x X_t = \mathbb{E}^x X_0 = f(\boldsymbol{x})$, which implies that

$$(6.64) \qquad \mathbb{E}^x f(\boldsymbol{B}_t) = f(\boldsymbol{x}) + \int_0^t \mathbb{E}^x L f(\boldsymbol{B}_s) \, ds,$$

where the interchange of expectation and integral is allowed by Fubini's theorem.

■ **Example 6.65 (Exit Probability)** For any $q \geq 0$, let T_q be the first time that a d-dimensional standard Brownian motion hits the sphere centered at $\boldsymbol{0}$ with radius q. We are interested in the probability

$$p := \mathbb{P}^x(T_l < T_r),$$

when $\boldsymbol{x} \in D := \{\boldsymbol{x} \in \mathbb{R}^d : l \leq \|\boldsymbol{x}\| \leq r\}$, where $0 < l < r$. That is, the probability that the Brownian motion hits the inner sphere before it hits the outer sphere. We can calculate p via (6.64), by selecting a function f for which $\Delta f = 0$ for all $\boldsymbol{x}$ (here, on the annulus D). Such functions are said to be *harmonic*; they have been widely studied in mathematical analysis and have many interesting properties. If f is harmonic on D and twice differentiable and bounded on $\mathbb{R}^d$, then (6.64) implies that $\mathbb{E}^x f(\boldsymbol{B}_t) = f(\boldsymbol{x})$. In particular, for spherically symmetric functions f (that is, $f(\boldsymbol{x}) = g(\|\boldsymbol{x}\|)$ for some function g), we then have $\mathbb{E}^x g(\|\boldsymbol{B}_t\|) = g(\|\boldsymbol{x}\|)$. Moreover, if the martingale (6.63) is uniformly integrable, we may, by Doob's stopping theorem, replace the fixed time t with any stopping time T. Taking $T := T_l \wedge T_r$, now gives

$$\mathbb{E}^x g(\|\boldsymbol{B}_T\|) = g(l)p + g(r)(1 - p) = g(\|\boldsymbol{x}\|),$$

whence,

$$p = \frac{g(r) - g(\|\boldsymbol{x}\|)}{g(r) - g(l)}.$$

In particular, the following functions:

$$(6.66) \qquad f(\boldsymbol{x}) := \begin{cases} |x| & \text{for } d = 1, \\ \ln \|\boldsymbol{x}\| & \text{for } d = 2, \\ \|\boldsymbol{x}\|^{2-d} & \text{for } d \geq 2, \end{cases}$$

are spherically symmetric and harmonic on D. We can extend f outside D in such a way that the function satisfies the conditions of Theorem 6.46. Moreover, $(f(\boldsymbol{B}_t))$ is uniformly integrable; see Exercise 10. Thus, we have found explicit expressions for the exit probability p for each dimension d. ∎

In Section 6.3 we established that $\boldsymbol{B}$ is a time-homogeneous Markov process with transition function $P_t = P_{0,t}$, whose transition density is given in (6.36). The left of (6.64) is exactly $(P_t f)(\boldsymbol{x})$; that is, the value at $\boldsymbol{x}$ of the function $P_t f$. It follows that

$$(6.67) \qquad \lim_{t \downarrow 0} \frac{(P_t f)(\boldsymbol{x}) - f(\boldsymbol{x})}{t} = \lim_{t \downarrow 0} \frac{\mathbb{E}^x f(\boldsymbol{B}_t) - f(\boldsymbol{x})}{t} = \mathbb{E}^x L f(\boldsymbol{B}_0) = L f(\boldsymbol{x}).$$

The limit in (6.67) defines the *infinitesimal generator* of the Markov process. Its domain consists of all bounded measurable functions for which the limit exists — this includes the domain of L, hence the infinitesimal generator extends the latter operator.

By (6.64), $(P_t f)(\boldsymbol{x})$ as a function of t has derivative $\mathbb{E}^x L f(\boldsymbol{B}_t) = (P_t L f)(\boldsymbol{x})$, which gives the *Kolmogorov forward equations*:

$$(6.68) \qquad\qquad\qquad P'_t f = P_t L f.$$

Also, by the Chapman–Kolmogorov equations (4.43), we have

$$P_{t+s} f(\boldsymbol{x}) = P_s P_t f(\boldsymbol{x}) = \mathbb{E}^x P_t f(\boldsymbol{B}_s),$$

and therefore

$$\frac{1}{s} \{ P_{t+s} f(\boldsymbol{x}) - P_t f(\boldsymbol{x}) \} = \frac{1}{s} \{ \mathbb{E}^x P_t f(\boldsymbol{B}_s) - P_t f(\boldsymbol{x}) \}.$$

Letting $s \downarrow 0$, we obtain the *Kolmogorov backward equations*:

$$(6.69) \qquad\qquad\qquad P'_t f = L P_t f.$$

In terms of the transition density $p_t(\boldsymbol{y} \mid \boldsymbol{x})$ this gives the partial differential equations

$$\frac{\partial}{\partial t} p_t(\boldsymbol{y} \mid \boldsymbol{x}) = \frac{1}{2} \Delta_y p_t(\boldsymbol{y} \mid \boldsymbol{x}) \quad \text{and} \quad \frac{\partial}{\partial t} p_t(\boldsymbol{y} \mid \boldsymbol{x}) = \frac{1}{2} \Delta_x p_t(\boldsymbol{y} \mid \boldsymbol{x}),$$

which we recognize as Laplace's *heat equation* (6.38) in $\boldsymbol{y}$ (with fixed $\boldsymbol{x}$) and $\boldsymbol{x}$ (with fixed $\boldsymbol{y}$), respectively.

The last point illustrates the important relation between partial differential equations of the form $u'_t = L u_t$ and the standard Brownian motion. Indeed, given the operator L, the pdf of $\boldsymbol{B}_t$ gives the fundamental solution (Green's function) of the differential operator $\partial t - L$. This idea can be extended to solve more general elliptical and higher-dimensional PDEs via Itô diffusion processes; see Section 7.3.

Another type of partial differential equation that can be solved via the Brownian motion process is the *boundary value* or *Dirichlet* problem. A typical application is found in electrostatics. Suppose a charge $g(\boldsymbol{x})$ is placed at every point $\boldsymbol{x}$ of the boundary ∂U of a bounded domain (open set) $U \subset \mathbb{R}^d$. This creates an electric field with potential (i.e., voltage) $v(\boldsymbol{x})$ for every $\boldsymbol{x}$ in the closure $\overline{U}$ of U, with

$$\Delta v(\boldsymbol{x}) = 0, \qquad \boldsymbol{x} \in U,$$
$$v(\boldsymbol{x}) = g(\boldsymbol{x}), \quad \boldsymbol{x} \in \partial U.$$

To solve the above problem (i.e., finding v) via simulation, we can employ once again the martingale (6.63), with $f = v$. Starting a Brownian motion from $\boldsymbol{x} \in U$, let T be the first time that ∂U is hit. Using the same reasoning as in Example 6.65, we may use Doob's stopping theorem to conclude that

$$v(\boldsymbol{x}) = \mathbb{E}^{\boldsymbol{x}} v(\boldsymbol{B}_0) = \mathbb{E}^{\boldsymbol{x}} v(\boldsymbol{B}_T) = \mathbb{E}^{\boldsymbol{x}} g(\boldsymbol{B}_T).$$

To estimate $v(\boldsymbol{x})$, simply run many standard Brownian motions starting at $\boldsymbol{x}$ until they hit the boundary, and take the average of their values at the boundary. It should be mentioned that some regularity conditions should be put on the domain U and on the function g. It suffices that g is continuous on U and that U satisfies an "exterior sphere" condition, meaning that it should be possible to roll a small enough sphere along the boundary such that sphere touches all the points at the boundary.

6.7   Path Properties

Recall the definition of a *segmentation* $\Pi_n = \{s_k, k = 0, \dots, n\}$ of $[0, t]$ given in (5.19), and that its *mesh* $\|\Pi_n\|$ is the maximum distance between consecutive points in the segmentation. We will frequently consider sequences of *nested* segmentations; that is, $\Pi_1 \subseteq \Pi_2 \subseteq \cdots$. For example, the following sequence of segmentations of $[0, 1]$ is nested: $\Pi_1 := \{0, 1\}, \Pi_2 := \{0, 1/2, 1\}, \Pi_3 := \{0, 1/4, 1/2, 1\}, \dots$. In this case, we say that Π_2 is a *refinement* of Π_1, and Π_3 is a refinement of Π_2.

We begin with a fundamental property of the *quadratic variation* of W on $[0, t]$ defined as the limit (in L^2 norm) of:

$$\langle W^{(n)} \rangle_t := \sum_{k=0}^{n-1} (W_{s_{k+1}} - W_{s_k})^2, \quad n \in \mathbb{N}.$$

Theorem 6.70: Quadratic Variation of the Wiener Process on $[0, t]$

For any (Π_n) such that $\|\Pi_n\| \to 0$, we have that $\langle W^{(n)} \rangle_t \xrightarrow{L^2} t$. If, additionally, the sequence of segmentations is nested, then $\langle W^{(n)} \rangle_t \xrightarrow{\text{a.s.}} t$.

Proof. To simplify notation, write $V_n := \langle W^{(n)} \rangle_t$ for a fixed t. The random variable V_n has expectation t and variance

$$\operatorname{Var} V_n = \sum_{k=0}^{n-1} \underbrace{\operatorname{Var} (W_{s_{k+1}} - W_{s_k})^2}_{= 3(s_{k+1} - s_k)^2} \leq 3t \, \|\Pi_n\| \to 0.$$

Hence, $\mathbb{E}(V_n - t)^2 = \operatorname{Var} V_n \to 0$ as $n \to \infty$, showing L^2-norm convergence.

Assume further that (Π_n) is nested and, without loss of generality, that Π_n and Π_{n-1} differ only by the addition of one point in the interval $[0, t]$, say $r \in (s_k, s_{k+1})$ for some k. Then,

$$V_n - V_{n-1} = X^2 + Y^2 - (X + Y)^2,$$

where $X := W_r - W_{s_k}$ and $Y := W_{s_{k+1}} - W_r$ are independent zero-mean Gaussian random variables. Let $\mathcal{G}_n$ be the σ-algebra generated by the sequence of random variables $(V_k, k \geq n)$, and note that

$$\mathcal{G}_1 \supseteq \mathcal{G}_2 \supseteq \cdots \supseteq \cap_{k=1}^{\infty} \mathcal{G}_k =: \mathcal{G}_\infty.$$

Then, for $n \geq 2$ we have that

$$\mathbb{E}[V_n - V_{n-1} \mid \mathcal{G}_n] = \mathbb{E}[X^2 + Y^2 - (X + Y)^2 \mid X^2 + Y^2] = -2 \, \mathbb{E}[XY \mid X^2 + Y^2].$$

Since (X, Y) has the same distribution as $(\pm X, \mp Y)$, and $(\pm X)^2 + (\mp Y)^2 = X^2 + Y^2$, it must be true that

$$\mathbb{E}[XY \mid X^2 + Y^2] = -\mathbb{E}[XY \mid X^2 + Y^2] = 0.$$

In other words, if $Z_n := V_{-n}$ and $\mathcal{F}_n := \mathcal{G}_{-n}$ for all $n \in \mathbb{T} = \{\dots, -2, -1\}$, then the process $Z := (Z_n, n \in \mathbb{T})$ is a square-integrable reversed-time martingale with respect to the filtration $(\mathcal{F}_n, n \in \mathbb{T})$. By Theorem 5.58 the reversed-time martingale converges almost surely to

$$\mathbb{E}[Z_{-1} \mid \mathcal{F}_{-\infty}] = \mathbb{E}V_1 = \mathbb{E}(W_t - W_0)^2 = t.$$

$\square$

Recall from Section 5.3.1 that a function $x : \mathbb{R}_+ \to \mathbb{R}$ is said to be of *bounded variation* over an interval $[0, t]$ if its *total variation* is finite. When the function x is right-continuous and the sequence (Π_n) is nested with $\|\Pi_n\| \to 0$, then the total variation can be written as

$$\sup_{(\Pi_n)} \sum_{k=0}^{n-1} |x_{s_{k+1}} - x_{s_k}| = \lim_{n \to \infty} \sum_{k=0}^{n-1} |x_{s_{k+1}} - x_{s_k}|.$$

This is because by adding a segmentation point to Π_n, the triangle inequality yields:

$$\sum_{k=0}^{n-1} |x_{s_{k+1}} - x_{s_k}| \le \sum_{k=0}^{n} |x_{s_{k+1}} - x_{s_k}|.$$

If x_s represents the position of a particle at time s, then the total variation over $[0,t]$ represents the total amount of vertical distance that the particle has traveled over the time interval $[0,t]$. Recall from Section 5.3.1 that for such functions we can define integrals

$$\int_0^t f_s \, dx_s$$

in the Lebesgue–Stieltjes sense. Our aim in Section 7.1 is to define stochastic integrals of the form

$$\int_0^t F_s \, dW_s$$

with respect to a Wiener process. If the paths of W had bounded variation, we would be able to define these integrals pathwise, i.e., for (almost) every function $w := W(\cdot, \omega)$, $\omega \in \Omega$. But, in fact, we have the following result:

Theorem 6.71: Infinite Variation of the Wiener Process

Suppose that the sequence (Π_n) of segmentations of $[0,t]$ is nested and $\|\Pi_n\| \to 0$. Then, for the Wiener process W,

$$\lim_{n \to \infty} \sum_{k=0}^{n-1} |W_{s_{k+1}} - W_{s_k}| = \infty \quad \text{almost surely.}$$

That is, the sample paths of W do not have bounded variation on $[0,t]$.

Proof. From Theorem 6.70 we know that there is an almost certain event Ω_0 such that $\langle W^{(n)}(\omega) \rangle_t \to t$ for all $\omega \in \Omega_0$. Now, consider a path $w := W(\omega)$ of the Wiener process, where $\omega \in \Omega_0$. Let its total variation on $[0,t]$ be v^*. We have

$$\sum_{k=0}^{n-1} (w_{s_{k+1}} - w_{s_k})^2 \le \sup_k |w_{s_{k+1}} - w_{s_k}| \sum_{k=0}^{n-1} |w_{s_{k+1}} - w_{s_k}| \le \sup_{s \in [0,t]} |w_{s+\varepsilon_n} - w_s| v^*,$$

where $\varepsilon_n := \|\Pi_n\| \to 0$. Letting $n \to \infty$, the left-most term goes to t, and in the right-most term, the supremum goes to 0, by the uniform continuity of each path w; see Theorem 6.21. It follows that v^* cannot be finite. $\square$

While the total and quadratic variation of the Wiener process on $[0, t]$ are ∞ and t, respectively, these can be random variables for other stochastic processes. For example, the total and quadratic variation of the Poisson counting process $(N_s, s \geq 0)$ on the interval $[0, t]$ are both equal to N_t; see Exercise 26.

By the existence constructions in Sections 6.2, the Wiener process has almost surely uniformly continuous paths on any interval $[0, t]$. In fact, the following stronger result holds; see also Theorem 4.34.

Theorem 6.72: Lévy's Modulus of Continuity

There exists a constant $c > 0$ such that with probability 1:

$$\limsup_{\delta \downarrow 0} \ \sup_{0 \leq t \leq 1-\delta} \frac{|W_{t+\delta} - W_t|}{\sqrt{c\,\delta \ln(1/\delta)}} \leq 1.$$

The function $\delta \mapsto \sqrt{c\,\delta \ln(1/\delta)}$ is called the *modulus of continuity* of the Wiener process.

Proof. We recycle the arguments in the proof of Theorem 6.21. First, applying the mean-value theorem to the Schauder functions in (6.18), we have that

$$|c_{k,j}(t+\delta) - c_{k,j}(t)| \leq \delta \sup_t |h_{k,j}(t)| \leq \delta\, 2^{k/2+1} \times \mathbb{1}_{[j/2^k,\,(j+1)/2^k]}(t).$$

It follows from the Haar basis expansion (6.19) that

$$|W_{t+\delta}^{(n)} - W_t^{(n)}| \leq \delta\,|Z_0| + \sum_{k=0}^{n-1} \sum_{j=0}^{2^k-1} |c_{k,j}(t+\delta) - c_{k,j}(t)|\,|Z_{j+2^k}| \leq 2\delta\, 2^{n/2}\, M_n,$$

where M_n is the maximum of the absolute value of 2^n iid $\mathcal{N}(0,1)$ random variables. Second, define the event $A_n := \{M_n > 2\sqrt{n}\}$. Since (see Exercise 6)

$$\sum_n \mathbb{P}(A_n) \leq \sum_n e^{-(4n+\ln(4n))/2 + n\ln 2} < \infty,$$

the Borel–Cantelli Lemma 3.14 implies that with probability 1 only finitely many of the events (A_n) occur. In other words, there exists a finite (random) N such that almost surely $M_n \leq 2\sqrt{n}$ for all $n \geq N$.

Third, using the triangle inequality (B.14) and the bound (6.22), we obtain:

$$|W_{t+\delta} - W_t| \leq 2\|W^{(n)} - W\| + |W_{t+\delta}^{(n)} - W_t^{(n)}|$$
$$\leq \gamma_1 \sqrt{n} 2^{-n/2} + 2\delta\, 2^{n/2}\, M_n,$$

where $\gamma_1 > 0$ is a constant. By setting $n := \lceil -\ln \delta / \ln 2 \rceil$ and dividing by $\sqrt{\delta \ln(1/\delta)}$, we obtain:

$$\sup_{0 \le t \le 1-\delta} \frac{|W_{t+\delta} - W_t|}{\sqrt{\delta \ln(1/\delta)}} \le \gamma_2 + \gamma_3 \frac{M_n}{\sqrt{\ln(1/\delta)}}$$

for some positive constants γ_2 and γ_3. The proof is completed by observing that a sufficiently small δ ensures that $n = \lceil -\ln \delta / \ln 2 \rceil \ge N$, so that with probability 1:

$$\frac{M_n}{\sqrt{\ln(1/\delta)}} \le \sqrt{\frac{2n}{\ln(1/\delta)}} \le \sqrt{\frac{2}{\ln 2} + \frac{2}{\ln(1/\delta)}} \le \gamma_4,$$

where $\gamma_4 > 0$ is some constant.   □

Lévy's result above suggests that the maximum amount of "smoothness" of a Wiener path is Hölder continuity with an exponent $\alpha < 1/2$; see Definition 4.33. For any $\alpha > 1/2$ Hölder continuity does *not* hold. In particular, differentiability ($\alpha = 1$) does not hold.

Theorem 6.73: Nowhere Differentiability of Wiener Paths

Almost surely, for $\alpha > 1/2$ the Wiener process is not Hölder continuous at any point. In particular, almost surely every path of a Wiener process is nowhere differentiable.

Proof. We argue by contradiction: assume that there is a point, say $s_0 \in [0,1]$, at which the Wiener process is Hölder continuous with $\alpha > 1/2$. In other words, we assume that there exists a constant $c < \infty$ and an $\alpha > 1/2$ such that almost surely:

(6.74) $$\sup_{\varepsilon \in [0,1]} \frac{|W_{s_0+\varepsilon} - W_{s_0}|}{\varepsilon^\alpha} \le c.$$

We now proceed to show that if Ω_0 is the event that there exists an $s_0 \in [0,1]$ satisfying (6.74), then $\mathbb{P}(\Omega_0) = 0$.

First, note that there exists a large enough integer m such that $1/m \in (0, \alpha - 1/2)$. For a given integer n, there exists a $k \in \{1, \ldots, 2^n\}$ such that $s_0 \in [(k-1)/2^n, k/2^n]$, and we can choose n large enough so that $2^n - k \ge m$.

Second, by assumption (6.74) and the inequality $|x|^\alpha + |y|^\alpha \le 2^{(1-\alpha)^+}(|x|+|y|)^\alpha$, we have for $l = 1, 2, \ldots, m$:

$$
\begin{aligned}
|W_{(k+l)/2^n} - W_{(k-1+l)/2^n}| &\le |W_{(k+l)/2^n} - W_{s_0}| + |W_{s_0} - W_{(k-1+l)/2^n}| \\
&\le c([l+1]/2^n)^\alpha + c(l/2^n)^\alpha \\
&\le \underbrace{c2^{(1-\alpha)^+}(2l+1)^\alpha}_{=: \, c_{l,\alpha}} 2^{-\alpha n} \le c_{m,\alpha} 2^{-\alpha n}.
\end{aligned}
$$

Third, if

$$A_{k,n} := \cap_{l=1}^{m}\{|W_{(k+l)/2^n} - W_{(k-1+l)/2^n}| \le c_{m,\alpha}2^{-\alpha n}\},$$

then the event Ω_0 implies that $H_n := \cup_{k=1}^{2^n-m}A_{k,n}$ occurs for infinitely many n. In other words,

$$\mathbb{P}(\Omega_0) \le \mathbb{P}\left(\sum_{n}\mathbb{1}_{H_n} = \infty\right),$$

and the proof will be complete if we can show that $\mathbb{P}(\sum_n \mathbb{1}_{H_n} = \infty) = 0$. By the independence of nonoverlapping increments of the Wiener process, we have

$$\mathbb{P}(A_{k,n}) = [\mathbb{P}(|W_{1/2^n}| \le c_{m,\alpha}2^{-\alpha n})]^m = [\mathbb{P}(|W_1| \le c_{m,\alpha}2^{-n(\alpha-1/2)})]^m.$$

Since $\int_{-|a|}^{|a|} \frac{dx}{\sqrt{2\pi}}\exp(-x^2/2) \le |a|$, it follows that $\mathbb{P}(A_{k,n}) \le c_{m,\alpha}^m 2^{-nm(\alpha-1/2)}$. Finally, by the countable subadditivity property in Theorem 2.2:

$$\sum_{n}\mathbb{P}(H_n) = \sum_{n}\mathbb{P}(\cup_{k=1}^{2^n-m}A_{k,n}) \le c_{m,\alpha}^m\sum_{n}2^{-nm(\alpha-1/2-1/m)} < \infty,$$

where the finiteness of the sum follows from $1/m \in (0, \alpha - 1/2)$. Therefore, from the Borel–Cantelli Lemma 3.14, the probability that the events (H_n) occur infinitely many times is 0. This implies that there is no point in $[0, 1]$ such that the Wiener process is almost surely Hölder continuous with $\alpha > 1/2$. The nondifferentiability follows as the special case with $\alpha = 1$.   □

Exercises

1. Simulate a two-dimensional Wiener process $(\boldsymbol{W}_t, t \ge 0)$ and show a typical realization.

2. Let $(\boldsymbol{W}_t, t \ge 0)$ be a d-dimensional Wiener process. Derive the probability distribution of $\|\boldsymbol{W}_t\|^2/t$. The process $(\|\boldsymbol{W}_t\|, t \ge 0)$ is called the d-dimensional *Bessel process*. Show that

$$\mathbb{E}\|\boldsymbol{W}_t\| = \sqrt{2t}\,\frac{\Gamma(\frac{d+1}{2})}{\Gamma(\frac{d}{2})}.$$

3. Let $W^{(n)}$ be defined as in (6.19). Show that $W_t^{(n+k)} = W_t^{(n)}, t \in D_n$ for $k = 1, 2, \dots$, and $\mathbb{E}\, W_s^{(n)}W_t^{(n)} = s \wedge t$ for all $s, t \in D_n$.

4.* To finish the proof of Theorem 6.21, show that for any choice of d and $0 \le t_1 < t_2 < \cdots < t_d \le 1$, the d-dimensional vector $[W_{t_1}, \dots, W_{t_d}]^\top$ is multivariate Gaussian with $\mathbb{E}W_{t_i} = 0$ and $\mathbb{Cov}(W_{t_i}, W_{t_j}) = t_{i \wedge j}$ for all i and j.

5. Given a sequence $(W_{t,k}, t \in [0, 1])$, $k = 0, 1, 2, \ldots$ of independent Wiener processes on $[0, 1]$, show that W defined via (6.23) is a Wiener process on $[0, \infty)$.

6. For $Z \sim \mathcal{N}(0, 1)$ and $z > 0$, show that

$$\mathbb{P}(|Z| > z) \le e^{-z^2/2}/z.$$

7. Explain how the construction of the Wiener process on $[0, 1]$ implies the existence of a Wiener process on $\mathbb{R}$.

8. Verify, using the transformation rule (2.39), that (6.45) is the pdf of x^2/Z^2, where $Z \sim \mathcal{N}(0, 1)$.

9.* Consider a standard Brownian motion (B_t) starting from $x \in (l, r)$. Using the fact that (B_t) is a martingale, show that the process exits through r rather than l with probability $(x - l)/(r - l)$. Use the fact that $(B_t^2 - t)$ is a martingale to show that the expected time to exit the interval $[l, r]$ is $(x - l)(b - r)$.

10. Show that the functions f in (6.66) are harmonic on the annulus D and that the process $(f(B_t))$ is uniformly integrable.

11. Example 6.65 gives a simple expression for the probability that a d-dimensional standard Brownian motion (B_t), starting at x, with $l \le \|x\| \le r$, hits the 0-centered sphere with radius l before it hits the 0-centered sphere with radius r, where $0 < l < r$. For $l \ge 0$, let $T_l := \inf\{t \ge 0 : \|B_t\| = l\}$.

   (a) Show that for $d = 2$ and any $l > 0$,

$$\mathbb{P}^x(T_l < \infty) = 1 \quad \text{for all } \|x\| > l.$$

   Show that, however, for the case $d = 2$ and $l = 0$, we have

$$\mathbb{P}^x(T_0 < \infty) = 0 \quad \text{for all } \|x\| > 0.$$

   Thus, starting from anywhere outside a 0-centered sphere of any radius $l > 0$, the process (B_t) will hit the sphere with certainty, but will never exactly hit the origin 0.

   (b) For the case $d \ge 2$, show that

$$\mathbb{P}^x(T_l < \infty) = \left(\frac{l}{\|x\|}\right)^{d-2} \quad \text{for all } \|x\| > l.$$

   Thus, any standard Brownian motion in dimension 3 or greater is *transient*, meaning that there is a strictly positive probability that the process will never hit a sphere of radius l, starting outside it, no matter how large l is.

12. With the hitting time T_x distributed according to (6.55), show that $\mathbb{P}(T_x < \infty) = 1$ and $\mathbb{E}T_x = \infty$ for all $x > 0$.

13.* Let $T_x^{(1)}$ and $T_x^{(2)}$ be independent copies of the hitting time T_x, as defined in (6.48). Show that for any $a, b > 0$ the random variable $aT_x^{(1)} + bT_x^{(2)}$ has the same distribution as cT_1 for some c.

14. Let $U := \{(x_1, x_2) \in \mathbb{R}^2 : x_1^2 + x_2^2 < 1\}$. On the unit circle ∂U, let

$$g(x) := \sin(4x_1) \cos(x_2).$$

Solve the Dirichlet problem $\Delta v(x) = 0, x \in U, v(x) = g(x), x \in \partial U$ via simulation, and make a density plot of v.

15. Let $(W_t, t \geq 0)$ be a Wiener process and let $\mathcal{F}^+$ be the right-continuous filtration defined by

$$\mathcal{F}_t^+ := \bigcap_{u > t} \mathcal{F}_u, \quad \text{where} \quad \mathcal{F}_u := \sigma(W_s : 0 \leq s \leq u).$$

Let $T_c := \min\{t : W_t = c\}$ and define

$$T^* := \begin{cases} \min\{T_{10}, T_{-1}\} & \text{if } \min\{T_{10}, T_{-1}\} \leq 1, \\ \min\{T_{20}, T_{-1}\} & \text{if } \min\{T_{10}, T_{-1}\} > 1. \end{cases}$$

(a) Show that T^* is a stopping time with respect to the filtration $\mathcal{F}^+$.

(b) Show $\mathbb{P}(W_{T^*} = 10) \leq 0.04$.

(c) Show $\mathbb{P}(W_{T^*} = 20) \geq 0.026$.

16. Let $W^{(1)}$ and $W^{(2)}$ be two independent Wiener processes. Define

$$M_n := \max_{1 \leq k \leq n} \left| \sum_{j=1}^{k} \left(W_{j/n}^{(1)} - W_{(j-1)/n}^{(1)} \right) \left(W_{j/n}^{(2)} - W_{(j-1)/n}^{(2)} \right) \right|.$$

Show that $M_n \xrightarrow{\mathbb{P}} 0$ as $n \to \infty$.

17. Let W be a Wiener process. Define $T := \inf\{t \geq 0 : W_t^2 = 1 - t\}$. Compute $\mathbb{E}T$.

18. Let (B_t) be a standard Brownian motion. Show that, for any $x > 0$ and measurable set $A \subset [0, \infty)$,

$$\mathbb{P}^x(B_s \geq 0 \text{ for all } 0 \leq s \leq t \text{ and } B_t \in A) = \mathbb{P}^x(B_t \in A) - \mathbb{P}^{-x}(B_t \in A).$$

19. Define $g(w, r, t) := \exp(rw - \frac{1}{2}r^2 t)$ for $w \in \mathbb{R}, r \in \mathbb{R}$, and $t \geq 0$. Let (W_t) be a Wiener process and let $g(W_t, r, t), t \geq 0$ be the exponential martingale defined in (6.40). Show that the process

$$X_t := \left. \frac{\partial^n g(W_t, r, t)}{\partial r^n} \right|_{r=0}, \quad t \geq 0$$

is a martingale with respect to the natural filtration of (W_t) for every $n \in \{1, 2, \ldots\}$. Show, in particular, that each of the processes

$$W_t^2 - t, \quad t \geq 0,$$
$$W_t^3 - 3t\, W_t, \quad t \geq 0,$$
$$W_t^4 - 6t\, W_t^2 + 3\, t^2, \quad t \geq 0,$$

is a martingale.

20. Find $\mathbb{E}T^2$ for $T := \min\{t \geq 0 : W_t \notin [-a, b]\}$ and $-a < 0 < b$, using the martingales in Exercise 19. You may assume that $\mathbb{E}T^2 < \infty$.

21. (a) Use the optional stopping theorem for the exponential martingale to show that, with $T_x := \inf\{t \geq 0 : W_t = x\}$,

$$\mathbb{E}\,e^{-sT_x} = e^{-x\sqrt{2s}}, \qquad \text{for all } s, x > 0.$$

(b) Show that, with $T_{-x} = \inf\{t \geq 0 : W_t = -x\}$, we have

$$\mathbb{E}\,e^{-sT_x} = \mathbb{E}[e^{-sT_x}\mathbb{1}(T_x < T_{-x})] + \mathbb{E}[e^{-sT_{-x}}\mathbb{1}(T_{-x} < T_x)]\,e^{-2x\sqrt{2s}}.$$

(c) Deduce that $T = T_x \wedge T_{-x}$ satisfies

$$\mathbb{E}\,e^{-sT} = \operatorname{sech}(x\sqrt{2s}).$$

22.* Let W be a Wiener process and define $T_z := \inf\{t : W_t = z\}$ where $z \in \mathbb{R}$. For any three constants $-a < 0 < b < c$, determine $\mathbb{P}(T_b < T_{-a} < T_c)$.

23. Show that with probability 1 for every $N < \infty$ there is a $t > N$ with $W_t = 0$.

24. Show that with probability 1 for every $\varepsilon > 0$ there is a $t \in (0, \varepsilon)$ with $W_t = 0$.

25. A continuous function f is said to have a *local maximum* at t^* if there exists an $\varepsilon > 0$ such that
$$f(t^*) \geq f(s) \quad \text{for all } s \in (t^* - \varepsilon, t^* + \varepsilon).$$

We know that with probability 1 the Wiener process $W := (W_t, t \geq 0)$ is not monotone on any interval $[a, b]$. Use this fact to show that the set of local maxima of W is dense in $[0, \infty)$ almost surely.

26.* The quadratic variation of the Poisson counting process $(N_s, s \geq 0)$ on the interval $[0, t]$ is defined as the almost sure limit

$$\langle N \rangle_t := \lim_{n \to \infty} \sum_{k=0}^{n-1} (N_{s_{k+1}} - N_{s_k})^2,$$

where $\{s_k, k = 0, \ldots, n\} =: \Pi_n$ is a segmentation of $[0, t]$ such that its mesh size $\|\Pi_n\| \to 0$ as $n \to \infty$. Show that the total and quadratic variation of the Poisson counting process $(N_s, s \geq 0)$ on the interval $[0, t]$ are both equal to N_t.

27. Let $W := (W_t, t \geq 0)$ be a Wiener process, and let $f : \mathbb{R} \to [0, 1]$ be a Hölder continuous function of order $\alpha = 1$; see Definition 4.33. Define $X_t := W_t + f_t, t \in [0, 1]$ and define the quadratic variation of $(X_t, t \in [0, 1])$ as the almost sure limit

$$\langle X \rangle_1 := \lim_{n \to \infty} \sum_{k=0}^{n-1} (X_{s_{k+1}} - X_{s_k})^2,$$

where $\{s_k, k = 0, \ldots, n\} =: \Pi_n$ is a nested segmentation of $[0, 1]$ such that its mesh size $\|\Pi_n\| \to 0$ as $n \to \infty$. Prove that $\langle X \rangle_1 = 1$.

28. Let $T := \min\{t : W_t \in \{-1, 1\}\}$. By considering the sequence of events

$$\{W_{n+1} - W_n > 2\}_{n=0}^{\infty},$$

show that $\mathbb{P}(T > n) \leq C\lambda^n$ for some $C > 0$ and $\lambda \in (0, 1)$.

29. Let W be a Wiener process and define $T := \min\{t : W_t = 1 - t\}$. Use the exponential martingale (6.40) to determine the Laplace transform of the distribution of T.

CHAPTER 7

ITÔ CALCULUS

In this chapter we introduce the framework for stochastic integration with respect to the Wiener process. The resulting *Itô integral* provides the fundamental example for stochastic integration with respect to integrators with unbounded variation. We then prove *Itô's formula* — the stochastic analogue of the chain rule in calculus. The important class of *Itô processes*, which are constructed from Itô integrals, forms the basis for the theory of diffusion processes and stochastic differential equations.

An important class of stochastic processes — that of *Itô processes* — is constructed from the Wiener process via the notion of the Itô integral. The Itô integral provides the mathematical justification of integrals of the form

$$\int_0^t F_s \, dX_s,$$

where the *integrator* $X := (X_s)$ and *integrand* $F := (F_s)$ are stochastic processes. The most important case is where X is a Wiener process — we will then use W instead of X. We already encountered stochastic integration in Section 5.3.1 and saw that integrals of the form above can be well-defined if suitable restrictions are placed on X and F. In particular, when X is of bounded variation, the integral can be defined pathwise. Unfortunately, for processes of unbounded variation, like the Wiener process, this is not possible in general, and a different approach is needed. The clue as to what to do in this case is provided by the discrete integral in (5.23), where F is a *predictable* process. In that case, the integral-transformed process (Z_n) in (5.24) is a martingale; see Theorem 5.26. Recall that, in this setting, X_n can be thought of as a share price and F_n as the number of shares owned at time n, so that Z_n represents the total capital at time n.

For consistency of the notation, in this chapter we will use the notation $\int_0^t F_s \, ds$ for Lebesgue integrals, rather than $\int_0^t ds \, F_s$.

7.1   Itô Integral

Throughout the rest of this chapter we will work with a Wiener process W on a complete probability space $(\Omega, \mathcal{H}, \mathbb{P})$ with a right-continuous filtration $\mathcal{F} := (\mathcal{F}_s, s \geq 0)$ and such that W is adapted to $\mathcal{F}$. We also use the following definition:

Definition 7.1: Simple Processes

A process $F := (F_s, s \in [0, t])$ is called *simple* if it is of the form:

$$(7.2) \qquad F_s := \xi_0 \mathbb{1}_{\{0\}}(s) + \sum_{k=0}^{n-1} \xi_k \mathbb{1}_{(s_k, s_{k+1}]}(s), \quad s \in [0, t],$$

where $\{s_0, \ldots, s_n : 0 = s_0 < \cdots < s_n = t\} =: \Pi_n$, $n \in \mathbb{N}$, is a segmentation of $[0, t]$ and $\xi_k \in \mathcal{F}_{s_k}$ for all k.

Note that the simple process in (7.2) is right-continuous at $s = 0$ and is left-continuous at $s \in (0, t]$. Figure 7.3 shows a realization of a simple process on $[0, 1]$ for the segmentation

$$\{s_0, \ldots, s_{10}\} = \{0, 0.07, 0.13, 0.19, 0.41, 0.49, 0.52, 0.67, 0.71, 0.76, 1\}.$$

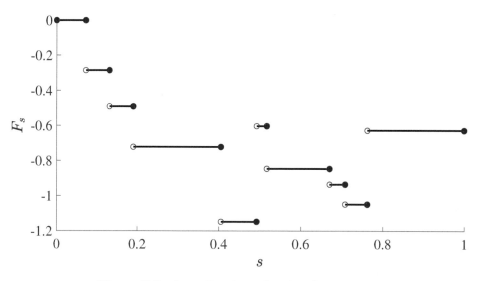

Figure 7.3: A realization of a simple process.

7.1.1   Itô Integral for Simple Processes

To introduce the Itô integral with respect to the Wiener process W, we need to revisit the generalization of predictability to continuous-time processes. Recall that a real-valued stochastic process is said to be $\mathcal{F}$-*predictable*, if it is measurable with respect to the σ-algebra in Definition 5.22, namely,

$$\mathcal{F}^p := \sigma((a,b] \times H : a,b \in \mathbb{R}_+, H \in \mathcal{F}_a) \vee \sigma(\{0\} \times H : H \in \mathcal{F}_0).$$

■ **Example 7.4 (Simple Processes are Predictable)** We now argue that for each $t \geq 0$ the stochastic process $F : [0,t] \times \Omega \to \mathbb{R}$ in (7.2) is not only $\mathcal{F}$-adapted, but also $\mathcal{F}^p$-measurable. Since the sum of measurable functions is measurable, it is enough to show that an arbitrarily chosen summand of (7.2), say $\xi_0 \mathbb{1}_{\{0\}}$, is $\mathcal{F}^p$-measurable. To this end, recall from Theorems 1.27 and 1.31 that $\xi_0 \in \mathcal{F}_0$ can be viewed as the pointwise limit of linear combinations of indicator functions. In other words,

$$\xi_0(\omega) = \lim_m \sum_{i=1}^{m} a_i \mathbb{1}_{H_i}(\omega)$$

for some real sequence (a_i) and subsets $H_1, H_2, \ldots \in \mathcal{F}_0$. As a result, we can write

$$\mathbb{1}_{\{0\}}(s)\,\xi_0(\omega) = \lim_m \sum_{i=1}^{m} a_i \mathbb{1}_{\{0\}}(s) \mathbb{1}_{H_i}(\omega),$$

and since the limits of measurable functions are themselves measurable (see Proposition 1.26), we only need to show that for an arbitrary i:

$$\mathbb{1}_{\{0\}}(s) \mathbb{1}_{H_i}(\omega) = \mathbb{1}_{\{0\} \times H_i}(s, \omega) \in \mathcal{F}^p.$$

Since $\{0\} \times H_i \in \mathcal{F}^p$, the indicator function of this set is $\mathcal{F}^p$-measurable (see Example 1.23), and therefore we can conclude that the process F defined in (7.2) is $\mathcal{F}$-predictable.

Note that the measurability of the mapping $F : [0,t] \times \Omega \to \mathbb{R}$ with respect to the product σ-algebra $\mathcal{B}_{[0,t]} \otimes \mathcal{F}_t$ confirms that F_t is indeed a random variable for each t, or equivalently (see Theorem 2.17) that $(F_s, s \in [0,t])$ is a stochastic process. If, in addition to $\xi_k \in \mathcal{F}_{s_k}$, we also have that

$$\sum_{k=0}^{n-1} \mathbb{E}\,\xi_k^2 (s_{k+1} - s_k) < \infty,$$

then F is square-integrable with respect to the product measure $\mathrm{Leb}_{[0,t]} \otimes \mathbb{P}$. In that case, Fubini's Theorem 1.67 allows us to write

$$\mathbb{E}\int_0^t F_s^2 \, \mathrm{d}s = \int_0^t \mathbb{E}\,F_s^2 \, \mathrm{d}s = \sum_{k=0}^{n-1} \mathbb{E}\,\xi_k^2(s_{k+1} - s_k) < \infty.$$

We are now ready to define the Itô Integral for simple processes.

Definition 7.5: Itô Integral For Simple Processes

Suppose that F is a simple process of the form (7.2), with $\int_0^t \mathbb{E}F_s^2 \, ds < \infty$. Then, the *Itô integral* of F with respect to W over $[0, r]$ is defined as

$$\int_0^r F_s \, dW_s := \sum_{k=0}^{n-1} \xi_k (W_{s_{k+1} \wedge r} - W_{s_k \wedge r}), \quad r \in [0, t].$$

The Itô integral from r to t is naturally defined as the difference of two integrals:

$$\int_r^t F_s \, dW_s := \int_0^t F_s \, dW_s - \int_0^r F_s \, dW_s.$$

The Itô integral has the important *isometry* property:

$$\mathbb{E}\left[\int_0^t F_s \, dW_s\right]^2 = \int_0^t \mathbb{E}F_s^2 \, ds,$$

which allows us to compute the variance of the Itô integral. In fact, the Itô integral satisfies the following four important properties:

Theorem 7.6: Properties of the Itô Integral

Assume that the Itô integral is well-defined. Then, for $\alpha, \beta \in \mathbb{R}$ and $r \in [0, t]$:

1. *(Linearity)*: $\int_0^r [\alpha F_s + \beta G_s] \, dW_s = \alpha \int_0^r F_s \, dW_s + \beta \int_0^r G_s \, dW_s$.

2. *(Zero mean)*: $\mathbb{E} \int_0^r F_s \, dW_s = 0$.

3. *(Isometry)*: $\mathbb{E}\left[\int_0^r F_s \, dW_s \times \int_0^r G_s \, dW_s\right] = \int_0^r \mathbb{E}[F_s G_s] \, ds$.

4. *(Martingale)*: $\left(\int_0^r F_s \, dW_s, r \in [0, t]\right)$ is an L^2-martingale.

We provide a proof for simple processes first. A proof for more general processes is given on page 253 after we establish a relevant existence result in Theorem 7.20.

Proof for Simple Processes. Let F be a simple process as in (7.2) and let G be the simple process:

$$G_s := \gamma_0 \mathbb{1}_{\{0\}}(s) + \sum_{k=0}^{n-1} \gamma_k \mathbb{1}_{(s_k, s_{k+1}]}(s), \quad s \in [0, t],$$

where $\gamma_k \in \mathcal{F}_{s_k}$ for all k. Linearity follows directly from Definition 7.5. Define $\Delta W_{s_k} := W_{s_{k+1} \wedge r} - W_{s_k \wedge r}$, so that $\int_0^r F_s\, dW_s = \sum_{k=0}^{n-1} \xi_k \Delta W_{s_k}$. By the martingale property of the Wiener process, combined with the fact that $\xi_k \in \mathcal{F}_{s_k}$, we have for each $k = 0, \ldots, n-1$:

$$\mathbb{E}\xi_k \Delta W_{s_k} = \mathbb{E}\,\mathbb{E}_{s_k}\xi_k \Delta W_{s_k} = \mathbb{E}\xi_k\,\mathbb{E}_{s_k} \Delta W_{s_k} = 0,$$

which implies the zero-mean property. For the isometry property, write the expectation $\mathbb{E}\left[\int_0^r F_s\, dW_s \times \int_0^r G_s\, dW_s\right]$ as the double sum:

$$(7.7)\quad \mathbb{E}\sum_{k=0}^{n-1}\sum_{j=0}^{n-1} \xi_k \Delta W_{s_k} \gamma_j \Delta W_{s_j} = \sum_{k=0}^{n-1} \mathbb{E}\xi_k \gamma_k [\Delta W_{s_k}]^2 + 2\sum_{k=1}^{n-1}\sum_{j=0}^{k-1} \mathbb{E}\xi_k \gamma_j \Delta W_{s_k} \Delta W_{s_j}.$$

The double sum on the right-hand side in (7.7) is 0, since for each $j < k$,

$$\mathbb{E}\left[\xi_k \gamma_j \Delta W_{s_k} \Delta W_{s_j}\right] = \mathbb{E}\left[\xi_k \gamma_j \Delta W_{s_j} \mathbb{E}_{s_k} \Delta W_{s_k}\right] = 0,$$

by the martingale property of W. Moreover, for the first sum on the right-hand side in (7.7) we have, again by conditioning on $\mathcal{F}_{s_k}$, that

$$\sum_{k=0}^{n-1} \mathbb{E}\xi_k \gamma_k [\Delta W_{s_k}]^2 = \sum_{k=0}^{n-1} \mathbb{E}[\xi_k \gamma_k \mathbb{E}_{s_k}[\Delta W_{s_k}]^2] = \sum_{k=0}^{n-1} \mathbb{E}\xi_k \gamma_k (s_{k+1} \wedge r - s_k \wedge r).$$

The latter sum is exactly $\int_0^r \mathbb{E}[F_s G_s]\, ds$, as $F_s = \xi_k$ and $G_s = \gamma_k$ on each interval $(s_k, s_{k+1}]$.

Finally, to show the martingale property, note that for $r \in [0, t)$ we can write:

$$(7.8)\quad \begin{aligned}\int_r^t F_s\, dW_s &= \sum_{k=0}^{n-1} \xi_k(W_{s_{k+1}} - W_{s_{k+1} \wedge r} + W_{s_k \wedge r} - W_{s_k}) \\ &= \sum_{k:s_{k+1}>r} \xi_k(W_{s_{k+1}} - W_{s_k \vee r}),\end{aligned}$$

where we used the fact that for any $r \in [0, t)$:

$$W_{s_{k+1}} - W_{s_{k+1} \wedge r} + W_{s_k \wedge r} - W_{s_k} = \begin{cases} W_{s_{k+1}} - W_{s_k \vee r}, & s_{k+1} > r, \\ 0, & s_{k+1} \le r. \end{cases}$$

By the repeated conditioning property in (4.5), we have:

$$\mathbb{E}_r \int_r^t F_s\, dW_s = \sum_{k:s_{k+1}>r} \mathbb{E}_r \xi_k \underbrace{\mathbb{E}_{s_k \vee r}[W_{s_{k+1}} - W_{s_k \vee r}]}_{= 0} = 0.$$

By construction, $\int_0^r F_s\, dW_s$ is $\mathcal{F}_r$-measurable for any arbitrary $r \in [0, t]$. In addition, it is square-integrable due to the isometry: $\mathbb{E}\left(\int_0^r F_s\, dW_s\right)^2 = \int_0^r \mathbb{E}F_s^2\, ds < \infty$. $\square$

■ **Example 7.9 (Deterministic Integrands and Gaussian Processes)** Suppose
that f is a simple process of the form (7.2) such that $\xi_0, \xi_1, \ldots$ are all deterministic.
Then, the random variable

$$Z_r := \int_0^r f_s \, dW_s$$

has a Gaussian distribution for each $r \in [0,t]$, because it is a linear combination
of the independent Gaussian random variables $W_{s_{k+1} \wedge r} - W_{s_k \wedge r}, k = 0, 1, \ldots, n-1$.
From Theorem 7.6 we deduce that $\mathbb{E} Z_r = 0$ and

$$\mathbb{V}\text{ar}\, Z_r = \int_0^r f_s^2 \, ds = \sum_{k=0}^{n-1} \xi_k^2 (s_{k+1} \wedge r - s_k \wedge r).$$

In fact, we can go further and show that $Z := (Z_t, t \geq 0)$ is a zero-mean Gaussian
process with independent increments. Namely, for any $0 \leq t_1 < t_2 < t_3 < t_4 \leq t$,
the increment $Z_{t_4} - Z_{t_3} = \int_{t_3}^{t_4} f_s \, dW_s$ is determined completely by the process $(W_s -
W_{t_3}, t_3 \leq s \leq t_4)$, whereas $Z_{t_2} - Z_{t_1} = \int_{t_1}^{t_2} f_s \, dW_s$ is determined by the process
$(W_s - W_{t_1}, t_1 \leq s \leq t_2)$. Since these two processes are independent of each other,
so are $Z_{t_4} - Z_{t_3}$ and $Z_{t_2} - Z_{t_1}$. This establishes the independence of the increments
of the process Z. This property of the increments combined with the fact that
$Z_r \sim \mathcal{N}(0, \int_0^r f_s^2 \, ds)$ for each $r \in [0,t]$ implies that Z is a Gaussian process with
covariance function

$$\mathbb{E} Z_{t_1} Z_{t_2} = \int_0^{t_1 \wedge t_2} f_s^2 \, ds, \quad t_1, t_2 \geq 0.$$

■

The main utility of simple processes of the form (7.2) is in approximating more
general processes. In this regard, we have the following definition:

Definition 7.10: Canonical Approximation

For any $\mathcal{F}$-adapted process $F := (F_s, s \in [0,t])$, we define its *canonical
approximation* as

(7.11) $$F_s^{(n)} := F_0 \mathbb{1}_{\{0\}}(s) + \sum_{k=0}^{n-1} F_{s_k} \mathbb{1}_{(s_k, s_{k+1}]}(s), \quad s \in [0,t].$$

Clearly, the canonical approximation is of the form (7.2), where $\xi_k := F_{s_k}$ for
$k = 0, \ldots, n-1$. Figure 7.12 shows the path of a Wiener process $(W_s, s \in [0,1])$
and a canonical approximation with $n = 10$, which coincides with the simple process
on Figure 7.3. Further to this, Figure 7.12 depicts the path of the corresponding Itô
integral process $Z^{(n)} = (\int_0^r F_s^{(n)} \, dW_s, r \in [0,1])$.

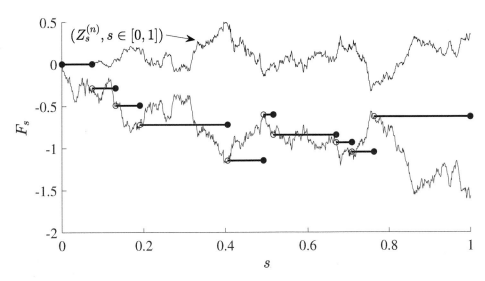

Figure 7.12: A Wiener process $F := (W_s, s \in [0, 1])$, its canonical approximation $F^{(n)}$, and the Itô integral process $Z^{(n)} = (\int_0^r F_s^{(n)} \, dW_s, r \in [0, 1])$.

Having defined the Itô integral for simple processes, we are now ready to extend the definition to more general predictable processes.

7.1.2   Itô Integral for Predictable Processes

While simple processes of the form (7.2) are $\mathcal{F}$-predictable, it is not immediately clear if more general processes are also $\mathcal{F}$-predictable. The following result shows that any process $F := (F_s, s \in [0, t])$ adapted to $\mathcal{F}$ and left-continuous on $(0, t]$ is not only $\mathcal{F}$-predictable, but also *indistinguishable* from any other left-continuous modification $\widetilde{F}$ of F; that is, if $\mathbb{P}(F_s = \widetilde{F}_s) = 1$ for all $s \in [0, t]$, then $\mathbb{P}(F_s = \widetilde{F}_s$ for all $s \in [0, t]) = 1$. Moreover, F is *progressively measurable*; that is, the mapping $F : [0, t] \times \Omega \to \mathbb{R}$ is measurable with respect to the product σ-algebra $\mathcal{B}_{[0,t]} \otimes \mathcal{F}_t$.

Theorem 7.13: Properties of Left-continuous Adapted Processes

Suppose that $F := (F_s, s \in [0, t])$ is a process on $(\Omega, \mathcal{H}, \mathbb{P})$ that is adapted to the filtration $\mathcal{F}$ and is left-continuous on $(0, t]$. Then, the following hold:

1. *(Predictable)*: The process F is $\mathcal{F}$-predictable (i.e., $\mathcal{F}^p$-measurable).

2. *(Progressively Measurable)*: For each t, the mapping $F : [0, t] \times \Omega \to \mathbb{R}$ is jointly measurable with respect to $\mathcal{B}_{[0,t]} \otimes \mathcal{F}_t$.

3. *(Indistinguishable)*: If $\widetilde{F}$ is a modification of F and almost surely left-continuous, then $\mathbb{P}(\cap_{s \leq t} \{F_s = \widetilde{F}_s\}) = 1$.

Proof. Consider the canonical approximation $F^{(n)}$ of F, given in (7.11), where the segmentation Π_n is such that mesh$(\Pi_n) \to 0$. For $s = 0$ and $\omega \in \Omega$, we have $F_0(\omega) = F_0^{(n)}(\omega)$ by construction. For any $s \in (0, t]$, we can always find a k (depending on n and s) such that $s_k < s \leq s_{k+1}$. For these k and s, define $\varepsilon_{n,s} := s - s_k$, so that $0 < \varepsilon_{n,s} \leq$ mesh(Π_n). Importantly,

$$F_s - F_s^{(n)} = F_s - F_{s_k} + \underbrace{F_{s_k} - F_s^{(n)}}_{= \, 0} = F_s - F_{s-\varepsilon_{n,s}},$$

where the last difference $F_s - F_{s-\varepsilon_{n,s}}$ vanishes as $n \uparrow \infty$, because F is left-continuous. In other words, the simple process approximates F arbitrarily well, in the sense that $F_s^{(n)}(\omega) \to F_s(\omega)$ for all $s \in [0, t]$ and $\omega \in \Omega$. Since F is adapted to $\mathcal{F}$, Example 7.4 implies that the simple process $F^{(n)}$ is $\mathcal{F}$-predictable. As a consequence, for each t, the mapping $F^{(n)} : [0, t] \times \Omega \to \mathbb{R}$ is jointly measurable with respect to $\mathcal{B}_{[0,t]} \otimes \mathcal{F}_t$. Since the limit of measurable functions is measurable, the same properties carry through to the limit F of $F^{(n)}$. This establishes the properties of $\mathcal{F}$-predictability and progressive measurability for F.

Next, assume that $\widetilde{F}$ is a modification of F; that is, $\mathbb{P}(\widetilde{F}_s \neq F_s) = 0$ for all $s \in [0, t]$. Let B be the event that both F and $\widetilde{F}$ are left-continuous, so that $\mathbb{P}(B) = 1$. Let $\mathbb{Q}_t := [0, t] \cap \mathbb{Q}$ and define $A := \cap_{s \in \mathbb{Q}_t} \{\widetilde{F}_s = F_s\}$ as the event that F and $\widetilde{F}$ coincide on $\mathbb{Q}_t$. Since the set $\mathbb{Q}_t$ is countable, the countable subadditivity property in Theorem 2.2 yields

$$\mathbb{P}(A^c) = \mathbb{P}(\cup_{s \in \mathbb{Q}_t} \{\widetilde{F}_s \neq F_s\}) \leq \sum_{s \in \mathbb{Q}_t} \mathbb{P}(\widetilde{F}_s \neq F_s) = 0.$$

In other words, $\mathbb{P}(A) = 1$. An important property of the set of rational numbers $\mathbb{Q}_t$, called *denseness*, is that any $s \in [0, t]$ can be approximated arbitrarily well with a sequence of rational numbers $s_1, s_2, \ldots \in \mathbb{Q}_t$. We can choose this sequence such that $s_n < s$ for all n; that is, $s_n \uparrow s$ as $n \uparrow \infty$. Therefore, for all $\omega \in A \cap B$:

$$F_s(\omega) - \widetilde{F}_s(\omega) = \lim_{n \uparrow \infty}(F_{s_n}(\omega) - \widetilde{F}_{s_n}(\omega)) = 0 \quad \text{for all } s \in [0, t].$$

Since $\mathbb{P}(A \cap B) = 1 - \mathbb{P}(A^c \cup B^c) = 1$, we conclude that F and $\widetilde{F}$ are indistinguishable. □

We have so far considered stochastic processes that are left-continuous and adapted to the filtration $\mathcal{F}$. However, in order to define the Itô integral for these more general processes, additional regularity conditions are needed.

Definition 7.14: The Class of Processes $\mathcal{P}_t$

Let $\mathcal{P}_t$ be the class of stochastic processes $F := (F_s, s \in [0, t])$ that are left-continuous on $(0, t]$, $\mathcal{F}$-adapted, and satisfy the norm condition:

$$\|F\|_{\mathcal{P}_t}^2 := \mathbb{E} \int_0^t F_s^2 \, ds < \infty.$$

Note that the progressively measurable property in Theorem 7.13 and Fubini's Theorem 1.67 together justify swapping the integral and expectation:

$$(7.15) \qquad \|F\|_{\mathcal{P}_t}^2 = \mathbb{E} \int_0^t F_s^2 \, ds = \int_0^t \mathbb{E} F_s^2 \, ds < \infty.$$

In addition, the processes in $\mathcal{P}_t$ are not only $\mathcal{F}^p$-measurable (by Theorem 7.13), but also belong to the Hilbert space $L^2([0, t] \times \Omega, \mathcal{F}^p, \text{Leb}_{[0,t]} \otimes \mathbb{P})$ equipped with the inner product that is induced by the norm $\| \cdot \|_{\mathcal{P}_t}$, namely,

$$\langle F, G \rangle_{\mathcal{P}_t} := \frac{\|F + G\|_{\mathcal{P}_t}^2 - \|F\|_{\mathcal{P}_t}^2 - \|G\|_{\mathcal{P}_t}^2}{2}.$$

We are now ready to begin extending the definition of the Itô integral to the class $\mathcal{P}_t$. A key step in this direction is to note that processes in $\mathcal{P}_t$ can be approximated arbitrarily well in the norm $\| \cdot \|_{\mathcal{P}_t}$ via the simple processes in Definition 7.1.

Theorem 7.16: Approximating the Class $\mathcal{P}_t$ via Simple Processes

Suppose that $F \in \mathcal{P}_t$. Then, there exists a sequence $(F^{(n)})$ of simple processes of the form (7.2) such that $\text{mesh}(\Pi_n) \to 0$ implies that $F_s^{(n)} \xrightarrow{\text{a.s.}} F_s$ for all $s \in [0, t]$ and

$$(7.17) \qquad \lim_{n \to \infty} \|F^{(n)} - F\|_{\mathcal{P}_t} = 0.$$

Proof. Let $\widetilde{\mathbb{P}} := \text{Leb}_{[0,t]} \otimes \mathbb{P}/t$ be a probability measure on $([0, t] \times \Omega, \mathcal{B}_{[0,t]} \otimes \mathcal{H})$ and let X, with $X(s, \omega) := F_s(\omega)$, be a numerical random variable on this probability space. With $\widetilde{\mathbb{E}}$ the expectation corresponding to $\widetilde{\mathbb{P}}$, and $\|X\|_2^2 := \widetilde{\mathbb{E}} X^2$ the squared L^2 norm of X, we have $t\|X\|_2 = \|F\|_{\mathcal{P}_t} < \infty$. Let $X_m := X \mathbb{1}_{\{|X| \leq m\}}$ for $m \in \mathbb{N}$ be a truncated version of X. Then,

$$\|X_m - X\|_2^2 = \widetilde{\mathbb{E}}(X_m - X)^2 = \widetilde{\mathbb{E}} X^2 \mathbb{1}_{\{|X| > m\}} = \|X\|_2^2 - \widetilde{\mathbb{E}} X^2 \mathbb{1}_{\{|X| \leq m\}} < \infty.$$

Since $X^2 \mathbb{1}_{\{|X| \leq m\}} \uparrow X^2$ as $m \to \infty$, the Monotone Convergence Theorem 2.34 yields $\widetilde{\mathbb{E}} X^2 \mathbb{1}_{\{|X| \leq m\}} \uparrow \|X\|_2^2 < \infty$ and hence $\lim_{m \to \infty} \|X_m - X\|_2 = 0$.

Next, we define the random variable

$$X_{m,n}(s,\omega) := F_{s_k}(\omega)\mathbb{1}_{\{|F_{s_k}(\omega)|\le m\}}, \quad s \in (s_k, s_{k+1}],$$

where $0 = s_0 < \cdots < s_n = t$ is a segmentation of $[0,t]$ such that $\max_k(s_{k+1}-s_k) \to 0$ as $n \to \infty$. In other words, $X_{m,n}$ is a simple process of the form (7.2):

$$X_{m,n}(s,\omega) = X_m(0,\omega)\mathbb{1}_{\{0\}}(s) + \sum_{k=0}^{n-1} X_m(s_k,\omega)\mathbb{1}_{(s_k,s_{k+1}]}(s), \quad s \in [0,t].$$

By the same arguments as in Theorem 7.13, the condition $\max_k(s_{k+1} - s_k) \to 0$ implies that $X_{m,n}(s,\omega) \xrightarrow{\text{a.s.}} X_m(s,\omega)$ as $n \to \infty$ for $(s,\omega) \in [0,t] \times \Omega$. Since $|X_{m,n}| \le m$ for each $m \in \mathbb{N}$, the Bounded Convergence Theorem 2.36 yields

$$\lim_{n\to\infty} \|X_{m,n} - X_m\|_2 = 0.$$

For each m, let n_m be the smallest integer such that $\|X_{m,n_m} - X_m\|_2 < 1/m$ and define

$$F_s^{(m)}(\omega) := X_{m,n_m}(s,\omega), \quad m \in \mathbb{N}.$$

Then, $(F^{(m)}, m \in \mathbb{N})$ corresponds to a sequence of simple processes of the form (7.2) that converges almost surely to X as $m \to \infty$ and satisfies

$$\|F^{(m)} - X\|_2 \le \|X_m - X\|_2 + 1/m \to 0.$$

This completes the proof.   □

■ **Example 7.18 (Uniform Integrability and Canonical Approximations)** Theorem 7.16 asserts that there exists a sequence of simple processes that can approximate any $F \in \mathcal{P}_t$ in the norm on $\mathcal{P}_t$, but it does not tell us how to construct such a sequence explicitly. An explicit approximating sequence for $n \in \mathbb{N}$ is the canonical approximation (7.11), provided that $(F_s^2, s \in [0,t])$ is uniformly integrable, in the sense that

$$(7.19) \qquad\qquad \lim_{m\to\infty} \sup_{s\in[0,t]} \mathbb{E}F_s^2 \mathbb{1}_{\{F_s^2 > m\}} = 0.$$

For example, this condition is met when $(F_s, s \in [0,t]) = (W_s, s \in [0,t])$ is the Wiener process, because $\mathbb{E}W_s^2 = s \le t < \infty$ is bounded.

To prove that, under condition (7.19), the canonical approximation $F^{(n)}$ satisfies $\|F^{(n)} - F\|_{\mathcal{P}_t} \to 0$, we can modify the proof in Theorem 7.16 as follows. For each $n \ge 1$, define the random variable

$$Y_n(s,\omega) := F_0(\omega)\mathbb{1}_{\{0\}}(s) + \sum_{k=0}^{n-1} F_{s_k}(\omega)\mathbb{1}_{(s_k,s_{k+1}]}(s), \quad s \in [0,t].$$

Then, as in the proof in Theorem 7.16, we have that $Y_n(s, \omega) \overset{\text{a.s.}}{\to} F_s(\omega) =: Y(s, \omega)$ as $n \to \infty$ for $(s, \omega) \in [0, t] \times \Omega$. Using the triangle inequality, we have:

$$\|Y_n - Y\|_2$$
$$\leq \|Y - Y\mathbb{1}_{\{|Y| \leq m\}}\|_2 + \|Y\mathbb{1}_{\{|Y| \leq m\}} - Y_n\mathbb{1}_{\{|Y_n| \leq m\}}\|_2 + \|Y_n\mathbb{1}_{\{|Y_n| \leq m\}} - Y_n\|_2$$
$$\leq \|Y\mathbb{1}_{\{|Y| > m\}}\|_2 + \|Y\mathbb{1}_{\{|Y| \leq m\}} - Y_n\mathbb{1}_{\{|Y_n| \leq m\}}\|_2 + \sup_n \|Y_n\mathbb{1}_{\{|Y_n| > m\}}\|_2.$$

Since $(Y\mathbb{1}_{\{|Y| \leq m\}} - Y_n\mathbb{1}_{\{|Y_n| \leq m\}})^2 \leq 4m^2$, the Dominated Convergence Theorem 2.36 gives:

$$\lim_{n \to \infty} \|Y\mathbb{1}_{\{|Y| \leq m\}} - Y_n\mathbb{1}_{\{|Y_n| \leq m\}}\|_2 = 0.$$

Hence, taking a limit as $n \to \infty$, we obtain

$$\lim_n \|Y_n - Y\|_2 \leq \|Y\mathbb{1}_{\{|Y| > m\}}\|_2 + \sup_n \|Y_n\mathbb{1}_{\{|Y_n| > m\}}\|_2.$$

The first term $\|Y\mathbb{1}_{\{|Y| > m\}}\|_2$ vanishes by the Monotone Convergence Theorem 2.34 as $m \to \infty$. For the second term we have, by the isometry property:

$$\|Y_n\mathbb{1}_{\{|Y_n| > m\}}\|_2^2 = \sum_{k=0}^{n-1}(s_{k+1} - s_k)\mathbb{E}F_{s_k}^2\mathbb{1}_{\{|F_{s_k}| > m\}} \leq t \sup_{s \in [0,t]} \mathbb{E}F_s^2\mathbb{1}_{\{|F_s| > m\}}.$$

Therefore, $\sup_n \|Y_n\mathbb{1}_{\{|Y_n| > m\}}\|_2 \to 0$ as $m \to \infty$, and since $\|F^{(n)} - F\|_{\mathcal{P}_t} = t\|Y_n - Y\|_2$, it follows that $\|F^{(n)} - F\|_{\mathcal{P}_t} \to 0$, as desired. $\blacksquare$

Theorem 7.16 permits us to define the Itô integral for any process in the class $\mathcal{P}_t$ as the limit of a suitable sequence of integrals of simple processes. For this definition to make sense, we need to ensure that such a limit exists.

Theorem 7.20: Existence of the Itô Integral with Integrand $F \in \mathcal{P}_t$

Suppose that $F \in \mathcal{P}_t$ and $(F^{(n)})$ is a sequence of simple processes satisfying (7.17). Then, the following limit exists in L^2 norm:

$$\int_r^t F_s\, dW_s := \lim_{n \to \infty} \int_r^t F_s^{(n)}\, dW_s, \quad r \in [0, t],$$

and it defines the *Itô integral* of F with respect to W on $[r, t]$.

Proof. Since $\int_r^t F_s^{(n)}\, dW_s = \int_0^t F_s^{(n)}\, dW_s - \int_0^r F_s^{(n)}\, dW_s$, it suffices to consider the existence of the limit of the integral $J^{(n)} := \int_0^r F_s^{(n)}\, dW_s$. Consider the Hilbert space $L^2(\Omega, \mathcal{H}, \mathbb{P})$, equipped with the L^2 norm: $\|X\|_2 = \sqrt{\mathbb{E}X^2}$. We will show that $(J^{(n)})$

is a Cauchy sequence in L^2, so that, by the completeness of L^2, there exists a random variable $J \in L^2$ such that $\mathbb{E}(J^{(n)} - J)^2 \to 0$. Thus, setting $\int_0^r F_s \, dW_s := J$, justifies the definition of the Itô integral.

That $(J^{(n)})$ is a Cauchy sequence in L^2 follows from the following bound:

$$\|J^{(n)} - J^{(m)}\|_2 = \sqrt{\mathbb{E}\left(\int_0^r (F_s^{(n)} - F_s^{(m)}) \, dW_s\right)^2}$$

$$= \|F^{(n)} - F^{(m)}\|_{\mathcal{P}_r} \leq \|F^{(m)} - F\|_{\mathcal{P}_r} + \|F^{(n)} - F\|_{\mathcal{P}_r},$$

where in the last line we used the linearity and isometry for the simple process $F^{(n)} - F^{(m)}$, as well as the triangle inequality applied to the norm $\|\cdot\|_{\mathcal{P}_r}$.

Finally, from (7.17) and $\|F\|_{\mathcal{P}_r} \leq \|F\|_{\mathcal{P}_t}$ we have that as $m, n \to \infty$:

$$\|J^{(n)} - J^{(m)}\|_2 \leq \|F^{(m)} - F\|_{\mathcal{P}_t} + \|F^{(n)} - F\|_{\mathcal{P}_t} \to 0.$$

Hence, $(J^{(n)})$ is a Cauchy sequence in L^2, guaranteeing that the Itô integral $\int_0^r F_s \, dW_s$ is well-defined as the limit J of $(J^{(n)})$ in L^2. $\qquad\square$

■ **Example 7.21 (Itô Integral as a Riemann Sum)** Recall that if the process F^2 is uniformly integrable, see (7.19), then an explicit approximating sequence $(F^{(n)}, n \in \mathbb{N})$ of F is (7.11). For such uniformly integrable F^2, the Itô integral over $[0, t]$ can be defined as the limit in L^2 norm of a Riemann sum:

$$\int_0^t F_s \, dW_s := \lim_{n \to \infty} \sum_{k=0}^{n-1} F_{s_k} (W_{s_{k+1}} - W_{s_k}).$$

For example, since $\mathbb{E}W_t^2 = t < \infty$, the Wiener process satisfies (7.19) and we can define $\int_0^t W_s \, dW_s = \lim_{n \to \infty} \sum_{k=0}^{n-1} W_{s_k} (W_{s_{k+1}} - W_{s_k}) = (W_t^2 - t)/2$.

To see the latter equality, consider the telescoping sum $\sum_{k=0}^{n-1} (W_{s_{k+1}} + W_{s_k})(W_{s_{k+1}} - W_{s_k}) = \sum_{k=0}^{n-1} (W_{s_{k+1}}^2 - W_{s_k}^2) = W_t^2$, and note that by Theorem 6.70 the quadratic variation of the Wiener process over $[0, t]$ is t; that is, $\langle W^{(n)} \rangle_t = \sum_{k=0}^{n-1} (W_{s_{k+1}} - W_{s_k})^2 \xrightarrow{L^2} t$. Hence,

$$\frac{W_t^2 - t}{2} = \lim_{n \to \infty} \sum_{k=0}^{n-1} \left[\frac{(W_{s_{k+1}}^2 - W_{s_k}^2)}{2} - \frac{(W_{s_{k+1}} - W_{s_k})^2}{2} \right]$$

$$= \lim_{n \to \infty} \sum_{k=0}^{n-1} W_{s_k} (W_{s_{k+1}} - W_{s_k}).$$

Figure 7.22 shows a realization of the process $(\int_0^t W_s \, dW_s, t \in [0, 1])$ together with the corresponding Wiener process $W := (W_s, s \in [0, 1])$ — the Wiener path from Figure 7.12. Observe that here the Riemann sum $\sum_{k=0}^{n-1} W_{s_k} (W_{s_{k+1}} - W_{s_k})$ not only

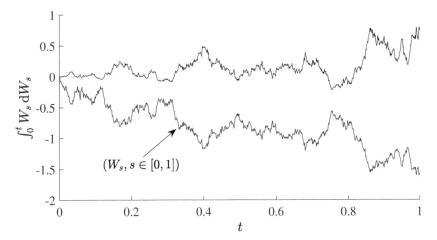

Figure 7.22: The Itô integral process $(\int_0^t W_s \, dW_s, t \in [0, 1])$ and its corresponding Wiener process W.

converges in L^2, but also almost surely, because $\langle W^{(n)} \rangle_t \xrightarrow{\text{a.s.}} t$ by Theorem 6.70. Similar calculations also show (see Exercise 1) that $\sum_{k=0}^{n-1} W_{s_{k+1}}(W_{s_{k+1}} - W_{s_k}) \xrightarrow{L^2} (W_t^2 + t)/2$. However, the limit of this latter Riemann sum does not define an Itô integral, because the simple process $\sum_{k=0}^{n-1} W_{s_{k+1}} \mathbb{1}_{[s_k, s_{k+1})}$ is not $\mathcal{F}$-adapted ($W_{s_{k+1}} \notin \mathcal{F}_{s_k}$). ∎

Next, we provide the proof of Theorem 7.6 for the more general class of integrands in $\mathcal{P}_t$.

Proof of Theorem 7.6 for Processes in $\mathcal{P}_t$. That the Itô integral for any $F \in \mathcal{P}_t$ has the same four properties as for simple processes follows by taking limits. In particular, let $(F^{(n)})$ and $(G^{(n)})$ be the approximating sequences of simple processes for F and G, respectively. Then, $H^{(n)} := \alpha F^{(n)} + \beta G^{(n)}$ is a simple process approximating $H := \alpha F + \beta G$, in the sense that

$$\|H^{(n)} - H\|_{\mathcal{P}_t} \le \alpha \|F^{(n)} - F\|_{\mathcal{P}_t} + \beta \|G^{(n)} - G\|_{\mathcal{P}_t} \to 0.$$

Hence, the limit of $\int_0^r H_s^{(n)} \, dW_s$ exists in L^2. By the linearity property of the integral for simple functions, we have:

$$\int_0^r H_s^{(n)} \, dW_s = \alpha \int_0^r F_s^{(n)} \, dW_s + \beta \int_0^r G_s^{(n)} \, dW_s.$$

By taking limits on both sides, we find $\int_0^r H_s \, dW_s = \alpha \int_0^r F_s \, dW_s + \beta \int_0^r F_s \, dW_s$, showing the linearity property.

Next, let $J^{(n)} := \int_0^r F_s^{(n)} \, dW_s$ and denote its limit in L^2 by $J := \int_0^r F_s \, dW_s$. Recall that $J^{(n)}$ satisfies the isometry $\|J^{(n)}\|_2 = \|F^{(n)}\|_{\mathcal{P}_r}$. By the monotonicity property in Theorem 2.47, $|\mathbb{E}J| = |\mathbb{E}J - \mathbb{E}J^{(n)}| \le \mathbb{E}|J - J^{(n)}| \le \|J - J^{(n)}\|_2 \to 0$, proving the zero-mean property. Further, by the triangle inequality, we have $|\,\|J^{(n)}\|_2 - \|J\|_2\,| \leq \|J^{(n)} - J\|_2 \to 0$ and $|\,\|F^{(n)}\|_{\mathcal{P}_r} - \|F\|_{\mathcal{P}_r}\,| \leq \|F^{(n)} - F\|_{\mathcal{P}_r} \to 0$. Hence,

$$\|J\|_2 = \lim_{n \to \infty} \|J^{(n)}\|_2 = \lim_{n \to \infty} \|F^{(n)}\|_{\mathcal{P}_r} = \|F\|_{\mathcal{P}_r},$$

proving the isometry for the special case where $G = F$. The more general isometry property follows by applying the linearity and isometry (in the special case of $G = F$) to the right-hand side of the *polarization identity*:

$$\mathbb{E}\left[\int_0^r F_s\,dW_s \times \int_0^r G_s\,dW_s\right] = \frac{\mathbb{E}\left[\int_0^r (F_s + G_s)\,dW_s\right]^2 - \mathbb{E}\left[\int_0^r (F_s - G_s)\,dW_s\right]^2}{4}$$

$$= \frac{\|F + G\|_{\mathcal{P}_r}^2 - \|F - G\|_{\mathcal{P}_r}^2}{4} = \langle F, G\rangle_{\mathcal{P}_r}.$$

Finally, for the martingale property, we show that $\mathbb{E}_r \int_r^t F_s\,dW_s = 0$, where

$$\int_r^t F_s\,dW_s := \lim_{n \to \infty} \int_r^t F_s^{(n)}\,dW_s,$$

and $\int_r^t F_s^{(n)}\,dW_s$ is the integral of a simple process and hence of the form (7.8), but with r assumed to be a point in the segmentation of $[0, t]$; that is, $r \in \Pi_n$. Define $\widetilde{W}_s := W_{r+s} - W_r$, $\widetilde{F}_s^{(n)} := F_{s+r}^{(n)}$, and $\widetilde{\mathcal{F}} := (\mathcal{F}_{s+r}, s \geq 0)$. We know that $(\widetilde{W}_s, s \geq 0)$ is a Wiener process (see Theorem 6.11) and that by construction $\widetilde{F}^{(n)}$ is $\widetilde{\mathcal{F}}$-adapted. Hence, $\int_0^{t-r} \widetilde{F}_s^{(n)}\,d\widetilde{W}_s$ converges to a limit in L^2 norm, which is the Itô integral $\int_0^{t-r} \widetilde{F}_s\,d\widetilde{W}_s$ with 0 mean. From the construction, $\int_0^{t-r} \widetilde{F}_s^{(n)}\,d\widetilde{W}_s = \int_r^t F_s^{(n)}\,dW_s$, and so we can conclude that

$$\int_0^{t-r} F_{s+r}\,d\widetilde{W}_s = \int_r^t F_s\,dW_s.$$

Hence, $\mathbb{E}_r \int_r^t F_s\,dW_s = \mathbb{E}_0 \int_0^{t-r} \widetilde{F}_s\,d\widetilde{W}_s = 0$, proving the martingale property. The square-integrability follows from the isometry:

$$\mathbb{E}_r\left(\int_r^t F_s\,dW_s\right)^2 = \mathbb{E}_0\left(\int_0^{t-r} \widetilde{F}_s\,d\widetilde{W}_s\right)^2$$

$$= \mathbb{E}_0 \int_0^{t-r} F_{s+r}^2\,ds = \mathbb{E}_r \int_r^t F_s^2\,ds < \|F\|_{\mathcal{P}_t}^2 < \infty.$$

$\square$

If we view the Itô integral $J_{[0,t]}(F) := \int_0^t F_s\,dW_s$ of $F \in \mathcal{P}_t$ on $[0, t]$ as a linear mapping $J_{[0,t]} : [0, t] \times \Omega \to \Omega$ from the Hilbert space $L^2([0, t] \times \Omega, \mathcal{F}^p, \mathrm{Leb}_{[0,t]} \otimes \mathbb{P})$ to the Hilbert space $L^2(\Omega, \mathcal{H}, \mathbb{P})$, then the isometry property in Theorem 7.6 shows that this mapping is a *linear isometry*; see Example B.3.

■ **Example 7.23 (Quadratic Martingale Process)** Let $J_{[r,t]} := \int_r^t F_s\, dW_s$ be the Itô integral of F on $[r,t]$ and define

$$Z_r := J_{[0,r]}^2 - \int_0^r F_s^2\, ds.$$

Then, the process $Z := (Z_r, r \in [0,t])$ is a martingale. Indeed, for any $r \in [0,t]$, the martingale property in Theorem 7.6 yields:

$$\begin{aligned}
\mathbb{E}_r[J_{[0,t]}^2 - J_{[0,r]}^2] &= \mathbb{E}_r(J_{[0,t]} - J_{[0,r]})(J_{[0,t]} + J_{[0,r]}) \\
&= \mathbb{E}_r J_{[r,t]}(J_{[r,t]} + 2J_{[0,r]}) \\
&= \mathbb{E}_r J_{[r,t]}^2 + 2J_{[0,r]}\mathbb{E}_r J_{[r,t]} \\
&= \mathbb{E}_r \int_r^t F_s^2\, ds + 0.
\end{aligned}$$

Finally, by the isometry and triangle inequality $\mathbb{E}|Z_t| \le 2\|F\|_{\mathcal{P}_t} < \infty$, and so Z is a martingale.   ■

■ **Example 7.24 (Stopping Times and Isometry)** Suppose that we are given two bounded stopping times $R \le T \le t$ adapted to the filtration $(\mathcal{F}_s, s \in [0,t])$. Then, an extension of the isometry property in Theorem 7.6 is the following:

$$\mathbb{E}_R\left[\int_R^T F_s\, dW_s \times \int_R^T G_s\, dW_s\right] = \mathbb{E}_R \int_R^T F_s G_s\, ds.$$

To prove this, we again first establish the identity for the special case of $G = F$. Using the notation in Example 7.23, recall that both $(J_{[0,r]}, r \in [0,t])$ and $(Z_r, r \in [0,t])$ are martingales, so that by Theorem 5.83 we have $\mathbb{E}_R[J_{[0,T]} - J_{[0,R]}] = 0$ and $\mathbb{E}_R[Z_T - Z_R] = 0$. In other words, we have $\mathbb{E}_R J_{[R,T]} = 0$ and $\mathbb{E}_R[J_{[0,T]}^2 - J_{[0,R]}^2] = \mathbb{E}_R \int_R^T F_s^2\, ds$. Hence, we obtain

$$\begin{aligned}
\mathbb{E}_R J_{[R,T]}^2 &= \mathbb{E}_R(J_{[0,T]} - J_{[0,R]})(J_{[0,T]} + J_{[0,R]} - 2J_{[0,R]}) \\
&= \mathbb{E}_R[J_{[0,T]}^2 - J_{[0,R]}^2] - 2J_{[0,R]}\mathbb{E}_R J_{[R,T]} \\
&= \mathbb{E}_R \int_R^T F_s^2\, ds - 0.
\end{aligned}$$

The general isometry identity follows by applying the linearity and isometry (in the case of $G \equiv F$) to the right-hand side of the polarization identity:

$$\mathbb{E}_R\left[\int_R^T F_s\, dW_s \times \int_R^T G_s\, dW_s\right] = \frac{\mathbb{E}_R\left[\int_R^T (F_s + G_s)\, dW_s\right]^2 - \mathbb{E}_R\left[\int_R^T (F_s - G_s)\, dW_s\right]^2}{4}.$$

■

We next generalize the results in Example 7.9 to the wider class of left-continuous deterministic integrands.

> ### Proposition 7.25: Deterministic Integrands and Gaussian Processes
>
> Let $f : \mathbb{R}_+ \times \mathbb{R}_+ \to \mathbb{R}$ be left-continuous in the first argument and such that $\|f(\cdot, t)\|_{\mathcal{P}_t}^2 = \int_0^t f^2(s, t)\, ds < \infty$. Then, the stochastic integral process
>
> $$Z_t := \int_0^t f(s, t)\, dW_s, \quad t \ge 0,$$
>
> is a zero-mean Gaussian process with covariance function
>
> $$\mathbb{C}\text{ov}(Z_{t_1}, Z_{t_2}) = \int_0^{t_1 \wedge t_2} f(s, t_1) f(s, t_2)\, ds, \quad t_1, t_2 \ge 0.$$

Proof. Take a sequence of segmentations of $[0, t]$ whose mesh is going to 0 as $n \to \infty$. Let $f^{(n)}(\cdot, r)$ be the corresponding canonical approximation of $f(\cdot, r)$ for $r \in [0, t]$, and define

$$Z_r^{(n)} := \int_0^r f^{(n)}(s, r)\, dW_s = \sum_{k=0}^{n-1} f(s_k, r)(W_{s_{k+1} \wedge r} - W_{s_k \wedge r}).$$

This is a zero-mean Gaussian random variable, with variance

$$\sum_{k=0}^{n-1} f^2(s_k, r)(s_{k+1} \wedge r - s_k \wedge r) = \int_0^r [f^{(n)}(s, r)]^2\, ds = \|f^{(n)}(\cdot, r)\|_{\mathcal{P}_r}^2.$$

Since $\big|\|f^{(n)}(\cdot, r)\|_{\mathcal{P}_r} - \|f(\cdot, r)\|_{\mathcal{P}_r}\big| \le \|f^{(n)}(\cdot, r) - f(\cdot, r)\|_{\mathcal{P}_r} \to 0$, we have that

$$\mathbb{V}\text{ar}\, Z_r^{(n)} = \|f^{(n)}(\cdot, r)\|_{\mathcal{P}_r}^2 \to \|f(\cdot, r)\|_{\mathcal{P}_r}^2 = \int_0^r f^2(s, r)\, ds.$$

Moreover, $Z_r^{(n)} \xrightarrow{d} Z_r$. By Theorem 3.24, the characteristic function of Z_r is given by the limit of the characteristic function of $Z_r^{(n)}$ as $n \to \infty$. Therefore, $Z_r \sim \mathcal{N}(0, \|f(\cdot, r)\|_{\mathcal{P}_r}^2)$. More generally, for any choice $t_1, \ldots, t_m \in [0, t]$ and real numbers $\alpha_1, \ldots, \alpha_m$, the linear combination

$$\sum_{i=1}^m \alpha_i Z_{t_i}^{(n)} = \sum_{i=1}^m \sum_{k=0}^{n-1} \alpha_i f(s_k, t_i)(W_{s_{k+1} \wedge t_i} - W_{s_k \wedge t_i})$$

is a Gaussian random variable, showing that $(Z_s^{(n)}, s \in [0, t])$ is a zero-mean Gaussian process for any n and, by taking the limit as $n \to \infty$, that $(Z_s, s \in [0, t])$ is a zero-mean Gaussian process as well. As t is arbitrary, the process $(Z_t, t \geq 0)$ is Gaussian. The covariance formula for $t_1 \leq t_2$ follows from

$$\mathbb{E}Z_{t_1}^{(n)}Z_{t_2}^{(n)} = \frac{\mathbb{V}\mathrm{ar}\left(Z_{t_1}^{(n)} + Z_{t_2}^{(n)}\right) - \mathbb{V}\mathrm{ar}\,Z_{t_1}^{(n)} - \mathbb{V}\mathrm{ar}\,Z_{t_2}^{(n)}}{2}$$

$$\rightarrow \frac{\mathbb{E}(Z_{t_1} + Z_{t_2})^2 - \mathbb{E}Z_{t_1}^2 - \mathbb{E}Z_{t_2}^2}{2} = \mathbb{E}Z_{t_1}Z_{t_2}$$

and an application of the isometry property in Theorem 7.6 gives:

$$\mathbb{E}Z_{t_1}Z_{t_2} = \mathbb{E}\left[\int_0^{t_1} f(s,t_1)\,\mathrm{d}W_s \times \int_{t_1}^{t_2} f(s,t_2)\,\mathrm{d}W_s\right]$$

$$+ \mathbb{E}\left[\int_0^{t_1} f(s,t_1)\,\mathrm{d}W_s \times \int_0^{t_1} f(s,t_2)\,\mathrm{d}W_s\right]$$

$$= 0 + \int_0^{t_1} f(s,t_1)f(s,t_2)\,\mathrm{d}s.$$

$\square$

■ **Example 7.26 (Time-changed Wiener Process)** Suppose that in Proposition 7.25 there is no dependence of f on t; that is, $f(s,t) = f_s$. Then, for $t_2 > t_1$:

$$\mathbb{C}\mathrm{ov}(Z_{t_1}, Z_{t_2} - Z_{t_1}) = \mathbb{C}\mathrm{ov}(Z_{t_1}, Z_{t_2}) - \mathbb{V}\mathrm{ar}\,Z_{t_1} = \int_0^{t_1 \wedge t_2} f_s^2\,\mathrm{d}s - \int_0^{t_1} f_s^2\,\mathrm{d}s = 0,$$

and therefore the Gaussian process $Z := (\int_0^t f_s\,\mathrm{d}W_s, t \geq 0)$ has *independent increments*, just like in Example 7.9. The variance of Z_t is in this case the continuous increasing function

$$(7.27) \qquad C(t) := \mathbb{V}\mathrm{ar}\,Z_t = \|f\|_{\mathcal{P}_t} = \int_0^t f_s^2\,\mathrm{d}s, \quad t \geq 0.$$

This increasing function can thus serve as a "clock" with which to measure time: showing the clock time $C(t)$ when t units of natural time have passed. In Exercise 2 we prove that the process $(W_{C(t)}, t \geq 0)$, which can be viewed as a *time change* of the Wiener process W, has the same statistical properties as Z. In other words, both Z and $(W_{C(t)}, t \geq 0)$ are Gaussian processes with the same mean and covariance functions. ■

We finish with a result that generalizes the quadratic variation of the Wiener process in Section 6.7 and that will be used many times in Section 7.2.

Lemma 7.28: Quadratic Variation Integral

Suppose that $F \in \mathcal{P}_t$ satisfies the condition (7.19). Then, as $\mathrm{mesh}(\Pi_n) \to 0$:

$$\sum_{k=0}^{n-1} F_{s_k}(W_{s_{k+1}} - W_{s_k})^2 \xrightarrow{L^2} \int_0^t F_s \, \mathrm{d}s.$$

Proof. Since $F \in \mathcal{P}_t$ and F^2 satisfies the uniform integrability condition, we can use the canonical approximation $F^{(n)}$ of the form (7.11). Set $\tilde{J}_n := \sum_{k=0}^{n-1} F_{s_k} \Delta s_k = \int_0^t F_s^{(n)} \, \mathrm{d}s$, and note that it converges almost surely and in mean squared error to the Lebesgue integral $J := \int_0^t F_s \, \mathrm{d}s$, defined in (5.17). In addition, if $J_n := \sum_{k=0}^{n-1} F_{s_k}(\Delta W_{s_k})^2$, then $\mathbb{E}[J_n - \tilde{J}_n] = \sum_{k=0}^{n-1} \mathbb{E}[F_{s_k} \mathbb{E}_{s_k}[(\Delta W_{s_k})^2 - \Delta s_k]] = 0$, and hence

$$\|J_n - \tilde{J}_n\|^2 = \sum_{k=0}^{n-1} \mathbb{E}[\underbrace{F_{s_k}^2 \mathbb{E}_{s_k}[(\Delta W_{s_k})^2 - \Delta s_k]^2}_{= 2\Delta s_k^2}] \leq 2 \underbrace{\max_k \Delta s_k}_{\to 0} \underbrace{\sum_{k=0}^{n-1} \mathbb{E} F_{s_k}^2 \Delta s_k}_{\to \|F\|_{\mathcal{P}_t}^2 < \infty}.$$

Therefore, by the triangle inequality: $\|J_n - J\| \leq \|J_n - \tilde{J}_n\| + \|\tilde{J}_n - J\| \to 0$, completing the proof. $\qquad\square$

7.1.3 Further Extensions of the Itô Integral

Recall that the Wiener process has continuous sample paths; see Definition 6.1 and Theorem 6.21. As a result of this, whenever the integrand F is a piecewise constant process of the form (7.2), the corresponding Itô integral in Definition 7.5 has continuous sample paths. A natural question then arises as to whether this continuity holds for a more general process $F \in \mathcal{P}_t$. While Theorem 7.20 ensures the existence of a unique L^2-norm limit when $F \in \mathcal{P}_t$, there is no guarantee that the Itô integral defined as this L^2-norm limit has almost surely continuous paths.

Fortunately, Theorem 7.29 below asserts that the Itô integral as defined by an L^2-norm limit in Theorem 7.20 has a continuous modification. Recall that a stochastic process $\tilde{Z} := (\tilde{Z}_r, r \in [0,t])$ is called a modification of $Z := (Z_r, r \in [0,t])$ if $\mathbb{P}(Z_r = \tilde{Z}_r) = 1$ for all $r \in [0,t]$.

Theorem 7.29: Continuous Modification of the Itô Integral

Suppose that $F \in \mathcal{P}_t$ and $(F^{(n)})$ is a sequence of simple processes satisfying (7.17). Denote the limit $\int_0^r F_s^{(n)} \, dW_s =: Z_r^{(n)} \xrightarrow{L^2} Z_r$ as the Itô integral on $[0, r]$ for $r \leq t$. Then, there exists a modification $\widetilde{Z}$ of $(Z_r, r \in [0, t])$ such that the mapping $r \mapsto \widetilde{Z}_r$ is almost surely continuous. Moreover, this modification is unique up to indistinguishability.

Proof. In what follows, we use similar arguments to the ones employed in the proof of the existence of the Wiener process; see Theorem 6.21. The condition (7.17) implies the existence of a subsequence $(F^{(n_k)})$ of $(F^{(n)})$ such that $(n_k, k \in \mathbb{N})$ is strictly increasing and $\|F^{(n_{k+1})} - F^{(n_k)}\|_{\mathcal{P}_t} \leq 2^{-2k}$. Define the processes $\widetilde{Z}^{(k)} := (Z_r^{(n_k)}, r \in [0, t])$ and the sequence of events with $\varepsilon_k := 2^{-k}$:

$$ H_k := \left\{ \sup_{0 \leq r \leq t} |\widetilde{Z}_r^{(k+1)} - \widetilde{Z}_r^{(k)}| \geq \varepsilon_k \right\}, \quad k = 1, 2, \ldots. $$

Recall from Theorem 7.6 that $Z^{(n)} := (Z_r^{(n)}, r \in [0, t])$ is an L^2-martingale with respect to $\mathcal{F}$ and, by construction, $Z^{(n)}$ has continuous sample paths, because $r \mapsto F_r^{(n)}(\omega)$ is piecewise constant. It follows that the process $|\widetilde{Z}^{(k+1)} - \widetilde{Z}^{(k)}|$ is a positive continuous $\mathcal{F}$-submartingale, and hence by Doob's maximal inequality (Proposition 5.86), we have

$$ \mathbb{P}(H_k) \leq \varepsilon_k^{-2} \mathbb{E}|\widetilde{Z}_t^{(k+1)} - \widetilde{Z}_t^{(k)}|^2 = 2^{2k} \|F^{(n_{k+1})} - F^{(n_k)}\|_{\mathcal{P}_t}^2 \leq 2^{-2k}, $$

where we used the isometry property in the last equality. Since $\sum_k \mathbb{P}(H_k) < \infty$, the Borel–Cantelli Lemma 3.14 implies that the event $\Omega_0 := \{\sum_k \mathbb{1}_{H_k} < \infty\}$ occurs with probability 1. From Proposition 3.2 the sequence $(\widetilde{Z}^{(k)}(\omega))$, $\omega \in \Omega_0$ is Cauchy convergent in the uniform norm and its limit $\widetilde{Z}$ defines the Itô integral for all $\omega \in \Omega_0$ and every $r \in [0, t]$. Since $\widetilde{Z}$ (the Itô integral) is the almost sure limit (in the uniform norm) of the continuous process $\widetilde{Z}^{(k)}$, its sample paths are almost surely continuous on $[0, t]$ and, in fact, uniformly continuous, because $[0, t]$ is a closed and bounded set; see Example B.13. Finally, since both $Z_r^{(n_k)} \xrightarrow{L^2} \widetilde{Z}_r$ and $Z_r^{(k)} \xrightarrow{L^2} Z_r$ converge to the same L^2-norm limit for each $r \in [0, t]$ (that is, $\mathbb{E}(Z_r - \widetilde{Z}_r)^2 = 0$), we conclude that $\mathbb{P}(Z_r = \widetilde{Z}_r) = 1$ for each $r \in [0, t]$. This shows that $\widetilde{Z}$ is a continuous modification of Z. If $\widetilde{Z}^*$ is another continuous modification of Z, then part 3 of Theorem 7.13 implies that $\widetilde{Z}^*$ is indistinguishable from $\widetilde{Z}$. $\square$

For the rest of this chapter we will assume that we are working with the continuous modification of every Itô integral, which, as the proof of Theorem 7.29 demonstrates, is defined as an almost sure limit. As a result of this, henceforth all equations and identities involving the Itô integrals are assumed to hold not only in L^2 norm, but also almost surely.

■ **Example 7.30 (Time-changed Wiener Process Continued)** In Example 7.26, we can go one step further and construct a time-changed Wiener process that is almost surely identical to the continuous modification of the process $Z = (\int_0^t f_s \, dW_s, t \geq 0)$. Since the time-change clock C in (7.27) is a continuous and increasing function, it has an inverse C^{-1} such that $C(C^{-1}(t)) = t$. Define the process $\widetilde{W}$ as

(7.31) $$\widetilde{W}_t := Z_{C^{-1}(t)}, \quad t \geq 0.$$

Then, $\widetilde{W}_0 = Z_0 = 0$ and the continuity of C^{-1} and Z implies that $\widetilde{W}$ has almost surely continuous paths. Moreover, for any $t_1 < t_2 \leq t_3 < t_4$, the increments $\widetilde{W}_{t_4} - \widetilde{W}_{t_3} =$ and $\widetilde{W}_{t_2} - \widetilde{W}_{t_1}$ are independent, because Z has independent increments. Finally,

$$\widetilde{W}_{t_2} - \widetilde{W}_{t_1} = Z_{C^{-1}(t_2)} - Z_{C^{-1}(t_1)} \sim \mathcal{N}(0, t_2 - t_1).$$

Hence, $\widetilde{W}$ is a Wiener process by Theorem 6.2. Note that $\widetilde{W}$ is not the same Wiener process as W, but when time-changed by the clock C, it is (almost surely) identical to Z:

$$\widetilde{W}_{C(t)} = Z_t, \quad t \geq 0.$$

■

Let $Z_t := \int_0^t F_s \, dW_s$ for $t \geq 0$. For a positive stopping time $T \leq t$, we can define $\int_0^T F_s \, dW_s$ as the random variable Z_T. An important consequence of working with continuous modifications of Itô integrals is that the following intuitive relationship holds:

Proposition 7.32: Continuous Modification and Stopping Times

If $F \in \mathcal{P}_t$ and $T \leq t$ is a stopping time with respect to $\mathcal{F}$, then almost surely:

$$\int_0^T F_s \, dW_s = \int_0^t \mathbb{1}_{[0,T]}(s) \, F_s \, dW_s.$$

Proof. Consider a segmentation Π_n of $[0, t]$ such that $\max_{k \leq n}(s_{k+1} - s_k) \to 0$ and

$$T_n := \sum_{k=0}^{n-1} s_{k+1} \mathbb{1}_{[s_k, s_{k+1})}(T)$$

is an approximation of T from above; that is, $T_n > T$ for each n and $T_n \xrightarrow{a.s.} T$ as $n \to \infty$. We now write for every $t \geq T$:

(7.33) $$\int_0^T F_s \, dW_s - \int_0^t \mathbb{1}_{[0,T]}(s) \, F_s \, dW_s =: J_1^{(n)} + J_2^{(n)} + J_3^{(n)},$$

where we have defined the quantities:

$$J_1^{(n)} := \int_0^T F_s \, dW_s - \int_0^{T_n} F_s \, dW_s,$$

$$J_2^{(n)} := \int_0^{T_n} F_s \, dW_s - \int_0^t \mathbb{1}_{\{s \le T_n\}} F_s \, dW_s,$$

$$J_3^{(n)} := \int_0^t (\mathbb{1}_{\{s \le T_n\}} - \mathbb{1}_{\{s \le T\}}) F_s \, dW_s.$$

Since we are using the continuous modification of the Itô integral, we have $J_1^{(n)} \overset{\text{a.s.}}{\to} 0$. The second quantity, $J_2^{(n)}$, is 0 almost surely for all $n \ge 1$, because $\sum_{k=1}^n \mathbb{1}_{\{T_n = s_k\}} = 1$ implies that:

$$\int_0^{T_n} F_s \, dW_s = \sum_{k=1}^n \mathbb{1}_{\{T_n = s_k\}} \int_0^{s_k} F_s \, dW_s$$

$$= \int_0^t \sum_{k=1}^n \mathbb{1}_{\{T_n = s_k\}} \mathbb{1}_{\{s \le s_k\}} F_s \, dW_s = \int_0^t \mathbb{1}_{\{s \le T_n\}} F_s \, dW_s.$$

For $J_3^{(n)}$, define $D^{(n)} := (F_s \mathbb{1}_{\{T < s \le T_n\}}, s \in [0, t])$, which is left-continuous, $\mathcal{F}$-adapted, and such that $D_s^{(n)} \overset{\text{a.s.}}{\to} 0$ for all $s \in [0, t]$. Since $D_s^{(n)} \le F_s$ and $\|F\|_{\mathcal{P}_t} < \infty$, the Dominated Convergence Theorem 2.36 implies that $\|D^{(n)}\|_{\mathcal{P}_t} \to 0$. Therefore, by the isometry property, $J_3^{(n)} = \int_0^t D_s^{(n)} \, dW_s \overset{L^2}{\to} 0$. Since the left-hand side of (7.33) does not depend on n, we conclude that it must be almost surely 0; see Exercise 3.16.   □

In the general theory of stochastic integration, the aim is to define integrators and integrands in as wide a sense as possible. For example, in Section 7.2 we show how integration with respect to the Wiener process can be extended to integration with respect to Itô processes; see Definition 7.39. We mention that the most general integrators in stochastic integration are the *semimartingale* processes, which are defined as the sum of a local martingale and a process of locally bounded variation; see Definition 5.74.

As indicated above, we can not only extend the type of integrators used in stochastic integration, but also the type of integrands. There are two possible extensions of the class of integrands $\mathcal{P}_t$ in Definition 7.14. The first extension, which we do not pursue in this book, is to relax the requirement that F is *left-continuous* with the less stringent requirement that F is *progressively measurable*. The second extension, which is our focus for the rest of this section, is to relax the L^2-norm condition $\|F\|_{\mathcal{P}_t} < \infty$ to the less restrictive condition:

$$(7.34) \qquad \mathbb{P}\left(\int_0^t F_s^2 \, ds < \infty \right) = 1, \quad \forall t \ge 0.$$

This relaxation is accomplished using a *localization* argument — rather than defining a property of a stochastic process on $[0, t]$ (or $\mathbb{R}$), we instead define it locally on a (random) interval $[0, T_n]$, where (T_n) is a sequence of increasing bounded stopping times called a *localizing sequence*.

Theorem 7.35: Enlarging the Class $\mathcal{P}_t$ via Localization

Let $F := (F_s, s \in [0, t])$ be an $\mathcal{F}$-adapted process, left-continuous on $(0, t]$, and such that (7.34) holds. Then, there exists a localizing sequence (T_n) such that the stopped process $(\int_0^{r \wedge T_n} F_s \, dW_s, r \in [0, t])$ is a square-integrable martingale. In addition, the Itô integral of F with respect to W on $[0, t]$ is defined as the almost sure limit:

$$\int_0^t F_s \, dW_s := \lim_{n \to \infty} \int_0^{t \wedge T_n} F_s \, dW_s.$$

Proof. First, let $T_n := n \wedge \inf \left\{ r : \int_0^r F_s^2 \, ds \geq n \right\}$ be a stopping time for each $n = 1, 2, \ldots$. Then, the condition (7.34) implies that (T_n) is an increasing sequence of stopping times such that $T_n \xrightarrow{\text{a.s.}} \infty$.

Second, consider the truncated process $F^{(n)}$, where $F_s^{(n)} := \mathbb{1}_{[0, T_n]}(s) F_s$ for all s. Since $\|F^{(n)}\|_{\mathcal{P}_t} \leq n$ (that is, $F^{(n)} \in \mathcal{P}_t$), the continuous modification of the Itô integral $\int_0^t F_s^{(n)} \, dW_s$ is well-defined by Theorem 7.29.

Third, since $T_n \xrightarrow{\text{a.s.}} \infty$, there exists an almost surely finite N such that $t \wedge T_n = t$ for $n \geq N$, and from Proposition 7.32 we have for any $m \geq n$ and $t \in [0, T_n]$ that:

$$\int_0^t F_s^{(m)} \, dW_s = \int_0^{t \wedge T_m} F_s \, dW_s = \int_0^{t \wedge T_n} F_s \, dW_s = \int_0^t F_s^{(n)} \, dW_s.$$

The last consistency relation implies that $\lim_n \int_0^{t \wedge T_n} F_s \, dW_s = \int_0^t F_s^{(m)} \, dW_s$ for each m and all $t \in [0, T_m]$. In addition, if (τ_n) is another strictly increasing sequence of stopping times such that $\int_0^{t \wedge \tau_n} F_s \, dW_s$ is well-defined as an L^2-norm limit, then almost surely $\lim_n \int_0^{t \wedge T_n} F_s \, dW_s = \lim_n \int_0^{t \wedge \tau_n} F_s \, dW_s$; that is, the Itô integral definition is independent of the choice of localizing sequence. $\square$

The Itô integral process $Z := (\int_0^t F_s \, dW_s, t \geq 0)$ defined in Theorem 7.35 is a continuous local martingale, because there exists a localizing sequence of stopping times (T_n) such that $T_n \xrightarrow{\text{a.s.}} \infty$. This extended definition of the Itô integral does not necessarily satisfy the isometry property in Theorem 7.6, as illustrated in the following example:

■ **Example 7.36 (Enlarging the Class $\mathcal{P}_t$)** Although $W^n \in \mathcal{P}_t$ for any $n \in \mathbb{N}$, the process $\exp(W^2) \notin \mathcal{P}_t$ for all t, because $\int_0^t \mathbb{E} \exp(2W_s^2) \, ds = \infty$ for $t \geq 1/4$.

Therefore, for $t \geq 1/4$ the process $(\int_0^t \exp(W_s^2)\,dW_s, t \geq 0)$ is not a square-integrable martingale and the isometry property does not hold. However, $\exp(W^2)$ satisfies the relaxed condition (7.34), because almost surely:

$$\int_0^t \exp(2W_s^2)\,ds \leq t\exp(2M_t^2) < \infty,$$

where M is the running maximum process (6.49), which satisfies $\mathbb{P}(M_t = \infty) = \lim_{x \to \infty} \mathbb{P}(M_t \geq x) = 0$; see (6.53). Hence, $(\int_0^t \exp(W_s^2)\,dW_s, t \geq 0)$ is a *local martingale* according to Theorem 5.74. ∎

7.2   Itô Calculus

Let f be a differentiable real-valued function and a a function of bounded variation (for example, an increasing function). For ordinary (Lebesgue–Stieltjes) integration, application of Theorem 1.62 gives the change of variable formula

$$\int_0^t f'(a_s)\,da_s = f(a_t) - f(a_0).$$

This formula no longer holds when the integrator is the Wiener process, motivating the study of stochastic calculus and one of its most celebrated formulas.

7.2.1   Itô's Formula

Itô's formula is the correct generalization of the change of variable formula for stochastic integrals and has a wide variety of applications in mathematical finance and the theory of stochastic differential equations; see Section 7.3.

Theorem 7.37: Itô's Formula for the Wiener Process

Let W be a Wiener process and $f : \mathbb{R} \to \mathbb{R}$ be twice continuously differentiable with derivatives f' and f''. Then, almost surely

$$f(W_t) - f(W_0) = \int_0^t f'(W_s)\,dW_s + \frac{1}{2}\int_0^t f''(W_s)\,ds.$$

Proof. We first prove the result under the assumption that $f^2, (f')^2$, and $(f'')^2$ are uniformly integrable as in condition (7.19), which ensures that the canonical approximations of the form (7.11) satisfy the crucial condition (7.17). In addition, we assume that f'' is uniformly continuous; that is, $\sup_x \sup_{y \in \mathbb{B}(x,r)} |f''(y) - f''(x)| \to 0$ as $r \downarrow 0$, where $\mathbb{B}(x, r) = \{u : |u - x| < r\}$ is the Euclidean ball centered at x with radius r.

By the fundamental theorem of calculus, $f(y) - f(x) = \int_x^y f'(s)\,ds$. Let $\Delta x := y - x$ and apply integration by parts to obtain $f(y) - f(x) = f'(s)s\big|_{s=x}^{s=y} - \int_x^y s f''(s)\,ds = f'(x)\Delta x + \int_x^y (y-s)f''(s)\,ds$. Using the change of variable $y - s \to s$ gives a suitable version of Taylor's theorem:

$$f(y) - f(x) = f'(x)\Delta x + \int_0^{\Delta x} f''(y - s)s\,ds = f'(x)\Delta x + \frac{1}{2}f''(x)[\Delta x]^2 + r(x, y),$$

where $r(x, y) := \int_0^{\Delta x}[f''(y - s) - f''(x)]s\,ds$ is an integral with magnitude

$$|r(x, y)| \leq \frac{[\Delta x]^2}{2} \sup_{u \in \mathbb{B}(x,\Delta x)} |f''(u) - f''(x)|.$$

Next, we take the sequence of simple processes (7.11), and then apply the Taylor expansion on each term in the telescoping sum with increments $\Delta W_{s_k} := W_{s_{k+1}} - W_{s_k}$:

$$f(W_t) - f(0) = \sum_{k=0}^{n-1}(f(W_{s_{k+1}}) - f(W_{s_k}))$$

$$= \underbrace{\sum_{k=0}^{n-1} f'(W_{s_k})\,\Delta W_{s_k}}_{=: J_1^{(n)}} + \underbrace{\frac{1}{2}\sum_{k=0}^{n-1} f''(W_{s_k})[\Delta W_{s_k}]^2}_{=: J_2^{(n)}} + \underbrace{\sum_{k=0}^{n-1} r(W_{s_k}, W_{s_{k+1}})}_{=: J_3^{(n)}}.$$

As $n \to \infty$, the first term $J_1^{(n)}$ converges in L^2 norm to the Itô integral $\int_0^t f'(W_s)\,dW_s$ by Theorem 7.20. Further, $J_2^{(n)} \xrightarrow{L^2} \int_0^t f''(W_s)\,ds$ from Lemma 7.28. As for the third term, its magnitude is at most

$$|J_3^{(n)}| \leq \sum_{k=0}^{n-1}|r(W_{s_k}, W_{s_{k+1}})| \leq \underbrace{\sup_k \sup_{u \in \mathbb{B}(W_{s_k},\Delta W_{s_k})} |f''(u) - f''(W_{s_k})|}_{=: V_n \xrightarrow{a.s.} 0} \times \underbrace{\sum_{k=0}^{n-1}(\Delta W_{s_k})^2}_{\xrightarrow{L^2} t}.$$

By Theorem 6.70, the second term converges in mean squared error to t as $\varepsilon_n := \max_k \Delta s_k \to 0$. By Theorem 6.72, we know that there exists a finite constant $c > 0$ such that, almost surely, for every sufficiently small Δs_k we have:

$$|\Delta W_{s_k}| \leq \sqrt{c\,\Delta s_k \ln(1/\Delta s_k)} \leq \sqrt{c\,\varepsilon_n \ln(1/\varepsilon_n)} =: r \quad \text{for all } k.$$

Since f'' is uniformly continuous, it follows that for a small enough r the following bound holds almost surely: $V_n \leq \sup_w \sup_{u \in \mathbb{B}(w,r)} |f''(u) - f''(w)|$. The bound vanishes, because $r \to 0$ as $\varepsilon_n \to 0$. Thus, the bounded random variable V_n converges to 0 almost surely, and hence in mean squared error, implying that $J_3^{(n)} \xrightarrow{L^2} 0$

as $n \to \infty$, and completing the proof under the restriction that $f^2, (f')^2$, and $(f'')^2$ are uniformly integrable, and f'' is uniformly continuous.

We can drop the latter restrictions on f, f', and f'' in the following way. Define the stopping time

$$T_m := \inf\{s \geq 0 : |W_s| \geq m\},$$

where m is a large positive integer and $t \wedge T_m \overset{\text{a.s.}}{\to} t$ as $m \to \infty$. Then, the range of the process $W^{(m)} := (W_{s \wedge T_m}, s \in [0, t])$ is restricted to the closed and bounded set $[-m, m]$. The continuity of f, f', f'' implies that the processes $f(W^{(m)}), f'(W^{(m)}), f''(W^{(m)})$ are almost surely bounded on the range of $W^{(m)}$; see Exercise 3.33(b). Hence, $f^2, (f')^2$, and $(f'')^2$ are uniformly integrable. Moreover, by the *Heine–Cantor* theorem in Example B.6 the f'' is uniformly continuous on the range of $W^{(m)}$. Therefore, Itô's formula holds with W replaced by $W^{(m)}$, or equivalently, with t replaced by $t \wedge T_m$ in all integrals. We can now take $m \to \infty$ as in Theorem 7.35 to remove any reference to the localizing sequence (T_m).   □

■ **Example 7.38 (Enlarging the Class $\mathcal{P}_t$ Continued)** An application of Theorem 7.37 to the function $x \mapsto \exp(x^2)$ yields:

$$\exp(W_t^2) = 1 + 2 \int_0^t W_s \exp(W_s^2) \, dW_s + \int_0^t [1 + 2W_s^2] \exp(W_s^2) \, ds,$$

where we recall from Example 7.36 that $\exp(W^2)$ does not belong to $\mathcal{P}_t$ for all t. As another example, we can apply Theorem 7.37 to the function $x \mapsto x^n$ to obtain:

$$W_t^n = n \int_0^t W_s^{n-1} \, dW_s + \frac{n(n-1)}{2} \int_0^t W_s^{n-2} \, ds,$$

which for $n = 2$ confirms the result in Example 7.21, namely, $W_t^2 = 2 \int_0^t W_s \, dW_s + t$.   ■

The Itô integral defines a new process $Z := (Z_s, s \in [0, t])$ via the integral transform $Z_t := \int_0^t F_s \, dW_s$. This motivates our definition of the *Itô process* as follows:

Definition 7.39: Itô Process

An *Itô process* is a stochastic process $X := (X_t, t \geq 0)$ that can be written in the form

$$X_t = X_0 + \int_0^t \mu_s \, ds + \int_0^t \sigma_s \, dW_s,$$

where $(\mu_s, s \geq 0)$ and $(\sigma_s, s \geq 0)$ are left-continuous $\mathcal{F}$-adapted processes such that almost surely $\int_0^t (|\mu_s| + \sigma_s^2) \, ds < \infty$ for all t.

Note that in Definition 7.39 $\int_0^t \mu_s \, ds$ is a Lebesgue integral as defined in Section 1.4.1, and $\int_0^t \sigma_s \, dW_s$ is an Itô integral such that the corresponding process $(\int_0^t \sigma_s \, dW_s, t \geq 0)$ is a local martingale like the one in Theorem 7.35.

The integral equation in Definition 7.39 is usually written in the shorthand differential form

$$dX_s = \mu_s \, ds + \sigma_s \, dW_s,$$

where μ is commonly referred to as the *drift* process and σ as the *dispersion* process. We denote an Itô process with drift μ and dispersion σ by $\mathsf{It\hat{o}}(\mu, \sigma)$.

If X is the Itô process in Definition 7.39 and F is a simple process of the form (7.2), then, similar to Definition 7.5, we can define the stochastic integral of F with respect to X as:

$$\int_0^r F_s \, dX_s := \sum_{k=0}^{n-1} \xi_k (X_{s_{k+1} \wedge r} - X_{s_k \wedge r}), \quad r \in [0, t].$$

As in Theorem 7.20, we can extend this definition to a more general process F by using sequences of simple processes $(F^{(n)})$, $(\mu^{(n)})$, $(\sigma^{(n)})$ approximating F, μ, σ, respectively, and with jumps on a common segmentation $0 = s_0 < s_1 < \cdots < s_n = t$ of $[0, t]$. If, for simplicity of exposition, all approximations are of the form (7.11), and we define $X_t^{(n)} := X_0 + \int_0^t \mu_s^{(n)} ds + \int_0^t \sigma_s^{(n)} dW_s$, then:

$$\int_0^t F_s^{(n)} \, dX_s^{(n)} = \sum_{k=0}^{n-1} F_{s_k} \left[\mu_{s_k} (s_{k+1} - s_k) + \sigma_{s_k} (W_{s_{k+1}} - W_{s_k}) \right]$$

$$= \underbrace{\int_0^t F_s^{(n)} \mu_s^{(n)} \, ds}_{=: \, J_1^{(n)}} + \underbrace{\int_0^t F_s^{(n)} \sigma_s^{(n)} \, dW_s}_{=: \, J_2^{(n)}}.$$

The first integral $J_1^{(n)}$ converges almost surely to the Lebesgue integral $\int_0^t F_s \mu_s \, ds$, under the assumption that $\int_0^t |F_s \mu_s| ds < \infty$; see Section 1.4.1. Additionally, if $F\sigma \in \mathcal{P}_t$ and $\|F^{(n)} \sigma^{(n)} - F\sigma\|_{\mathcal{P}_t} \to 0$, then by Theorem 7.20 the sequence $(J_2^{(n)})$ has an L^2-norm limit — the Itô integral $\int_0^t F_s \sigma_s \, dW_s$.

As in Theorem 7.35, the condition that $F\sigma \in \mathcal{P}_t$ can be relaxed to the less restrictive (7.34) (that is, $\int_0^t (F_s \sigma_s)^2 ds < \infty$ for all t), whilst still ensuring the existence of the Itô integral as an almost sure limit. We can conclude that the almost sure limit of $\int_0^t F_s^{(n)} \, dX_s^{(n)}$ is again an Itô process, but with drift $F\mu$ and dispersion $F\sigma$. In other words, we can define the integral $\int_0^t F_s \, dX_s$ as follows:

Definition 7.40: Integral with Respect to an Itô Process

Let X be an $\mathsf{Itô}(\mu, \sigma)$ process and F a left-continuous $\mathcal{F}$-adapted process such that almost surely $\int_0^t (|F_s \mu_s| + (F_s \sigma_s)^2)\, ds < \infty$ for all t. Then, the integral of F with respect to X is defined as the $\mathsf{Itô}(F\mu, F\sigma)$ process with:

$$\int_0^t F_s \, dX_s := \int_0^t F_s \, \mu_s \, ds + \int_0^t F_s \, \sigma_s \, dW_s.$$

Recall from Theorem 6.70 that we define the quadratic variation of the Wiener process as the limit of $\langle W^{(n)} \rangle_t$ in L^2 norm. Similarly, we can define the quadratic variation over $[0, t]$ of any $\mathcal{F}$-adapted process X, as follows:

Definition 7.41: Quadratic Variation on $[0, t]$

Let $\Pi_n := \{s_k, k = 0, \ldots, n\}$ be a segmentation of $[0, t]$ and let $(X^{(n)})$ be a sequence of simple processes with almost sure limit $X_s := \lim_n X_s^{(n)}$ for all $s \in [0, t]$ as $\mathrm{mesh}(\Pi_n) \to 0$. The *quadratic variation* of $X := (X_s, s \in [0, t])$ over $[0, t]$, denoted as $\langle X \rangle_t$, is defined as the limit in probability of

$$\langle X^{(n)} \rangle_t := \sum_{k=0}^{n-1} (X_{s_{k+1}}^{(n)} - X_{s_k}^{(n)})^2.$$

Note that the limit in the definition of quadratic variation can often be in a stronger mode of convergence, such as in almost sure or L^2-norm convergence. In particular, we have L^2-norm convergence for the quadratic variation of an Itô process.

Theorem 7.42: Quadratic Variation of an Itô Process

If X is an $\mathsf{Itô}(\mu, \sigma)$ process, then

$$\langle X^{(n)} \rangle_t \xrightarrow{L^2} \langle X \rangle_t = \int_0^t \sigma_s^2 \, ds.$$

Furthermore, if $F \in \mathcal{P}_t$ satisfies the condition (7.19) and $\mathrm{mesh}(\Pi_n) \to 0$, then:

$$\sum_{k=0}^{n-1} F_{s_k} (X_{s_{k+1}} - X_{s_k})^2 \xrightarrow{L^2} \int_0^t F_s \, \sigma_s^2 \, ds.$$

Proof. Suppose that $\mu^{(n)}, \sigma^{(n)}$, and $F^{(n)}$ are simple processes with jumps on the same segmentation $0 = s_0 < s_1 < \cdots < s_n = t$ of $[0, t]$, and approximating μ, σ, and F. Without loss of generality we may assume that μ, σ, and F are bounded, so that $\mu^{(n)}, \sigma^{(n)}$, and $F^{(n)}$ are canonical approximations of the form (7.11). Then,

$$X_r^{(n)} := X_0 + \int_0^r \mu_s^{(n)} \, \mathrm{d}s + \int_0^r \sigma_s^{(n)} \, \mathrm{d}W_s$$

converges almost surely to a limit which we can take to be the Itô process X in Definition 7.39. Denoting $G_s := F_s \mu_s \sigma_s$, it follows that

$$\sum_{k=0}^{n-1} F_{s_k} (X_{s_{k+1}}^{(n)} - X_{s_k}^{(n)})^2 = \sum_{k=0}^{n-1} F_{s_k} (\mu_{s_k} \Delta s_k + \sigma_{s_k} \Delta W_{s_k})^2$$

$$= \underbrace{\sum_{k=0}^{n-1} F_{s_k} \mu_{s_k}^2 (\Delta s_k)^2}_{=: \, J_1^{(n)}} + \underbrace{\sum_{k=0}^{n-1} F_{s_k} \sigma_{s_k}^2 [\Delta W_{s_k}]^2}_{=: \, J_2^{(n)}} + 2 \underbrace{\sum_{k=0}^{n-1} G_{s_k} \Delta s_k \Delta W_{s_k}}_{=: \, J_3^{(n)}} .$$

If $\varepsilon_n := \max_k \Delta s_k$, then $|J_1^{(n)}| \leq \varepsilon_n \int_0^t F_s^{(n)} (\mu_s^{(n)})^2 \, \mathrm{d}s \to 0$. Furthermore, $J_3^{(n)}$ is a zero-mean Itô integral, so that, by the isometry property,

$$\mathbb{V}\mathrm{ar}(J_3^{(n)}) = \mathbb{E} \sum_{k=0}^{n-1} G_{s_k}^2 (\Delta s_k)^3 \leq \varepsilon_n^2 \, \mathbb{E} \int_0^t (F_s^{(n)} \mu_s^{(n)} \sigma_s^{(n)})^2 \, \mathrm{d}s \to 0,$$

proving that $J_3^{(n)} \xrightarrow{L^2} 0$. Finally, we also have that $J_2^{(n)} \xrightarrow{L^2} \int_0^t F_s \sigma_s^2 \, \mathrm{d}s$ by Lemma 7.28. $\square$

■ **Example 7.43 (Quadratic Variation of the Wiener Process)** Since the Wiener process $W_t = 0 + \int_0^t 0 \, \mathrm{d}s + \int_0^t 1 \, \mathrm{d}W_s$ is a special case of the Itô process with $\mu = 0$ and $\sigma = 1$, we have that $\langle W \rangle_t = \int_0^t 1 \, \mathrm{d}s = t$, in agreement with Theorem 6.70.   ■

■ **Example 7.44 (Quadratic Variation and the Integral $\int_0^t X_s \, \mathrm{d}X_s$)** It is possible to define the meaning of the integral $\int_0^t X_s \, \mathrm{d}X_s$ using the quadratic variation of X, namely,

$$\int_0^t X_s \, \mathrm{d}X_s := \frac{X_t^2 - X_0^2 - \langle X \rangle_t}{2}.$$

That this definition makes sense follows from the corresponding telescoping sum:

$$X_t^2 - X_0^2 = \sum_{k=0}^{n-1} (X_{s_{k+1}} - X_{s_k})^2 + 2 \sum_{k=0}^{n-1} X_{s_k} (X_{s_{k+1}} - X_{s_k}).$$

■

Using the shorthand notation $\mathrm{d}\langle X \rangle_t = \sigma_t^2 \, \mathrm{d}t$ for the result in Theorem 7.42, we can now generalize Itô's formula in Theorem 7.37.

Theorem 7.45: Itô's Formula for Itô Processes

Let X be an Itô process and let f be twice continuously differentiable. Then,

$$(7.46) \qquad f(X_t) = f(X_0) + \int_0^t f'(X_s) \, \mathrm{d}X_s + \frac{1}{2} \int_0^t f''(X_s) \, \mathrm{d}\langle X \rangle_s.$$

Proof. Repeating the arguments in Theorem 7.37, we obtain from Taylor's expansion:

$$f(X_t) - f(X_0) = \underbrace{\sum_{k=0}^{n-1} f'(X_{s_k}) \, \Delta X_{s_k}}_{=:\, J_1^{(n)}} + \underbrace{\frac{1}{2} \sum_{k=0}^{n-1} f''(X_{s_k}) [\Delta X_{s_k}]^2}_{=:\, J_2^{(n)}} + \underbrace{\sum_{k=0}^{n-1} r(X_{s_k}, X_{s_{k+1}})}_{=:\, J_3^{(n)}}.$$

As in the proof of Theorem 7.37, the function f and its derivatives are assumed to be bounded and continuous. Therefore, the first term $J_1^{(n)} \xrightarrow{L^2} \int_0^t f'(X_s) \, \mathrm{d}X_s$, and by Theorem 7.42, the second term $J_2^{(n)} \xrightarrow{L^2} \int_0^t f''(X_s) \, \mathrm{d}\langle X \rangle_s$. Finally, similar to the proof in Theorem 7.37, we have:

$$|J_3^{(n)}| \le \underbrace{\sup_k \sup_{u \in \mathbb{B}(X_{s_k}, \Delta X_{s_k})} |f''(u) - f''(X_{s_k})|}_{=:\, V_n \xrightarrow{a.s.} 0} \times \underbrace{\sum_{k=0}^{n-1} (\Delta X_{s_k})^2}_{\xrightarrow{L^2} \langle X \rangle_t \, < \, \infty}.$$

Here V_n is bounded and converges to 0 (almost surely and in L^2 norm), because:

$$|\Delta X_{s_k}| \le |\mu_{s_k} \, \Delta s_k| + |\sigma_{s_k} \, \Delta W_{s_k}| \le \varepsilon_n \max_{s \in [0,t]} \mu_s + \sqrt{c \varepsilon_n \ln(1/\varepsilon_n)} \max_{s \in [0,t]} \sigma_s \to 0.$$

As in Theorem 7.37, the assumption that f and its derivatives are bounded can be removed by using a suitably chosen localizing sequence of stopping times.   □

In differential form, (7.46) reads:

$$\mathrm{d}f(X_t) = f'(X_t) \, \mathrm{d}X_t + \frac{1}{2} f''(X_t) \, \mathrm{d}\langle X \rangle_t$$

$$= \left(f'(X_t)\mu_t + \frac{1}{2} f''(X_t)\sigma_t^2 \right) \mathrm{d}s + f'(X_t)\sigma_t \, \mathrm{d}W_t.$$

Compare this with the corresponding *chain rule* of ordinary calculus:

$$df(x(t)) = f'(x(t))\, dx(t).$$

If X is an Itô(μ, σ) process, then an immediate consequence of (7.46) is that $f(X)$ is also an Itô process:

$$\text{Itô}\left(f'(X)\mu + \frac{1}{2}f''(X)\sigma^2,\ f'(X)\sigma\right).$$

Another consequence of Itô's formula and Theorem 7.42 is that

$$\langle f(X)\rangle_t = \int_0^t [f'(X_s)]^2 \, d\langle X\rangle_s.$$

7.2.2 Multivariate Itô's Formula

In this section we extend the Itô calculus to multivariate Itô processes. Suppose that

$$\mu_s := \begin{bmatrix} \mu_{s,1} \\ \vdots \\ \mu_{s,d} \end{bmatrix} \quad \text{and} \quad \sigma_s := \begin{bmatrix} \sigma_{s,1,1} & \cdots & \sigma_{s,1,d} \\ \vdots & \ddots & \vdots \\ \sigma_{s,d,1} & \cdots & \sigma_{s,d,d} \end{bmatrix}$$

are processes such that $\mu_{s,i}, \sigma_{s,i,j} \in \mathcal{F}_s$ for all i, j, and both $\int_0^t |\mu_{s,i}|\, ds$ and $\int_0^t \sigma_{s,i}^2\, ds$ are almost surely finite for all i. If W is a d-dimensional Wiener process, see Definition 6.8, then in analogy to $dX_s = \mu_s\, ds + \sigma_s\, dW_s$, we define the *Itô process in $\mathbb{R}^d$* via:

$$(7.47) \qquad dX_{t,i} = \mu_{t,i}\, dt + \sum_{j=1}^d \sigma_{t,i,j}\, dW_{t,j}, \quad i = 1, \ldots, d,$$

or in matrix differential notation:

$$dX_t = \mu_t\, dt + \sigma_t\, dW_t.$$

Further, we denote a d-dimensional Itô process with drift μ and dispersion σ by Itô(μ, σ). The results of Theorem 7.42 can be extended (see Exercise 8) to evaluate the quadratic variation of $X_{.,i}$ for a given i:

$$\langle X_{.,i}\rangle_t = \sum_j \int_0^t \sigma_{s,i,j}^2 \, ds.$$

Using the above formula, we can extend the concept of a quadratic variation to the covariation of two processes.

Definition 7.48: Covariation of Two Itô Processes

Let X and Y be two Itô processes adapted to the same filtration $\mathcal{F}$. Then, the *covariation* between X and Y is defined as the quantity:

$$\langle X, Y \rangle_t := \frac{\langle X + Y \rangle_t - \langle X \rangle_t - \langle Y \rangle_t}{2}.$$

The special case $\langle X, X \rangle_t$ simplifies to $\langle X \rangle_t$, so that the covariation of X with respect to itself is the quadratic variation of X. Note that the covariation satisfies the same properties as the inner product in Definition B.17. In particular, the covariation satisfies the symmetry property $\langle X, Y \rangle_t = \langle Y, X \rangle_t$ and the bilinearity property $\langle \alpha X + \beta Y, Z \rangle_t = \alpha \langle X, Z \rangle_t + \beta \langle Y, Z \rangle_t$ for any $\alpha, \beta \in \mathbb{R}$. Finally, if $X^{(n)}, Y^{(n)}$ are simple processes with jumps on the same segmentation Π_n of $[0, t]$ and approximating X, Y, then (see Exercise 6) the covariation $\langle X, Y \rangle_t$ is the limiting value of:

$$\langle X^{(n)}, Y^{(n)} \rangle_t = \sum_{k=0}^{n-1} (X^{(n)}_{s_{k+1}} - X^{(n)}_{s_k})(Y^{(n)}_{s_{k+1}} - Y^{(n)}_{s_k}).$$

■ **Example 7.49 (Covariation and Dispersion Matrix)** Let $dX_t = \mu_t \, dt + \sigma_t \, dW_t$ and $dY_t = \nu_t \, dt + \varrho_t \, dW_t$ define two Itô processes with respect to the *same* Wiener process W. Then, from the definition of the covariation, we have:

$$\langle X, Y \rangle_t = \int_0^t \sigma_s \, \varrho_s \, ds,$$

or in shorthand differential form: $d\langle X, Y \rangle_t = \sigma_t \varrho_t$. More generally, for a d-dimensional Itô process $dX_t = \mu_t \, dt + \sigma_t \, dW_t$, we obtain:

$$(7.50) \qquad d\langle X_{\cdot, i}, X_{\cdot, j} \rangle_t = \sum_{k=1}^{d} \sigma_{t,i,k} \sigma_{t,j,k} \, dt, \quad \forall i, j,$$

or in shorthand matrix notation: $d\langle X, X \rangle_t = \sigma_t \sigma_t^\top \, dt$. ■

Note that all the results so far suggest the following heuristic Itô calculus rules for manipulating a multivariate Itô process X:

$$dX_{t,i} \times dX_{t,j} = \sum_{k=1}^{d} \sigma_{t,i,k} \sigma_{t,j,k} \, dt, \quad dt \times dX_{t,i} = 0, \quad (dt)^2 = 0.$$

In particular, we have the following Itô calculus rule for the increments of the d-dimensional Wiener process:

$$dW_{t,i} \times dW_{t,j} = \begin{cases} dt & \text{if } i = j, \\ 0 & \text{if } i \neq j. \end{cases}$$

The utility of the covariation concept in Definition 7.48 is that it permits us to state the multivariate version of Itô's formula in compact and intuitive notation.

Theorem 7.51: Itô's Formula in $\mathbb{R}^d$

Let (X_t) be an d-dimensional Itô process, and let $f : \mathbb{R}^d \to \mathbb{R}$ be twice continuously differentiable in all variables. Then,

$$df(X_t) = \sum_{i=1}^{d} \partial_i f(X_t)\, dX_{t,i} + \frac{1}{2} \sum_{i=1}^{d} \sum_{j=1}^{d} \partial_{ij} f(X_t)\, d\langle X_{\cdot,i}, X_{\cdot,j} \rangle_t.$$

Proof. We repeat the arguments in the proof of Theorem 7.45, assuming that X is an $\mathrm{It\hat{o}}(\boldsymbol{\mu}, \boldsymbol{\sigma})$ process. The main difference is that we combine a version of the multivariate Taylor theorem with the covariation identity (7.50). Below, ∂f and $\partial^2 f$ denote the gradient and Hessian matrix of f. Let $\Delta x := y - x$ and $g(t) := f(x + t\,\Delta x)$ for some $t \in \mathbb{R}$, and note that by the one-dimensional Taylor expansion in Theorem 7.37:

$$g(1) - g(0) = g'(0) + \frac{1}{2}g''(0) + \int_0^1 s[g''(1 - s) - g''(0)]\, ds.$$

Using the chain rule, the last identity is equivalent to

$$f(y) - f(x) = (\Delta x)^\top \partial f(x) + \frac{1}{2}(\Delta x)^\top \partial^2 f(x)\,\Delta x + r(x, y),$$

where the residual is given by

$$r(x, y) := \int_0^1 (1 - s)(\Delta x)^\top [\partial^2 f(x + s\Delta x) - \partial^2 f(x)]\,\Delta x\, ds.$$

It follows that

$$f(X_t) - f(X_0) = \sum_{k=0}^{n-1} (f(X_{s_{k+1}}) - f(X_{s_k}))$$

$$= \sum_i \sum_k \partial_i f(X_{s_k})\,\Delta X_{s_k,i} + \frac{1}{2} \sum_{i,j} \sum_k \partial_{ij} f(X_{s_k})\,\Delta X_{s_k,i}\,\Delta X_{s_k,j} + \sum_k r(X_{s_k}, X_{s_{k+1}}),$$

where $\Delta X_{s_k,i} := X_{s_{k+1},i} - X_{s_k,i}$. For a given i, let $G^{(n)}_{s,i} := \partial_i f(X_s)$ for $s \in [s_k, s_{k+1})$ and all $k = 0, 1 \ldots, n - 1$. Then, we have

$$\sum_k \partial_i f(X_{s_k})\,\Delta X_{s_k,i} = \int_0^t G^{(n)}_{s,i}\, dX_{s,i} \xrightarrow{L^2} \int_0^t \partial_i f(X_s)\, dX_{s,i}.$$

Therefore, $\sum_i \sum_k \partial_i f(X_{s_k}) \Delta X_{s_k,i} \xrightarrow{L^2} \sum_i \int_0^t \partial_i f(X_s) \, dX_{s,i}$, dealing with the first term in the Taylor expansion. Next, we use the fact that for any bounded $F \in \mathcal{P}_t$:

$$2 \sum_k F_{s_k} \Delta X_{s_k,i} \Delta X_{s_k,j} = \sum_k F_{s_k} (\Delta X_{s_k,i} + \Delta X_{s_k,j})^2$$
$$- \sum_k F_{s_k} (\Delta X_{s_k,i})^2 - \sum_k F_{s_k} (\Delta X_{s_k,j})^2.$$

Since $(X_{s,i} + X_{s,j}, s \in [0,1])$ is an Itô process with drift $(\mu_{s,i} + \mu_{s,j})$ and dispersion $(\sum_k (\sigma_{s,i,k} + \sigma_{s,j,k}))$, Theorem 7.42 yields

$$\sum_k F_{s_k} (\Delta X_{s_k,i} + \Delta X_{s_k,j})^2 \xrightarrow{L^2} \int_0^t F_s \sum_k (\sigma_{s,i,k} + \sigma_{s,j,k})^2 \, ds.$$

In addition, $\sum_k F_{s_k} (\Delta X_{s_k,i})^2 \xrightarrow{L^2} \int_0^t F_s \sum_k \sigma_{s,i,k}^2 \, ds$. Therefore,

$$2 \sum_k F_{s_k} \Delta X_{s_k,i} \Delta X_{s_k,j} \xrightarrow{L^2} \int_0^t F_s \sum_k \left((\sigma_{s,i,k} + \sigma_{s,j,k})^2 - \sigma_{s,i,k}^2 - \sigma_{s,j,k}^2 \right) ds.$$

In other words, for the second term in the Taylor approximation, (7.50) yields:

$$\sum_{i,j,k} \partial_{ij} f(X_{s_k}) \Delta X_{s_k,i} \Delta X_{s_k,j} \xrightarrow{L^2} \sum_{i,j} \int_0^t \partial_{ij} f(X_s) \, d\langle X_{\cdot,i}, X_{\cdot,j} \rangle_s.$$

Finally, denote $h_{ij}(u, x) := \partial_{ij} f(u) - \partial_{ij} f(x)$, and note that the matrix $\mathbf{H} = [h_{ij}]$ is a symmetric matrix, implying that its eigenvalues $\lambda_1(u, x), \ldots, \lambda_d(u, x)$ are real. Thus, the spectral radius of $\mathbf{H}(u, x)$:

$$\varrho(u, x) := \max_{\|v\|=1} |v^\top \mathbf{H}(u, x) v| = \max_k |\lambda_k(u, x)|$$

can be bounded as:

$$\varrho(u, x) \le \sqrt{\sum_k [\lambda_k(u, x)]^2} = \sqrt{\sum_{i,j} [h_{i,j}(u, x)]^2} \le d \max_{i,j} |h_{i,j}(u, x)|.$$

Hence, we have the following bound on the residual:

$$|r(x, y)| \le \|\Delta x\|^2 \int_0^1 (1 - s) \varrho(x + s\Delta x, x) \, ds$$
$$\le \frac{d \|\Delta x\|^2}{2} \sup_{u \in \mathbb{B}(x, \|\Delta x\|)} \max_{i,j} |h_{i,j}(u, x)|,$$

where $\mathbb{B}(x, r) = \{u \in \mathbb{R} : \|u - x\|_2 < r\}$ is the Euclidean ball centered at x with radius r. Therefore,

$$\sum_k |r(X_{s_k}, X_{s_{k+1}})| \leq \underbrace{\sup_k \sup_{u \in \mathbb{B}(X_{s_k}, \|\Delta X_{s_k}\|)} \max_{i,j} |h_{i,j}(u, X_{s_k})| \times d \sum_k \|\Delta X_{s_k}\|^2}_{=: V_n}.$$

By Theorem 7.42, we have $\sum_k (X_{s_{k+1},i} - X_{s_k,i})^2 \xrightarrow{L^2} \int_0^t \sigma_{s,i}^2 \, ds$ for all i, and hence $\sum_k \|\Delta X_{s_k}\|^2 \xrightarrow{L^2} \sum_i \int_0^t \sigma_{s,i}^2 \, ds$. Further, if we define $\varepsilon_n := \max_k (s_{k+1} - s_k)$, then Theorem 6.72 implies the following almost sure inequalities:

$$\|\Delta X_{s_k}\| \leq d \varepsilon_n \max_i \max_{s \in [0,t]} \mu_{s,i} + \sqrt{c \varepsilon_n \ln(1/\varepsilon_n)} \max_i \max_{s \in [0,t]} \sigma_{s,i} \to 0.$$

Hence, the uniform continuity of $(u, x) \mapsto \max_{i,j} |h_{i,j}(u, x)|$ and the boundedness of V_n together imply that the residual $\sum_k |r(X_{s_k}, X_{s_{k+1}})|$ converges to 0 (both in L^2 norm and almost surely). Similarly to Theorem 7.37, one can lift any boundedness restrictions on f and its derivatives using a localization argument. $\square$

A consequence of Itô's formula in Theorem 7.51 is that the covariation satisfies:

$$(7.52) \qquad d\langle f(X), X_{\cdot,i} \rangle_t = \sum_j \partial_j f(X_t) \, d\langle X_{\cdot,i}, X_{\cdot,j} \rangle_t \quad \text{for all } i.$$

Further, if X is an Itô(μ, σ) process, then:

$$df(X_t) = \left([\partial f(X_t)]^\top \mu_t + \frac{1}{2} \text{tr}\left(\sigma_t^\top \partial^2 f(X_t) \sigma_t \right) \right) dt + [\partial f(X_t)]^\top \sigma_t \, dW_t.$$

■ **Example 7.53 (Multivariate Itô Formula)** A special case of Theorem 7.51 is $X_t := [X_t, t]^\top$, where $t \, (\geq 0)$ is deterministic and the process (X_t) is governed by $dX_t = \mu_t \, dt + \sigma_t \, dW_t$. Then,

$$(7.54) \quad df(X_t, t) = \left(\frac{\partial f}{\partial t}(X_t) + \mu_t \frac{\partial f}{\partial x}(X_t) + \frac{\sigma_t^2}{2} \frac{\partial^2 f}{\partial x^2}(X_t) \right) dt + \sigma_t \frac{\partial f}{\partial x}(X_t) \, dW_t.$$

As an example, consider the exponential martingale $S_t := \exp(r W_t - t r^2/2), t \geq 0$ from (6.40), where (W_t) is a Wiener process. An application of (7.54) with $\mu_t = 0$ and $\sigma_t = 1$ yields

$$dS_t = \left(-\frac{r^2}{2} S_t + 0 + \frac{r^2}{2} S_t \right) dt + r S_t \, dW_t = r S_t \, dW_t,$$

or equivalently $S_t - S_0 = r \int_0^t S_s \, dW_s$.

■

■ **Example 7.55 (Theorem 6.46 Revisited)** Recall that in Theorem 6.46 the standard d-dimensional Brownian motion $\boldsymbol{B}$ is of the form $\boldsymbol{B}_t = \boldsymbol{B}_0 + \boldsymbol{W}_t$. In other words, $\boldsymbol{B}$ is an Itô process with drift vector $\boldsymbol{\mu} = \boldsymbol{0}$ and dispersion matrix $\boldsymbol{\sigma} = \boldsymbol{I}_d$, so that by (7.50) we have $\mathrm{d}\langle W_{.,i}, W_{.,j}\rangle_t = \mathbb{1}_{\{i=j\}} \, \mathrm{d}t$. An application of Theorem 7.51 yields:

$$\mathrm{d}f(\boldsymbol{B}_t) = \sum_{i=1}^{d} \partial_i f(\boldsymbol{B}_t) \, \mathrm{d}W_{t,i} + \frac{1}{2}\sum_{i=1}^{d} \partial_{ii} f(\boldsymbol{B}_t) \, \mathrm{d}\langle W_{.,i}\rangle_t,$$

or in integral form we can write:

$$X_t := f(\boldsymbol{B}_t) - \frac{1}{2}\sum_{i=1}^{d} \int_0^t \partial_{ii} f(\boldsymbol{B}_s) \, \mathrm{d}s = f(\boldsymbol{B}_0) + \sum_{i=1}^{d} \int_0^t \partial_i f(\boldsymbol{B}_s) \, \mathrm{d}W_{s,i}.$$

Since the sum of martingales adapted to the same filtration is another martingale, the process $\left(\sum_{i=1}^{d} \int_0^t \partial_i f(\boldsymbol{B}_s) \, \mathrm{d}W_{s,i}\right)$ is a martingale, which confirms the result of Theorem 6.46, namely, that (X_t) is a martingale.   ■

■ **Example 7.56 (Stochastic Product Rule and Integration by Parts)** A notable application of Theorem 7.51 on the function $(x, y) \mapsto x \times y$ yields the *product rule* for Itô processes:

(7.57) $$\mathrm{d}(X_t Y_t) = Y_t \, \mathrm{d}X_t + X_t \, \mathrm{d}Y_t + \mathrm{d}\langle X, Y\rangle_t.$$

The corresponding integral form,

$$X_t Y_t - X_0 Y_0 = \int_0^t Y_s \, \mathrm{d}X_s + \int_0^t X_s \, \mathrm{d}Y_s + \langle X, Y\rangle_t,$$

is the Itô *integration by parts* formula.   ■

■ **Example 7.58 (Integral of the Wiener Process)** Consider the integral process $Y := (Y_t, t \geq 0)$ of the Wiener process, defined as

$$Y_t := \int_0^t W_s \, \mathrm{d}s, \quad t \geq 0.$$

Using the Itô integration by parts formula for the process $(t\, W_t, t \geq 0)$, we obtain

$$Y_t = t\, W_t - \int_0^t s \, \mathrm{d}W_s = \int_0^t (t - s) \, \mathrm{d}W_s,$$

which shows, by Proposition 7.25, that Y is a zero-mean Gaussian process with covariance function

$$\mathbb{E}Y_t Y_{t+u} = \int_0^t (t - s)(t + u - s) \, \mathrm{d}s = \frac{t^3}{3} + u\frac{t^2}{2}.$$

In addition, if we define

$$\widetilde{W}_t := \int_0^{(3t)^{1/3}} s \, dW_s,$$

then the time change in (7.31) with $C(t) := t^3/3$ shows that $\widetilde{W}$ is a Wiener process. In other words,

$$Y_t = tW_t - \widetilde{W}_{C(t)}, \quad t \geq 0.$$

Figure 7.59 shows typical paths of the process (Y_t). Note that both $(t\,W_t)$ and $(\int_0^t s \, dW_s)$ have paths of *unbounded* variation, but the difference between these two processes, that is (Y_t), has paths of *bounded* variation. In fact, its total variation on $[0, t]$ is $\int_0^t |W_s| \, ds$.

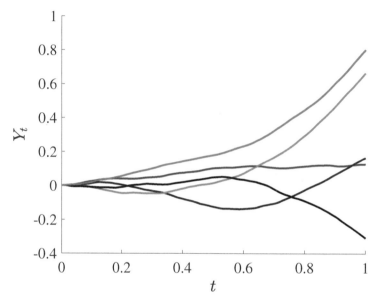

Figure 7.59: Typical paths of $(\int_0^t W_s \, ds, t \geq 0)$. The paths are of bounded variation.

We say that the Itô process (7.47) is a *local martingale* in $\mathbb{R}^d$ if the d components of the process are each local martingales.

Theorem 7.60: Itô Processes and Martingales

An Itô process in $\mathbb{R}^d$ is a local martingale if and only if its drift is 0.

Proof. Let the Itô process be $X_t = X_0 + \int_0^t \mu_s \, ds + \int_0^t \sigma_s \, dW_s$. Sufficiency of $\mu_t = 0$ follows from the fact that the d components of the process $X_t = X_0 + \int_0^t \sigma_s \, dW_s$ are all Itô integrals and hence local martingales. To prove the necessity, we assume that

X is a local martingale. Then, $(X_t - X_0 - \int_0^t \sigma_s \, dW_s)$ is also a local martingale. However, the identity $X_t - X_0 - \int_0^t \sigma_s \, dW_s = \int_0^t \mu_s \, ds$ implies that $(\int_0^t \mu_s \, ds)$ is a local martingale, which is continuous and of bounded variation, because

$$\sup_{\Pi_n} \sum_k \left| \int_{s_k}^{s_{k+1}} \mu_{s,i} \, ds \right| = \int_0^t |\mu_{s,i}| \, ds < \infty.$$

It follows from Lemma 5.90 that $(\int_0^t \mu_{s,i} \, ds)$ for each $i = 1, \ldots, d$ are almost surely constant. In other words, μ_s is almost surely 0.   □

■ **Example 7.61 (Lévy's Characterization of the Wiener Process)** Suppose that X is an Itô(μ, σ) process such that both $(X_t, t \geq 0)$ and $(X_t^2 - t, t \geq 0)$ are local martingales. Then, X must be the Wiener process governing the Itô process. To see this, we apply Theorem 7.60. First, X being a local martingale implies that $\mu = 0$. Second, by Itô's formula: $d(X_t^2 - t) = (-1 + \sigma_t^2) \, dt + 2\sigma_t X_t \, dW_t$, so that $(X_t^2 - t)$ is an Itô process with with drift $(-1 + \sigma_t^2)$. Hence, if $(X_t^2 - t)$ is a martingale, then we must have $\sigma_t^2 = 1$. In other words, X is an Itô process with drift 0 and dispersion $\sigma_t = \pm 1$, implying that X is a Wiener process.

This example gives a characterization of the Wiener process in terms of two martingales. We demonstrated it under the restriction that X is an Itô process. However, the restriction that the process X must be an Itô process can be removed, resulting in the celebrated *Lévy's characterization theorem* (Karatzas and Shreve, 1998, page 157).   ■

■ **Example 7.62 (Quadratic Variation and Martingales)** Let $X := (X_t)$ be an Itô process that is also a local martingale, and let $Z := (Z_t)$ be an increasing process with $Z_0 = 0$. Then, $(X_t^2 - Z_t)$ is a local martingale if and only if $Z = \langle X \rangle$. To see this, use Itô's formula to write

$$X_t^2 = 2 \int_0^t X_s \, dX_s + \langle X \rangle_t,$$

and note that by Theorem 7.60 the Itô process (X_t) must be of the form $X_t = X_0 + \int_0^t \sigma_s \, dW_s$, because $\mu_t = 0$ almost surely.

First, assume that $Z = \langle X \rangle$. Then,

$$X_t^2 - Z_t = \int_0^t X_s \, dX_s = \int_0^t X_s \sigma_s \, dW_s.$$

Hence, $(X_t^2 - Z_t)$ is a local martingale, either because the integrator process (X_t) in $\int_0^t X_s \, dX_s$ is a local martingale, or simply because $\int_0^t X_s \sigma_s \, dW_s$ is an Itô integral.

Conversely, suppose that $(X_t^2 - Z_t) =: (M_t)$ is a local martingale. Then,

$$Z_t - \langle X \rangle_t = 2 \int_0^t X_s \, dX_s - M_t$$

is a local martingale as well. Moreover, the process $(Z_t - \langle X \rangle_t)$ is the difference of two increasing processes, and so has bounded variation; see Exercise 6.11. It follows from Lemma 5.90 that $(Z_t - \langle X \rangle_t)$ is the 0 process. Hence, $Z_t = \langle X \rangle_t$ almost surely for all $t \geq 0$. ■

7.3   Stochastic Differential Equations

Stochastic differential equations are based on the same principle as ordinary differential equations, relating an unknown function to its derivatives, but with the additional feature that part of the unknown function is driven by randomness.

Definition 7.63: Stochastic Differential Equation

A *stochastic differential equation* (SDE) for a stochastic process $(X_t, t \geq 0)$ is an expression of the form

$$(7.64) \qquad dX_t = a(X_t, t)\, dt + b(X_t, t)\, dW_t,$$

where $(W_t, t \geq 0)$ is a Wiener process and a and b are deterministic functions, called the *drift* and *diffusion*[1] functions. The resulting process $(X_t, t \geq 0)$ is called an *(Itô) diffusion*.

Note that the stochastic process $(b(X_t, t))$ in Definition 7.63 is a *dispersion* process $\sigma := (\sigma_t)$ (as used in Definition 7.39) for which σ_t only depends on the values of X_t and t, rather than on the whole path $(X_s, s \in [0, t])$. Intuitively, (7.64) expresses that the infinitesimal change, dX_t, in X_t at time t, is the sum of an infinitesimal displacement $a(X_t, t)\, dt$ and an infinitesimal noise term $b(X_t, t)\, dW_t$. The precise mathematical meaning of (7.64) is that the stochastic process $(X_t, t \geq 0)$ satisfies the integral equation

$$(7.65) \qquad X_t = X_0 + \int_0^t a(X_s, s)\, ds + \int_0^t b(X_s, s)\, dW_s,$$

where the last integral is an Itô integral. The definition of such integrals is discussed in Section 7.1.

When a and b do not depend on t explicitly (that is, $a(x, t) = \widetilde{a}(x)$, and $b(x, t) = \widetilde{b}(x)$), the corresponding SDE is referred to as being *time-homogeneous* or *autonomous*. For simplicity, we will focus on such autonomous SDEs.

Multidimensional SDEs can be defined in a similar way as in (7.64). A stochastic differential equation in $\mathbb{R}^d$ is an expression of the form

$$(7.66) \qquad d\boldsymbol{X}_t = \boldsymbol{a}(\boldsymbol{X}_t, t)\, dt + \mathbf{B}(\boldsymbol{X}_t, t)\, d\boldsymbol{W}_t,$$

---

[1]Some authors refer to b^2 as the diffusion function (or diffusion coefficient).

where (W_t) is an m-dimensional Wiener process, $\boldsymbol{a}(\boldsymbol{x}, t)$ is a d-dimensional vector and $\mathbf{B}(\boldsymbol{x}, t)$ a $d \times m$ matrix, for each $\boldsymbol{x} \in \mathbb{R}^d$ and $t \geq 0$.

■ **Example 7.67 (Brownian Motion)** The solution of the simplest SDE

$$dX_t = a\, dt + b\, dW_t$$

is the Brownian motion process

$$X_t = X_0 + a\, t + b\, W_t, \quad t \geq 0,$$

as this satisfies (7.65) with $a(x, t) = a$ and $b(x, t) = b$.   ■

■ **Example 7.68 (Geometric Brownian Motion)** A fundamental stochastic process in finance is defined as the solution to the SDE

(7.69)
$$dS_t = \mu S_t\, dt + \sigma S_t\, dW_t,$$

which is called a *geometric Brownian motion*. In finance, S_t typically models the price of a risky financial asset at a time t, characterized by a *drift* parameter μ and a *volatility* parameter σ. If (B_t) is the Brownian motion process with $B_t = \mu t + \sigma W_t$, then the SDE can also be written as

$$dS_t = S_t\, dB_t,$$

which suggests an exponential/geometric growth of the process, hence the name. When $\mu = 0$, we obtain the *exponential martingale* $S_t = \exp(\sigma W_t - t\sigma^2/2), t \geq 0$; see Theorem 6.39.

We can solve the SDE (7.69) via a "separation of variables" argument, similar to the way certain ordinary differential equations are solved. The derivation below also uses Itô's formula (7.46).

First, note that $S_t = \int_0^t \mu S_s\, ds + \int_0^t \sigma S_s\, dW_s$, which shows that the diffusion function of (S_t) is $(\sigma^2 S_t^2)$. Next, a separation of variables for the SDE (7.69) yields

$$\frac{dS_t}{S_t} = \mu\, dt + \sigma\, dW_t.$$

Taking the integral from 0 to t on both sides results in

(7.70)
$$\int_0^t \frac{1}{S_s}\, dS_s = \mu t + \sigma W_t.$$

With $f(x) := \ln x$, $f'(x) = 1/x$, and $f''(x) = -1/x^2$, Itô's formula (7.46) gives

$$\ln S_t = \ln S_0 + \int_0^t \frac{1}{S_s}\, dS_s - \frac{1}{2}\int_0^t \frac{1}{S_s^2}\sigma^2 S_s^2\, ds = \int_0^t \frac{1}{S_s}\, dS_s - \frac{1}{2}\sigma^2 t.$$

Combining this with (7.70), we conclude that

$$\ln \frac{S_t}{S_0} = \sigma W_t + \left(\mu - \frac{1}{2}\sigma^2 \right) t,$$

so that

(7.71) $$S_t = S_0 \exp \left(\sigma W_t + \left(\mu - \frac{1}{2}\sigma^2 \right) t \right), \quad t \geq 0.$$

Figure 7.72 shows typical trajectories for the case $\mu = 1$, $\sigma = 0.2$, and $S_0 = 1$.

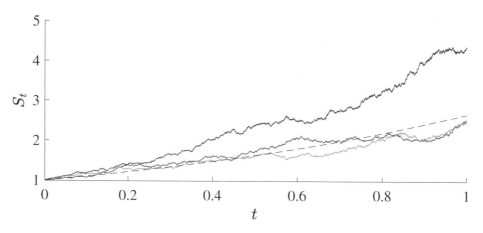

Figure 7.72: Typical paths of a geometric Brownian motion process with drift $\mu = 1$, volatility $\sigma = 0.2$, and initial state $S_0 = 1$.

■

■ **Example 7.73 (Ornstein–Uhlenbeck Process)** Another prominent SDE is

(7.74) $$dX_t = \theta(\nu - X_t)\, dt + \sigma\, dW_t,$$

with $\sigma > 0$, $\theta > 0$, and $\nu \in \mathbb{R}$. Its solution (X_t) is called the *Ornstein–Uhlenbeck* process or *mean-reverting* process. In finance, the process is used to describe price fluctuations around a mean price ν such that the process "reverts to the mean" — that is, when $X_t > \nu$ the drift is negative, and when $X_t < \nu$ the drift is positive, so that at all times the process tends to drift toward ν. In physics, X_t is used to describe the *velocity* of a Brownian particle. The term ν is then usually taken to be 0, and the resulting SDE is said to be of *Langevin* type.

We show that the Itô diffusion

(7.75) $$X_t = e^{-\theta t} X_0 + \nu(1 - e^{-\theta t}) + \sigma e^{-\theta t} \int_0^t e^{\theta s}\, dW_s, \quad t \geq 0$$

satisfies the SDE (7.74). Namely, define $f(y,t) := \sigma e^{-\theta t}y$ and $Y_t := \int_0^t e^{\theta s}\, dW_s$, so that $X_t = e^{-\theta t}X_0 + \nu(1 - e^{-\theta t}) + f(Y_t, t)$. Then, Itô's formula (7.54) yields:

$$dX_t = -\theta e^{-\theta t}X_0\, dt + \nu\theta e^{-\theta t}\, dt - \sigma\theta e^{-\theta t}Y_t\, dt + e^{\theta t}\sigma e^{-\theta t}dW_t$$
$$= \theta e^{-\theta t}(\nu - X_0)\, dt - \theta f(Y_t, t)\, dt + \sigma dW_t$$
$$= \theta(\nu - X_t)\, dt + \sigma\, dW_t.$$

Since the integrand in the Itô integral in (7.75) is deterministic (i.e., $e^{\theta s}$), it follows from Proposition 7.25 that (X_t) is a Gaussian process, with mean

$$\mathbb{E}X_t = e^{-\theta t}\,\mathbb{E}X_0 + \nu(1 - e^{-\theta t}),$$

variance $\mathbb{V}\mathrm{ar}\, X_t = \sigma^2(1 - e^{-2\theta t})/(2\theta)$, and covariance

$$\mathbb{C}\mathrm{ov}(X_s, X_t) = \frac{\sigma^2}{2\theta}e^{-\theta(s+t)}\left(e^{2\theta(s\wedge t)} - 1\right).$$

Note that (X_t) has independent increments, because the integrand in the Itô integral in (7.75) does not depend on t. This shows that X_t converges in distribution to an $\mathcal{N}(\nu, \sigma^2/(2\theta))$ random variable as $t \to \infty$. Also, by the time-change property (7.31), we have that

$$\widetilde{W}_t := \int_0^{C^{-1}(t)} e^{\theta s}\, dW_s, \quad t \geq 0$$

is a Wiener process, where $C(t) := (e^{2t\theta} - 1)/(2\theta), t \geq 0$. As a result of this, we can write

$$X_t = e^{-\theta t}X_0 + \nu(1 - e^{-\theta t}) + \sigma e^{-\theta t}\widetilde{W}_{C(t)}, \quad t \geq 0,$$

where the Wiener process $\widetilde{W}$ is time-changed using the clock function C. This provides a straightforward way to simulate the process from a realization of a Wiener process. Typical realizations are depicted in Figure 7.76.

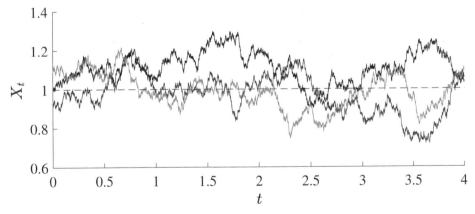

Figure 7.76: Three realizations of an Ornstein–Uhlenbeck process with parameters $\nu = 1, \sigma = 0.2$, and $\theta = 2$. The initial value X_0 is drawn from the $\mathcal{N}(\nu, \sigma^2/(2\theta))$ distribution.

7.3.1 Existence of Solutions to SDEs

Consider the autonomous SDE

(7.77) $$dX_t = a(X_t)\,dt + b(X_t)\,dW_t.$$

As already mentioned, the meaning of this SDE is that X is a stochastic process that satisfies the integral equation

(7.78) $$X_t = X_0 + \int_0^t a(X_s)\,dx + \int_0^t b(X_s)\,dW_s.$$

However, it is not clear that such a process indeed exists. We show in this section that under mild conditions on a and b we can obtain a unique solution X of the SDE via a method of successive approximations, similar to *Picard iteration* for ordinary differential equations.

The condition that is placed on a and b is that they be *Lipschitz continuous*; that is, for all $x \in \mathbb{R}$ there exists a constant c such that

(7.79) $$|a(y) - a(x)| + |b(y) - b(x)| \le c|y - x|.$$

The successive approximation idea is very simple. Suppose that $X_0 = x$. Define $X_t^{(0)} := x$ for all t, and let

(7.80) $$X_t^{(n+1)} := x + \int_0^t a(X_s^{(n)})\,ds + \int_0^t b(X_s^{(n)})\,dW_s$$

for $n \in \mathbb{N}$. The following is the main existence result for SDEs:

Theorem 7.81: Existence of SDE Solutions

For any $t \in [0, v]$, the sequence $(X^{(n)})$ in (7.80) converges almost surely in the uniform norm on $[0, t]$ to a process X that satisfies the integral equation (7.78).

For the proof, we first need the following lemma. Recall also the definition of the norm $\|\cdot\|_{\mathcal{P}_t}$ in Definition 7.14.

Lemma 7.82: Integral Bound

Let $X^{(n)}, n = 1, 2, \ldots$ be defined in (7.80), with $X^{(0)} = x$. Then, for every $v > 0$ there exists a constant α such that

$$\mathbb{E}\sup_{s \le t}\left(X_s^{(n+1)} - X_s^{(n)}\right)^2 \le \alpha\|X^{(n)} - X^{(n-1)}\|_{\mathcal{P}_t}^2, \quad t \le v.$$

Proof.

$$X_t^{(n+1)} - X_t^{(n)} = \int_0^t \left[a(X_s^{(n)}) - a(X_s^{(n-1)}) \right] ds + \int_0^t \left[b(X_s^{(n)}) - b(X_s^{(n-1)}) \right] dW_s$$

$$=: A_t + B_t.$$

By Jensen's inequality $t^2 (1/t \int_0^t f_s \, ds)^2 \leq t \int_0^t f_s^2 \, ds$, which implies that almost surely

$$A_t^2 \leq t \int_0^t \left[a(X_s^{(n)}) - a(X_s^{(n-1)}) \right]^2 ds \leq tc^2 \int_0^t \left[X_s^{(n)} - X_s^{(n-1)} \right]^2 ds,$$

using the Lipschitz continuity (7.79) of a in the last inequality. Taking expectations, we find

$$\mathbb{E} A_t^2 \leq tc^2 \| X^{(n)} - X^{(n-1)} \|_{\mathcal{P}_t}^2.$$

To bound $\mathbb{E} B_t^2$, we use the fact that

$$B_v^2 - \int_0^v \left[b(X_s^{(n)}) - b(X_s^{(n-1)}) \right]^2 ds, \quad v \in [0, t],$$

is a martingale; see Example 7.23. In particular,

$$\mathbb{E} B_t^2 = \| b(X^{(n)}) - b(X^{(n-1)}) \|_{\mathcal{P}_t}^2 \leq c^2 \| X^{(n)} - X^{(n-1)} \|_{\mathcal{P}_t}^2,$$

again using Lipschitz continuity (7.79) in the last inequality. By combining the latter result with Doob's norm inequality (5.89), we obtain

$$\sup_{s \leq t} \mathbb{E} B_s^2 \leq 2^2 \, \mathbb{E} B_t^2 \leq 4 \, c^2 \| X^{(n)} - X^{(n-1)} \|_{\mathcal{P}_t}^2.$$

Putting everything together, gives

$$\mathbb{E} \sup_{s \leq t} (A_s + B_s)^2 \leq 2 \left\{ \sup_{s \leq t} \mathbb{E} A_s^2 + \sup_{s \leq t} \mathbb{E} B_s^2 \right\} \leq (2t + 8) c^2 \| X^{(n)} - X^{(n-1)} \|_{\mathcal{P}_t}^2,$$

which proves the lemma for $\alpha := (2t + 8)c^2$. □

Proof of Theorem 7.81. The proof is similar to the existence result for the Wiener process in Theorem 6.21. We will show that for almost all ω, the sequence $(X^{(n)}(\omega))$ converges in the uniform norm $\sup_{s \leq t} |f(s)|$, and hence has a limit $X(\omega)$ in this norm. Take an arbitrary $v \geq 0$. For a fixed x and $t \in [0, v]$, there exists the following bound:

$$\mathbb{E}(X_t^{(1)} - x)^2 = \mathbb{E}(a(x)t + b(x)W_t)^2 = a^2(x)t^2 + b^2(x)t \leq \beta,$$

where $\beta \geq 0$ is a constant that depends on x and v. Consequently, by applying Lemma 7.82 iteratively, we have

$$\mathbb{E} \sup_{s \leq t}(X_s^{(2)} - X_s^{(1)})^2 \leq \alpha \int_0^t \beta \, ds = \beta \alpha t,$$

so that

$$\mathbb{E} \sup_{s \leq t}(X_s^{(3)} - X_s^{(2)})^2 \leq \alpha \int_0^t \alpha \beta s \, ds = \frac{\beta \alpha^2 t^2}{2},$$

and, in general, for every $t \leq v$ and $n \in \mathbb{N}$, we have

$$\mathbb{E} \sup_{s \leq t}(X_s^{(n+1)} - X_s^{(n)})^2 \leq \frac{\beta \alpha^n t^n}{n!}.$$

Hence, for the event $A_n := \{\sup_{s \leq t} |X_s^{(n+1)} - X_s^{(n)}| > 2^{-n}\}$ Markov's inequality yields:

$$\mathbb{P}(A_n) \leq \frac{\beta(4\alpha t)^n}{n!}.$$

By the first part of Borel–Cantelli's Lemma 3.14, we have that $\mathbb{P}(\Omega_0) = 1$, where Ω_0 is the event $\{\sum_k \mathbb{1}_{A_k} < \infty\}$. If $x_k := X_k(\omega)$ for an $\omega \in \Omega_0$, then (x_k) is a Cauchy sequence by part three of Proposition 3.2, where $\varepsilon_n := 2^{-n}$ and we use the result with the uniform norm. Hence, the sequence $(X^{(n)}(\omega))$ is Cauchy convergent in the uniform norm for every $\omega \in \Omega_0$, and hence has a limit $X(\omega)$ in this norm; see Proposition 3.7. We can set $X(\omega) = x$ for $\omega \notin \Omega_0$. Since all $X^{(n)}$ are continuous processes and convergence is in uniform norm, X is a continuous process as well, on any interval $[0, t]$. Moreover, taking limits on both sides of (7.80), we see that X satisfies the integral equation (7.78) with $X_0 = x$; so it is a solution of the SDE (7.77) with this initial condition. That it is the *only* solution follows from an application of *Gronwall's inequality*; see Exercise 11. Namely, suppose X and $\widetilde{X}$ are solutions to (7.78), with $X_0 = \widetilde{X}_0 = x$. Then, duplicating the proof of Lemma 7.82, with X and $\widetilde{X}$, we find for all $t \leq v$:

$$\mathbb{E}(X_t - \widetilde{X}_t)^2 \leq (2v + 8)c^2 \|X - \widetilde{X}\|_{\mathcal{P}_t}.$$

Application of (7.115), with $f(s) := \mathbb{E}(X_s - \widetilde{X}_s)^2$ and $g(t) := 0$, shows that $f(t) = 0$ for all $t \leq v$. Since v is arbitrary, X and $\widetilde{X}$ are modifications of each other and hence, by their continuity, indistinguishable; see Theorem 7.13.   □

7.3.2 Markov Property of Diffusion Processes

Let X be the unique solution to the SDE (7.77) obtained from the successive substitution procedure in (7.80). This solution is called the *strong solution* of the SDE, in the sense that almost surely every path of X is a deterministic function of the starting value of X and the path of W; that is, for almost every $\omega \in \Omega$ and all $t \geq 0$,

$$X_t(\omega) = \phi(X_0(\omega), (W_s(\omega), s \in [0,t]), t)$$

for some function ϕ. In contrast, a *weak solution* is one which satisfies the SDE for *some* Wiener process $\widetilde{W}$ different from W. Weak solutions are solutions in distribution, in the sense that while W and $\widetilde{W}$ may exist on different probability spaces, they still have the same statistical properties. For a brief exploration of the antipodal concept of a weak solution of an SDE; see Example 7.105, Exercise 22, and Exercise 27.

For a fixed $t \geq 0$, let $\widetilde{W} := (W_{t+u} - W_t, u \geq 0)$ and $\widetilde{X} := (X_{t+u} - X_t, u \geq 0)$. Then, from (7.78) we have

$$\widetilde{X}_u = X_t + \int_t^{t+u} a(X_s)\,ds + \int_t^{t+u} b(X_s)\,dW_s$$
$$= \widetilde{X}_0 + \int_0^u a(\widetilde{X}_s)\,ds + \int_0^u b(\widetilde{X}_s)\,d\widetilde{W}_s.$$

Thus, if $X = \phi(X_0, W)$, then $\widetilde{X} = \phi(\widetilde{X}_0, \widetilde{W})$. Let $(\mathcal{F}_t)$ the filtration to which W is adapted, and assume that X_0 is independent of $\mathcal{F}_0$ and W. Then, $\widetilde{X} = \phi(\widetilde{X}_0, \widetilde{W}) = \phi(X_t, \widetilde{W})$ is conditionally independent of $\mathcal{F}_t$ given X_t, showing that the Markov property holds. Moreover, the conditional distribution of $\widetilde{X}$ given $X_t = x$ is the same as the distribution of X starting at state x. Thus, the Itô diffusion X is a (time-homogeneous) Markov process with continuous sample paths. We will show that its infinitesimal generator is given by the differential operator L defined by

$$(7.83) \qquad Lf(x) := a(x)f'(x) + \frac{1}{2}b^2(x)f''(x),$$

for all twice continuously differentiable functions on compact sets.

First, by Itô's formula,

$$f(X_t) - f(X_0) = \int_0^t f'(X_s)\,dX_s + \int_0^t \frac{1}{2}f''(X_s)\,d\langle X\rangle_s$$
$$= \int_0^t \left[f'(X_s)a(X_s) + \frac{1}{2}f''(X_s)b^2(X_s)\right] ds + \int_0^t f'(X_s)b(X_s)\,dW_s$$
$$= \int_0^t Lf(X_s)\,ds + M_t,$$

where $(M_t, t \geq 0)$ is the martingale given by

$$(7.84) \qquad M_t := f(X_t) - f(X_0) - \int_0^t Lf(X_s)\,ds = \int_0^t f'(X_s)b(X_s)\,dW_s.$$

Hence, denoting by $\mathbb{E}^x$ the expectation operator under which the process starts at x, we have for all $t \geq 0$:

$$(7.85) \qquad \mathbb{E}^x f(X_t) = f(x) + \mathbb{E}^x \int_0^t Lf(X_s) \, ds.$$

Under mild conditions, we may also replace t in (7.85) with a stopping time T; that is,

$$(7.86) \qquad \mathbb{E}^x f(X_T) = f(x) + \mathbb{E}^x \int_0^T Lf(X_s) \, ds.$$

This is *Dynkin's formula*. A sufficient condition is that $\mathbb{E}^x T < \infty$ for all x. To prove (7.86), we need to show that $\mathbb{E}^x M_T = 0$, with M the martingale in (7.84). Consider the stopped martingale $\widehat{M} := (\widehat{M}_t, t \geq 0)$, with $\widehat{M}_t := M_{T \wedge t}, t \geq 0$. Since f' and b are continuous, $f'b$ is bounded on $[0, t]$ by some constant c. From Proposition 7.32, it follows that

$$\mathbb{E}^x \widehat{M}_t^2 = \mathbb{E}^x \left(\int_0^t f'(X_s) \, b(X_s) \mathbb{1}_{\{s \leq T\}} \, dW_s \right)^2$$

$$= \mathbb{E}^x \int_0^t (f'(X_s) \, b(X_s))^2 \mathbb{1}_{\{s \leq T\}} \, ds \leq c^2 \, \mathbb{E}^x T < \infty.$$

Consequently, $\widehat{M}$ is uniformly integrable, and thus it converges almost surely and in L^1 to M_T. By Theorem 5.95, $\widehat{M}$ can be extended to a Doob martingale on $\overline{\mathbb{R}}_+$ by defining $\widehat{M}_\infty := M_T$. In particular,

$$\mathbb{E} M_T = \mathbb{E} \widehat{M}_T = \mathbb{E} \widehat{M}_0 = \mathbb{E} M_0 = 0,$$

which had to be shown.

■ **Example 7.87 (Exit Probability)** We can generalize the methodology in Example 6.65 to calculate exit probabilities for general diffusion processes. In this example we consider the Brownian motion process $B_t = at + bW_t, t \geq 0$; but see also Exercise 12. As in Example 6.65, we are interested in the probability

$$p := \mathbb{P}^x(T_l < T_r),$$

that the Brownian motion exits the interval $[l, r]$ through l rather than r, when starting from $x \in [l, r]$. We can calculate p by finding a function f such that $Lf = 0$ on $[l, r]$, where

$$Lf(x) = af'(x) + \frac{1}{2}b^2 f''(x).$$

One particular solution (verify by differentiating the function twice) is

$$f(x) := e^{-2ax/b^2}.$$

Application of Dynkin's formula (7.86) with $T = T_l \wedge T_r$ now immediately gives:

$$p = \frac{e^{-2ar/b^2} - e^{-2ax/b^2}}{e^{-2ar/b^2} - e^{-2al/b^2}},$$

assuming that $\mathbb{E}^x T < \infty$. We leave the proof of the latter as an exercise. ∎

To show that L is the infinitesimal generator of the diffusion, we can reason in the same way as for the infinitesimal generator of the Wiener process in Section 6.6. Namely, from (7.85) it follows that

$$(7.88) \qquad\qquad Lf(x) = \lim_{t\downarrow 0} \frac{\mathbb{E}^x f(X_t) - f(x)}{t}.$$

The limit on the right in (7.88) defines the infinitesimal generator of the Markov process X. Its domain consists of all bounded measurable functions for which the limit exists — this includes the domain of L, hence the infinitesimal generator extends L.

Let P_t be the transition kernel of the Markov process and define the operator P_t by

$$P_t f(x) := \int P_t(x, \mathrm{d}y) f(y) = \mathbb{E}^x f(X_t).$$

Then, by (7.85), we obtain the *Kolmogorov forward equations*:

$$(7.89) \qquad\qquad P_t' f = P_t L f.$$

Moreover, by the Chapman–Kolmogorov equations (4.43), we have $P_{t+s} f(x) = P_s P_t f(x) = \mathbb{E}^x P_t f(X_s)$, and therefore

$$\frac{1}{s}\{P_{t+s}f(x) - P_t f(x)\} = \frac{1}{s}\{\mathbb{E}^x P_t f(X_s) - P_t f(x)\}.$$

Letting $s \downarrow 0$, we obtain the *Kolmogorov backward equations*:

$$(7.90) \qquad\qquad P_t' f = L P_t f.$$

If P_t has a transition density p_t, then we can write the last equation as

$$\frac{\mathrm{d}}{\mathrm{d}t} \int p_t(y \mid x) f(y)\,\mathrm{d}y = \int L p_t(y \mid x) f(y)\,\mathrm{d}y,$$

so that $p_t(y \mid x)$ for fixed y satisfies the Kolmogorov backward equations:

$$(7.91) \qquad \frac{\partial}{\partial t} p_t(y \mid x) = a(x) \frac{\partial}{\partial x} p_t(y \mid x) + \frac{1}{2} b^2(x) \frac{\partial^2}{\partial x^2} p_t(y \mid x).$$

Similarly, (7.89) can be written as $\frac{d}{dt} \int p_t(y \mid x) f(y) \, dy = \int p_t(y \mid x) L f(y) \, dy = \int f(y) L^* p_t(y \mid x) \, dy$, where L^* (acting here on y) is the adjoint operator of L defined by $\int g(y) L h(y) \, dy = \int h(y) L^* g(y) \, dy$. Hence, for fixed x the density $p_t(y \mid x)$ satisfies the Kolmogorov forward equations:

$$(7.92) \qquad \frac{\partial}{\partial t} p_t(y \mid x) = -\frac{\partial}{\partial y} \left(a(y) \, p_t(y \mid x) \right) + \frac{1}{2} \frac{\partial^2}{\partial y^2} \left(b^2(y) \, p_t(y \mid x) \right).$$

This illustrates the important connection between partial differential equations of the form $u_t' = L u_t$ and diffusion processes. Similarly, for multidimensional SDEs of the form (7.66) one can show that the infinitesimal generator extends the operator

$$L f(\boldsymbol{x}) = \sum_{i=1}^{m} a_i(\boldsymbol{x}) \frac{\partial}{\partial x_i} f(\boldsymbol{x}) + \frac{1}{2} \sum_{i=1}^{m} \sum_{j=1}^{m} c_{ij}(\boldsymbol{x}) \frac{\partial^2}{\partial x_i \, \partial x_j} f(\boldsymbol{x}),$$

where $\{a_i\}$ are the components of $\boldsymbol{a}$ and $\{c_{ij}\}$ the components of $\mathbf{C} := \mathbf{B} \mathbf{B}^\top$.

7.3.3 Methods for Solving Simple SDEs

If a strong solution of the Itô diffusion SDE (7.64) exists, then it can be expressed as $X_t = \phi(X_0, (W_s, s \in [0, t]), t)$ for some function ϕ. Finding an analytical formula for ϕ is usually impossible, except in a few special cases.

7.3.3.1 Linear Stochastic Differential Equations

One setting where the strong solution of an SDE can be expressed as a simple analytical formula is when both the drift and diffusion of the SDE (7.64) are linear functions. In particular, suppose that

$$a(x, t) = \alpha_t + \beta_t x,$$
$$b(x, t) = \delta_t + \gamma_t x.$$

An SDE of the form

$$(7.93) \qquad dX_t = (\alpha_t + \beta_t X_t) \, dt + (\delta_t + \gamma_t X_t) \, dW_t$$

is said to be *linear*.

■ **Example 7.94 (Stochastic Exponential)** Let Y be an Itô(β, γ) process. A special case of (7.93) is

$$dX_t = X_t \, dY_t = \beta_t X_t \, dt + \gamma_t X_t \, dW_t,$$

whose solution is called the *stochastic exponential* of Y. If we conjecture that the strong solution is of the form $X_t = f(Y_t, z_t)$ for some twice continuously differentiable deterministic functions f and z, then an application of Theorem 7.51 with $Y_t := (Y_t, z_t)^\top$ yields:

$$df(Y_t, z_t) = \partial_1 f(\boldsymbol{Y}_t)\, dY_t + \partial_2 f(\boldsymbol{Y}_t)\, dz_t + \frac{1}{2}\partial_{11} f(\boldsymbol{Y}_t)\, d\langle Y\rangle_t.$$

Since $df(\boldsymbol{Y}_t) = X_t\, dY_t = f(\boldsymbol{Y}_t)\, dY_t$, matching terms yields the system of equations for $f(\boldsymbol{y}) := f(y, z)$:

$$\partial_1 f(\boldsymbol{y}) - f(\boldsymbol{y}) = 0,$$

$$\partial_2 f(\boldsymbol{y})\, z_t' + \frac{1}{2}\partial_{11} f(\boldsymbol{y})\, \frac{d}{dt}\langle Y\rangle_t = 0.$$

The most general solution to the first equation is $f(y, z) = c_1(z)\exp(y)$, where $c_1(z)$ is an arbitrary function of z. Substituting this into the second equation and solving yields $\ln c_1(z_t) = -\frac{1}{2}\langle Y\rangle_t + c_2(t)$, where c_2 is an arbitrary function of t. Hence, we must have that $f(y, z) = \exp\left(y - \frac{1}{2}\langle Y\rangle_t + c_2(t)\right)$. In other words, from $X_0 = f(Y_0, z_0)$ we obtain that the stochastic exponential of Y is:

$$X_t = X_0 \exp\left(Y_t - Y_0 - \frac{1}{2}\langle Y\rangle_t\right).$$

∎

The solution to the general linear SDE in (7.93) is given next.

Proposition 7.95: Solution to a Linear Diffusion SDE

The solution to the linear SDE (7.93) is given by

$$X_t = Y_t\left(X_0 + \int_0^t \frac{\alpha_s - \delta_s\gamma_s}{Y_s}\, ds + \int_0^t \frac{\delta_s}{Y_s}\, dW_s\right),$$

where $Y_t = \exp\left(\int_0^t (\beta_s - \gamma_s^2/2)\, ds + \int_0^t \gamma_s\, dW_s\right)$ is the stochastic exponential of an Itô(β, γ) process.

Proof. Let Z be an Itô$((\alpha - \delta\gamma)/Y, \delta/Y)$ process such that $X_t = Y_t Z_t$. We verify that this formula satisfies (7.93) via direct computation of the differential of $Y_t Z_t$. Since Y is the stochastic exponential of an Itô(β, γ) process, it satisfies $dY_t = Y_t(\beta_t\, dt + \gamma_t\, dW_s)$. Using the stochastic product rule (7.57) and the covariation formula (7.50), we obtain:

$$\begin{aligned}
dX_t &= Y_t\, dZ_t + Z_t\, dY_t + \langle Y, Z\rangle_t \\
&= [(\alpha_t - \delta_t\gamma_t)\, dt + \delta_t\, dW_t] + [Z_t Y_t(\beta_t\, dt + \gamma_t\, dW_t)] + \delta_t\gamma_t\, dt \\
&= (\alpha_t + \beta_t X_t)\, dt + (\delta_t + \gamma_t X_t)\, dW_t,
\end{aligned}$$

verifying that $X_t = Y_t Z_t$ satisfies (7.93). □

■ **Example 7.96 (Brownian Bridge)** Consider the SDE for $t \in [0, 1)$:

$$dX_t = \frac{b - X_t}{1 - t} dt + dW_t, \quad X_0 = a.$$

This SDE is linear with $\alpha_t := b/(1 - t)$, $\beta_t := -1/(1 - t)$, $\delta := 1$, $\gamma_t := 0$, and so an application of Proposition 7.95 yields:

$$X_t = a + (b - a)t + \underbrace{\int_0^t \frac{1 - t}{1 - s} dW_s}_{=:Z_t}, \quad t \in [0, 1).$$

Moreover, applying Proposition 7.25 to $Z := (Z_t, t \in [0, 1))$ shows that $(X_t, t \in [0, 1))$ is a Gaussian process with mean $a + (b - a)t$ and covariance function:

$$\mathbb{C}\mathrm{ov}(X_{s_1}, X_{s_2}) = (1 - s_1)(1 - s_2) \left[\frac{1}{1 - s_1 \wedge s_2} - 1 \right] = s_1 \wedge s_2 - s_1 s_2.$$

Since $\mathbb{V}\mathrm{ar}\, X_{1-\varepsilon} = \varepsilon(1 - \varepsilon) \to 0$ as $\varepsilon \downarrow 0$, it is clear that $X_{1-\varepsilon} \overset{L^2}{\to} b$. In fact, Exercise 16 shows that $X_{1-\varepsilon} \overset{\mathrm{a.s.}}{\to} b$ as $\varepsilon \downarrow 0$. Hence, if we define $X_1 := b$, then $X := (X_t, t \in [0, 1])$ is a Gaussian process with continuous paths. As a matter of fact, X is the *Brownian bridge* discussed in Example 6.5.   ■

7.3.3.2 Itô–Stratonovich Method

Recall that the ordinary chain rule

$$df(\boldsymbol{x}(t)) = \sum_{i=1}^{d} \partial_i f(\boldsymbol{x}(t))\, dx_i(t)$$

is not in agreement with the multivariate Itô's formula in Theorem 7.51. Nevertheless, it is possible to recover the ordinary form of the chain rule in stochastic settings by using the following modification of the Itô integral:

Definition 7.97: Itô–Stratonovich Integral

Let X and Y be Itô processes such that $\int_0^t Y_s\, dX_s$ and $\int_0^t X_s\, dY_s$ are well-defined stochastic integrals. Then, the *Itô–Stratonovich* integral of Y with respect to X is defined as

$$\int_0^t Y_s \circ dX_s := \int_0^t Y_s\, dX_s + \frac{1}{2} \langle X, Y \rangle_t.$$

The differential form of the identity in Definition 7.97 is

$$Y_t \circ dX_t = Y_t \, dX_t + \frac{1}{2} d\langle X, Y \rangle_t.$$

With this differential identity we can write the stochastic product rule (7.57) as:

$$d(X_t \, Y_t) = X_t \circ dY_t + Y_t \circ dY_t.$$

More importantly, the corresponding chain rule in Theorem 7.51 is now formally in agreement with the chain rule in ordinary calculus.

Proposition 7.98: Itô–Stratonovich Formula

Let (X_t) be a d-dimensional Itô process, and let $f : \mathbb{R}^d \to \mathbb{R}$ be three times continuously differentiable in all variables. Then,

$$df(X_t) = \sum_{i=1}^{d} \partial_i f(X_t) \circ dX_{t,i}.$$

Proof. Define $g_i(x) := \partial_i f(x)$ for all i and note that each g_i is twice continuously differentiable. Since $g_i(X_t) \circ dX_{t,i} = g_i(X_t) \, dX_{t,i} + \frac{1}{2} d\langle g_i(X), X_{\cdot,i} \rangle_t$, the proof will be complete if we can show that for all i:

$$\langle g_i(X), X_{\cdot,i} \rangle_t = \sum_j \partial_j g_i(X_t) \, d\langle X_{\cdot,i}, X_{\cdot,j} \rangle_t.$$

To see this, apply Theorem 7.51 to obtain:

$$dg_i(X_t) = \sum_j \partial_j g_i(X_t) \, dX_{t,j} + \frac{1}{2} \sum_{k,j} \partial_{kj} g_i(X_t) \, d\langle X_{\cdot,k}, X_{\cdot,j} \rangle_t,$$

and then use formula (7.52) to conclude that the covariation is $d\langle g_i(X), X_{\cdot,i} \rangle_t = \sum_j \partial_j g_i(X_t) \, d\langle X_{\cdot,i}, X_{\cdot,j} \rangle_t$.   □

Since the definition of the Itô–Stratonovich integral is such that both the chain rule and product rule of ordinary calculus are formally the same, this suggests yet another method for solving diffusion SDEs. Namely, we convert all Itô integrals to their Stratonovich equivalents, making it possible to use our experience with (systems of) ODEs to solve SDEs. An example will illustrate this point.

■ **Example 7.99 (Itô–Stratonovich Method)** Consider solving the nonlinear SDE:

$$dX_t = \frac{n}{2} X_t^{2n-1} \, dt + X_t^n \, dW_t, \qquad X_0 = a$$

for some constant $a > 0$. By Itô's formula $d(X_t^n) = nX_t^{n-1}dX_t + \frac{n(n-1)}{2}X_t^{n-2}d\langle X\rangle_t$, and therefore

$$d\langle X^n, W\rangle_t = nX_t^{n-1}X_t^n\,dt = nX_t^{2n-1}\,dt.$$

Hence, $X_t^n\,dW_t = X_t^n \circ dW_t - \frac{n}{2}X_t^{2n-1}\,dt$, and the SDE can be written as in its Itô–Stratonovich format:

$$dX_t = X_t^n \circ dW_t.$$

The differential equation $dx(t) = [x(t)]^n dw(t)$ implies that

$$\int \frac{dx(t)}{[x(t)]^n} = \int dw(t) = w(t) + c$$

for some constant c. Hence, the strong solution to the SDE is

$$X_t = [(1-n)(W_t + c)]^{\frac{1}{1-n}}, \quad c = a^{1-n}/(1-n).$$

Note that this may not be a solution for all $t \geq 0$. For example, when $n = 3$, the expression $[-2(W_t + c)]^{-1/2}$ is only well-defined until the exit time $\inf\{t \geq 0 : W_t + c > 0\}$.

∎

7.3.3.3 Girsanov's Method

Girsanov's method explores how random variables and stochastic processes (e.g., determined by an SDE) behave when the underlying probability measure is changed. Let $(\Omega, \mathcal{H}, \mathbb{P})$ be a probability space and let M be a real-valued random variable with expectation 1. We can use M to construct a *change of measure* $\widetilde{\mathbb{P}}$ on $(\Omega, \mathcal{H})$ by defining

$$\widetilde{\mathbb{P}}(A) := \int_A M\,d\mathbb{P} = \mathbb{E}M\mathbb{1}_A, \quad A \in \mathcal{H}.$$

Thus, $\widetilde{\mathbb{P}}$ is the *indefinite integral* of M with respect to $\mathbb{P}$; see Section 1.4.3. It is a probability measure, because $\widetilde{\mathbb{P}}(\Omega) = \mathbb{E}M = 1$.

As an example, suppose X is a numerical random variable with pdf g. Let f be another pdf, with $f(x) = 0$ whenever $g(x) = 0$. Define $M := f(X)/g(X)$. Then, $\mathbb{E}M = 1$, and we have for any measurable function h:

$$\widetilde{\mathbb{E}}h(X) = \mathbb{E}Mh(X) = \int \frac{f(x)}{g(x)}h(x)\,g(x)\,dx = \int h(x)f(x)\,dx.$$

Thus, under $\widetilde{\mathbb{P}}$, X has pdf f. More generally, we have by Theorem 1.58 that for any random variable V:

$$\mathbb{E}MV = \widetilde{\mathbb{E}}V.$$

We can generalize the change of measure idea to stochastic processes, as follows. Suppose $\mathcal{F} := (\mathcal{F}_t)$ is a filtration on the probability space $(\Omega, \mathcal{H}, \mathbb{P})$ and $M := (M_t)$

is an $\mathcal{F}$-martingale with $\mathbb{E} M_t = 1$ for all t. Then, for each t we can define a new measure $\widetilde{\mathbb{P}}^t$ on $(\Omega, \mathcal{F}_t)$ by

(7.100) $$\widetilde{\mathbb{P}}^t(A) := \mathbb{E} M_t \mathbb{1}_A, \quad A \in \mathcal{F}_t.$$

We say that $\widetilde{\mathbb{P}}$ is the change of measure *induced* by M.

We now come to the connection with Itô processes, because here a natural way arises to create a mean-1 martingale, and hence a change of measure. Let X be an Itô$(\mu, 1)$ process under probability measure $\mathbb{P}$, with respect to a filtration $\mathcal{F} := (\mathcal{F}_t, t \geq 0)$. Suppose that (M_t) is the solution to the SDE

$$dM_t = M_t \mu_t \, dW_t,$$

ensuring that (M_t) is a continuous local martingale. We can solve the SDE in the same way as in Example 7.68. In particular, Itô's formula gives

$$d \ln M_t = M_t^{-1} \, dM_t - \frac{1}{2} \mu_t^2 \, dt = \mu_t \, dW_t - \frac{1}{2} \mu_t^2 \, dt,$$

so that

(7.101) $$M_t = \exp\left(\int_0^t \mu_s \, dW_s - \frac{1}{2} \int_0^t \mu_s^2 \, ds \right) = e^{Z_t - \frac{1}{2}\langle Z \rangle_t}, \quad t \geq 0,$$

where $Z := (Z_t)$ is the martingale defined by $Z_t := \int_0^t \mu_s \, dW_s$. Exercises 19 and 20 show that a sufficient condition for (M_t) to be a proper martingale (rather than a local one) is that $\langle Z \rangle$ satisfies the *Krylov condition*:

(7.102) $$\lim_{\varepsilon \downarrow 0} \varepsilon \ln \mathbb{E} \exp((1 - \varepsilon)\langle Z \rangle_t / 2) = 0.$$

Note that the Krylov condition is implied by the better known *Novikov condition*:

$$\mathbb{E} \exp\left(\langle Z \rangle_t / 2 \right) < \infty;$$

see Exercises 19 and 20. Thus, assuming the Krylov condition (7.102), we have $\mathbb{E} M_t = \mathbb{E} M_0 = 1$ for all t and (M_t) is the *stochastic exponential* of Z. As a consequence of this, we have the following important result:

Theorem 7.103: Girsanov's Change of Measure

Suppose that μ satisfies the Krylov condition (7.102) and M is the exponential martingale (7.101). Then, under the change of measure (7.100) induced by M, the Itô process

$$\widetilde{W}_t := - \int_0^t \mu_s \, ds + W_t, \quad t \geq 0$$

is a Wiener process.

Proof. We first show that $\widetilde{W}$ is a martingale under $\widetilde{\mathbb{P}}$. Noting that

$$\mathrm{d}M_t = M_t\,\mu_t\,\mathrm{d}W_t,$$
$$\mathrm{d}\widetilde{W}_t = -\mu_t\,\mathrm{d}t + \mathrm{d}W_t,$$
$$\mathrm{d}\langle M, \widetilde{W}\rangle_t = M_t\,\mu_t\,\mathrm{d}t,$$

we find, by using the product rule (7.57), that:

$$
\begin{aligned}
\mathrm{d}(M_t\widetilde{W}_t) &= M_t\,\mathrm{d}\widetilde{W}_t + \widetilde{W}_t\,\mathrm{d}M_t + M_t\,\mu_t\,\mathrm{d}t \\
&= -\mu_t\,M_t\,\mathrm{d}t + M_t\,\mathrm{d}W_t + \widetilde{W}_t\,M_t\,\mu_t\,\mathrm{d}W_t + M_t\,\mu_t\,\mathrm{d}t \\
&= M_t(1 + \mu_t\widetilde{W}_t)\,\mathrm{d}W_t,
\end{aligned}
$$

which shows that $(M_t\widetilde{W}_t)$ is a martingale under $\mathbb{P}$, and so $(\widetilde{W}_t)$ is a martingale under $\widetilde{\mathbb{P}}$. Similarly, applying the product rule to $(M_t(\widetilde{W}_t^2 - t))$ gives:

$$
\begin{aligned}
\mathrm{d}(M_t(\widetilde{W}_t^2 - t)) &= M_t\,\mathrm{d}(\widetilde{W}_t^2 - t) + (\widetilde{W}_t^2 - t)\,\mathrm{d}M_t + 2M_t\,\mu_t\widetilde{W}_t\,\mathrm{d}t \\
&= M_t 2\widetilde{W}_t\,\mathrm{d}\widetilde{W}_t + (\widetilde{W}_t^2 - t)\,M_t\,\mu_t\,\mathrm{d}W_t + 2M_t\,\mu_t\widetilde{W}_t\,\mathrm{d}t \\
&= M_t(2\widetilde{W}_t + (\widetilde{W}_t^2 - t)\mu_t)\,\mathrm{d}W_t,
\end{aligned}
$$

which shows that $(\widetilde{W}_t^2 - t)$ is martingale under $\widetilde{\mathbb{P}}$. *Lévy's characterization theorem* then proves that $\widetilde{W}$ is a Wiener process under $\widetilde{\mathbb{P}}$; see Example 7.61. $\qquad\square$

■ **Example 7.104 (Girsanov's Method I)** Suppose that W is a Wiener process under $\mathbb{P}$. Girsanov's Theorem 7.103 can be used to solve SDEs of the form

$$\mathrm{d}X_t = a(X_t, t)\,\mathrm{d}t + \gamma_t X_t\,\mathrm{d}W_t,$$

where a and γ are continuous functions. Suppose that μ satisfies the condition (7.102) and $\widetilde{\mathbb{P}}$ is the change of measure (7.100) induced by the exponential martingale (7.101). Then, Girsanov's theorem states that $\widetilde{W}$, which satisfies $\mathrm{d}\widetilde{W}_t = -\mu_t\,\mathrm{d}t + \mathrm{d}W_t$ with $\widetilde{W}_0 = 0$, is a Wiener process under $\widetilde{\mathbb{P}}$. Hence, under the change of measure, the SDE can be written as:

$$\mathrm{d}X_t = (a(X_t, t) + \mu_t\gamma_t X_t)\,\mathrm{d}t + \gamma_t X_t\,\mathrm{d}\widetilde{W}_t.$$

Taking μ such that $a(X_t, t) + \mu_t\gamma_t X_t = 0$, the solution under $\widetilde{\mathbb{P}}$ is the stochastic exponential $X_t = X_0 \exp\left(\int_0^t \gamma_s\,\mathrm{d}\widetilde{W}_s - \frac{1}{2}\int_0^t \gamma_s^2\,\mathrm{d}s\right)$. Substituting with $\mathrm{d}\widetilde{W}_t = -\mu_t\,\mathrm{d}t + \mathrm{d}W_t$, yields $X_t = Y_t Z_t$, where

$$Y_t := X_0 \exp\left(\int_0^t \gamma_s\,\mathrm{d}W_s - \frac{1}{2}\int_0^t \gamma_s^2\,\mathrm{d}s\right) \quad \text{and} \quad Z_t := \exp\left(-\int_0^t \mu_s\gamma_s\,\mathrm{d}s\right).$$

Since Y is uniquely determined by γ, we only need to solve for Z to complete the solution. This amounts to finding a function z that satisfies the ordinary differential equation:

$$a(y_t z_t, t) + \mu_t \gamma_t y_t z_t = a(y_t z_t, t) - y_t \frac{dz}{dt} = 0, \quad z_0 = 1.$$

As a particular example, suppose that we wish to find the strong solution of

$$dX_t = X_t^n \, dt + \gamma X_t \, dW_t, \quad n \leq 1.$$

Girsanov's method calls for the solution of the differential equation: $y_t^n z_t^n = y_t \frac{dz}{dt}$. The SDE solution is thus:

$$X_t = Y_t \left(1 + (1 - n) \int_0^t Y_s^{n-1} ds \right)^{\frac{1}{1-n}},$$

where $Y_t = X_0 \exp \left(\gamma W_t - \gamma^2 t / 2 \right)$.   ∎

■ **Example 7.105 (Girsanov's Method II)** Suppose we wish to solve SDEs of the form

$$dX_t = a(X_t, t) \, dt + \delta_t \, dW_t, \quad X_0 = x.$$

Under the measure $\widetilde{\mathbb{P}}$, the SDE is written as: $dX_t = (a(X_t, t) + \mu_t) \, dt + \delta_t \, d\widetilde{W}_t$. Taking μ such that $a(X_t, t) + \mu_t = 0$, we obtain:

$$X_t - x = \int_0^t \delta_s \, d\widetilde{W}_s = \underbrace{\int_0^t \delta_s \, dW_s}_{=: Y_t} \underbrace{- \int_0^t \delta_s \mu_s \, ds}_{=: Z_t}.$$

Hence, the solution is given by $X_t - x = Y_t + Z_t$, where each realization (z_t) of (Z_t) satisfies the nonlinear differential equation: $dz_t / dt = \delta_t \, a(x + y_t + z_t, t)$ with $z_0 = 0$. For instance, if $X_0 = x \geq 0$, $\delta_t := 1$, and $a(x, t) := 1/x$, then $X_t - x = W_t + Z_t$, such that each path (z_t) satisfies the ODE:

$$(7.106) \qquad \frac{dz_t}{dt} = \frac{1}{x + w_t + z_t}, \quad z_0 = 0.$$

This ODE can be easily solved numerically using standard methods; see Exercise 27. When $x = 0$, the process $X := (W_t + Z_t, t \geq 0)$ is the strong solution to the d-dimensional *Bessel SDE*

$$(7.107) \qquad dX_t = \frac{d-1}{2X_t} \, dt + dW_t, \quad X_0 = 0$$

for the case $d = 3$. In Exercise 22 we show that the Euclidean norm $\|W\|$ of a d-dimensional Wiener process W is a weak solution to the d-dimensional Bessel SDE. In the next example we highlight an important link between Brownian excursions and the three-dimensional Bessel SDE.   ∎

■ **Example 7.108 (Brownian Excursions)** This example deals with the continuous version of the gambler's ruin problem in Example 5.33; see also Example 7.87. Let $B := (B_t, t \geq 0)$ be a standard Brownian motion, starting at some x with $0 < x < r$ under probability measure $\mathbb{P}$. Denote the natural filtration of B by $\mathcal{F} := (\mathcal{F}_t)$ and define

$$T := \min\{t : B_t = r \text{ or } 0\}$$

as the first time that the process hits either r or 0. What do the trajectories of B look like that hit r before 0? We are thus interested in the conditional distribution of B given that the event $\{B_T = r\}$ occurs. The corresponding probability measure, $\widetilde{\mathbb{P}}$, is given by

$$\widetilde{\mathbb{P}}(A) := \frac{\mathbb{P}(A \cap \{B_t = r\})}{\mathbb{P}(B_t = r)} = \frac{r}{x}\mathbb{P}(A \cap \{B_t = r\}), \quad A \in \mathcal{F}_T,$$

where we used the fact that $\mathbb{P}(B_t = r) = x/r$, by Example 7.87. It follows that the Radon–Nikodym derivative $d\widetilde{\mathbb{P}}/d\mathbb{P}$ is equal to the random variable

$$Z := \frac{r}{x}\mathbb{1}_{\{B_T = r\}}.$$

Let $\widetilde{\mathbb{P}}^t$ be the restriction of $\widetilde{\mathbb{P}}$ to $(\Omega, \mathcal{F}_{t \wedge T})$. For $A \in \mathcal{F}_{t \wedge T}$, we have

$$\widetilde{\mathbb{P}}^t(A) = \mathbb{E}Z\mathbb{1}_A = \mathbb{E}\mathbb{E}_{t \wedge T}[Z\mathbb{1}_A] = \mathbb{E}[\mathbb{1}_A\mathbb{E}_{t \wedge T}Z] = \mathbb{E}[M_{t \wedge T}\mathbb{1}_A],$$

where $M_{t \wedge T} := \mathbb{E}_{t \wedge T}Z$ and $\mathbb{E}_{t \wedge T}$ denotes the conditional expectation given $\mathcal{F}_{t \wedge T}$. By Example 5.11, we see that, under $\widetilde{\mathbb{P}}$, the stopped process $(M_{t \wedge T})$ is a uniformly integrable martingale. Moreover, $M_{t \wedge T} = (r/x)\mathbb{P}_{t \wedge T}(B_T = r) = (r/x)B_{t \wedge T}/r = B_{t \wedge T}/x$. Writing

$$M_{t \wedge T} = \exp(\ln(B_{t \wedge T}/x)) = \exp\left(\int_0^{t \wedge T} \frac{1}{B_s}\,dW_s - \frac{1}{2}\int_0^{t \wedge T} \frac{1}{B_s^2}\,ds\right),$$

we see that $(M_{t \wedge T})$ is of the form (7.101), with $\mu_s = B_s^{-1}$. Girsanov's Theorem 7.103, shows that the process

$$\widetilde{B}_t := B_t - x - \int_0^t \frac{1}{B_s}\,ds, \quad t \leq T$$

is a Wiener process under $\widetilde{\mathbb{P}}$. In other words, under $\widetilde{\mathbb{P}}$ the process (B_t) satisfies the SDE

$$dB_t = \frac{1}{B_t}\,dt + d\widetilde{B}_t, \quad t \leq T,$$

with $B_0 = x > 0$. Hence, excursions of a standard Brownian motion from x to r behave according to an Itô diffusion with drift function $a : x \mapsto 1/x$. While the strong solution of this SDE can be readily computed numerically, as discussed in Example 7.105 and Exercise 27, a weak solution to the SDE is the process $X := \|x + W\|$, where $x \in \mathbb{R}^3$ is a vector with Euclidean norm $\|x\| = x$ and W is a three-dimensional Wiener process; see Exercise 22. Recall from Example 7.105 that in the special case of $x = 0$, the process X is a weak solution of the three-dimensional Bessel SDE (7.107).

Figure 7.109 shows a typical excursion of a standard Brownian motion that starts at a position x, arbitrarily close to 0, and reaches the maximal position $r = 1$ before dropping back to x.

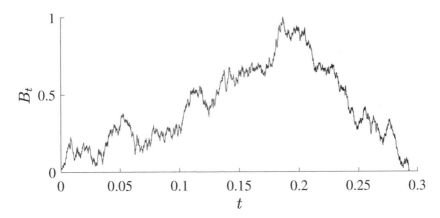

Figure 7.109: A typical excursion of a standard Brownian motion starting arbitrarily close to 0 and reaching a maximal position of 1.

The path of the excursion in Figure 7.109 consists of two parts. The first (depicted in blue) corresponds to an Itô process with drift (B_t^{-1}); i.e., the path that the Brownian motion follows conditional on reaching 1 before returning to x. The second part (depicted in red) corresponds to the path of a Brownian motion starting at 1, conditional on hitting x before returning to 1. This has the same distribution as 1 minus the path of a Brownian motion starting at x and hitting 1 before returning to x, and can be thus simulated in the same way as the first path. By concatenating the two paths, we obtain the path of the excursion depicted in Figure 7.109. ∎

7.3.4 Euler's Method for Numerically Solving SDEs

Let $(X_t, t \geq 0)$ be a diffusion process defined by the autonomous SDE

$$(7.110) \qquad dX_t = a(X_t)\, dt + b(X_t)\, dW_t, \quad t \geq 0,$$

where X_0 has a known distribution.

The *Euler* or *Euler–Maruyama* method for solving SDEs is a simple generalization of Euler's method for solving ordinary differential equations. The idea is to replace the SDE with the stochastic difference equation

$$(7.111) \qquad Y_{k+1} = Y_k + a(Y_k)\, h + b(Y_k)\, V_{k+1},$$

where $V_1, V_2, \ldots \sim_{\text{iid}} \mathcal{N}(0, h)$ and $Y_0 = X_0$. For a small step size h, the sequence $(Y_k, k \in \mathbb{N})$ approximates the process $(X_t, t \geq 0)$; that is, $Y_k \approx X_{kh}, k \in \mathbb{N}$. The generalization to non-autonomous and multidimensional SDEs is straightforward; see Exercises 26 and 23. *Milstein's* method is similar to Euler's, but uses an additional correction term, which gives slightly better approximations in low-dimensional settings; see Exercise 25.

■ **Example 7.112 (Euler Method for Geometric Brownian Motion)** We illustrate Euler's method via the geometric Brownian motion SDE (7.69). The (strong) solution is given by (7.71). To compare the Euler solution with the exact solution, we use the *same* random variables $V_1, V_2, \ldots \sim_{\text{iid}} \mathcal{N}(0, h)$ for a given step size h. The exact solution at time kh can be written as

$$S_{kh} = S_0 \exp\left(\left(\mu - \frac{\sigma^2}{2}\right) kh + \sigma \sum_{i=1}^{k} V_i \right).$$

Euler's method gives the approximation

(7.113) $Y_k = Y_{k-1}\left(1 + \mu h + \sigma V_k\right), \quad k = 1, 2, \ldots.$

For values not at the approximation points, we use linear interpolation.

We compute the approximation for two different step sizes, while ensuring that the appropriate common random numbers are used. In particular, we use step sizes $h := 2^{-8}$ and $\widetilde{h} := 2^{-4} = mh$, with $m := 2^4$. Then, set

$$\widetilde{V}_i := \sum_{j=1}^{m} V_{(i-1)m+j}.$$

The exact solution at time $k\widetilde{h}$ on the $\widetilde{h}$ time scale is generated as

$$S_{k\widetilde{h}} = S_0 \exp\left(\left(\mu - \frac{\sigma^2}{2}\right) k\widetilde{h} + \sigma \sum_{i=1}^{k} \widetilde{V}_i \right) = S_0 \exp\left(\left(\mu - \frac{\sigma^2}{2}\right) k\widetilde{h} + \sigma \sum_{j=1}^{mk} V_j \right).$$

Euler's approximation on this time scale is computed as

$$\widetilde{Y}_k = \widetilde{Y}_{k-1}\left(1 + \mu \widetilde{h} + \sigma \widetilde{V}_k\right) = \widetilde{Y}_{k-1}\left(1 + \mu \widetilde{h} + \sigma \sum_{j=1}^{m} V_{(k-1)m+j}\right).$$

Figure 7.114 depicts the output of the exact and Euler schemes for the case $\mu = 2$, $\sigma = 1$, initial value $S_0 = 1$, and step sizes $\widetilde{h}$ and h. For the smaller step size, we see that Euler's approximation is very close to the exact solution.

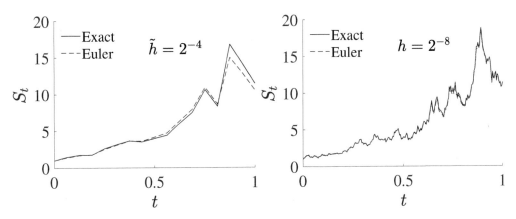

Figure 7.114: Approximation schemes and exact solution for a geometric Brownian motion trajectory.

Exercises

1. Let W be a Wiener process on $[0, t]$. Show that

$$\sum_{k=0}^{n-1} W_{s_{k+1}}(W_{s_{k+1}} - W_{s_k}) \overset{L^2}{\to} (W_t^2 + t)/2,$$

where $s_0, \ldots, s_n$ belong to a segmentation such that $\max_k (s_{k+1} - s_k) \to 0$.

2.* Suppose that W is a Wiener process and $f : \mathbb{R}_+ \to \mathbb{R}$ is a left-continuous deterministic function. Prove that the process $(W_{C(t)}, t \geq 0)$, where the time change C is defined in (7.27), has the same distribution as $Z := (\int_0^t f_s \, dW_s, t \geq 0)$.

3. Suppose we are given the Itô integral $\int_T^t F_s \, dW_s$, where $T \leq t$ is a stopping time with respect to the filtration $\mathcal{F}$. If $X \in \mathcal{F}_T$ and $\mathbb{E}X^2 < \infty$, show that

$$X \int_T^t F_s \, dW_s = \int_0^t X \mathbb{1}_{[T,t]}(s) F_s \, dW_s.$$

4. Consider the martingale process $S_t = \exp(rW_t - tr^2/2)$ from (6.40), where (W_t) is a Wiener process. Using the telescoping sum $S_t - S_0 = \sum_{k=0}^{n-1} S_{s_k}(S_{s_{k+1}}/S_{s_k} - 1)$ and without calling upon Itô's formula, prove that

$$\int_0^t rS_s \, dW_s = S_t - S_0.$$

5. Show that the covariation process $\langle X, Y \rangle_t$ is of finite variation and is at least right-continuous.

6. Let X, Y be two Itô processes on the same probability space, adapted to the same filtration, and with dispersion processes $\sigma, \gamma \in \mathcal{P}_t$, respectively. If $(X^{(n)}), (Y^{(n)})$ are sequences of simple processes with jumps on the same segmentation Π_n of $[0, t]$ and approximating X, Y, then show that

$$\sum_{k=0}^{n-1} (X_{s_{k+1}}^{(n)} - X_{s_k}^{(n)})(Y_{s_{k+1}}^{(n)} - Y_{s_k}^{(n)}) \xrightarrow{L^2} \langle X, Y \rangle_t.$$

7. Suppose that X and Y are Itô processes and define the process $Z := (Z_t, t \geq 0)$, with $Z_t := \langle X, Y \rangle_t$. Prove that

$$(Z_t)^2 \leq \langle X \rangle_t \langle Y \rangle_t.$$

Hence, deduce the *Cauchy–Schwarz* inequality:

$$(V_Z(t))^2 \leq \langle X \rangle_t \langle Y \rangle_t,$$

where $V_Z(t)$ is the total variation of Z in $[0, t]$; see (5.18).

8. Suppose that

$$X_r := X_0 + \int_0^r \mu_s \, \mathrm{d}s + \sum_j \int_0^r \sigma_{s,j} \, \mathrm{d}W_{s,j}, \quad r \in [0, t]$$

is a multivariate Itô process. Show that

$$\mathrm{d}\langle X \rangle_t = \sum_j \sigma_{t,j}^2 \, \mathrm{d}t.$$

9. Define

$$H_k(x, t) := \frac{t^{k/2}}{k!} \hbar_k(x/\sqrt{t}), \quad k \in \mathbb{N},$$

to be a scaled Hermite polynomial; see Section B.5 for the properties of these polynomials. Show that

$$\mathrm{d}H_k(W_t, t) = H_{k-1}(W_t, t)\mathrm{d}W_t,$$

that is, $H_k(x, t)$ is differentiated with respect to the Wiener process in the same way that the monomial $x^k/k!$ is differentiated in ordinary calculus.

10. Prove or disprove: if $X := (X_v, v \in [0, t])$ is a zero-mean Itô process and $\langle X \rangle_v = t, v \in [0, t]$, then X is a Wiener process on $[0, t]$.

11.* Suppose f and g are positive continuous functions on $\mathbb{R}_+$. Suppose that there exists a constant $c \in \mathbb{R}$ such that

$$f(t) \le g(t) + c \int_0^t f(s)\,\mathrm{d}s.$$

Then,

(7.115)
$$f(t) \le g(t) + c \int_0^t e^{c(t-s)} g(s)\,\mathrm{d}s.$$

This is *Gronwall's inequality*. Prove this by first finding an upper bound for

$$h(t) := c\mathrm{e}^{-ct} \int_0^t f(s)\,\mathrm{d}s,$$

in terms of the function g and the constant c.

12. For an Itô diffusion process of the form (7.77), show that the function

$$h(x) := \int_1^x \mathrm{d}y\, \exp\left(\int_1^y \mathrm{d}z\, R(z)\right), \quad \text{with} \quad R(z) := -\frac{2a(z)}{[b(z)]^2}$$

satisfies $Lh = 0$. Use it to derive the probability that the process exits the interval $[l, r]$ through l rather than r.

13. For an Itô diffusion process of the form (7.77), show that the expected exit time $s(x) := \mathbb{E}^x T$ of the interval $[l, r]$ satisfies the differential equation $Ls = -1$ with $s(l) = 0, s(r) = 0$.

14. Show that $\mathbb{E}^x T < \infty$ in Example 7.87.

15. Solve the SDE (7.74) via a separation of variables, similar to Example 7.68.

16. Using stochastic integration by parts (that is, the integral version of (7.57)), show that as $t \uparrow 1$:

$$\int_0^t \frac{1-t}{1-s}\,\mathrm{d}W_s \xrightarrow{\text{a.s.}} 0.$$

17. Let X, Y be two Itô processes on the same probability space, adapted to the same filtration, and with dispersion processes $\sigma, \gamma \in \mathcal{P}_t$, respectively. If $(X^{(n)}), (Y^{(n)})$ are sequences of simple processes with jumps on the same segmentation Π_n of $[0, t]$ and approximating X, Y, then show that

$$\frac{1}{2} \sum_{k=0}^{n-1} (Y_{s_k}^{(n)} + Y_{s_{k+1}}^{(n)})(X_{s_{k+1}}^{(n)} - X_{s_k}^{(n)}) \xrightarrow{L^2} \int_0^t Y_s \circ \mathrm{d}X_s.$$

In other words, the *Itô–Stratonovich* integral can be defined as the limit of the Riemann sum above.

18. Suppose that f is a three times continuously differentiable and invertible function such that $\int_a^b \frac{dx}{f(x)} = g(b) - g(a)$. Use the Itô–Stratonovich method to find the strong solution to the SDE:

$$dX_t = (cf(X_t) + f'(X_t)f(X_t)/2)\, dt + f(X_t)\, dW_t,$$

where c is a constant.

19. Let $Z := (\int_0^t \mu_s\, dW_s, t \geq 0)$ be a continuous local martingale, and let $f(Z) := \exp(Z - \langle Z \rangle/2)$ be its stochastic exponential. Assume that there exists an $\varepsilon > 0$ such that Z satisfies the *weak Novikov condition*:

$$\mathbb{E}\exp((1+\varepsilon)\langle Z \rangle_t/2) < \infty.$$

(a) Show that $(f(Z_s), s \leq t)$ is a local martingale.

(b) Let (T_n) be a localizing sequence for $f(Z)$, so that $(f(Z_{s \wedge T_n}), s \leq t)$ is a martingale. Use Hölder's inequality in Theorem 2.47 to prove that for $p, r, q \geq 1$ and $\frac{1}{p} + \frac{1}{q} = \frac{1}{r}$:

$$\|f(Z_{t \wedge T_n})\|_r \leq [\mathbb{E}f(prZ_{t \wedge T_n})]^{1/p}\left[\mathbb{E}\exp\left(\frac{qr(pr-1)}{2}\langle Z \rangle_{t \wedge T_n}\right)\right]^{1/q}.$$

(c) Use the inequality in (b) with $r = 1 + \varepsilon^4$ and $p = 1 + \varepsilon^2$ to prove that

$$\sup_n \mathbb{E}[f(Z_{t \wedge T_n})]^{1+\varepsilon^2} < \infty.$$

(d) Use the results above to show that $(f(Z_{t \wedge T_n}), n \in \mathbb{N})$ is UI. Hence, deduce that $(f(Z_s), s \leq t)$ is a martingale.

20. Let $Z := (\int_0^t \mu_s\, dW_s, t \geq 0)$ be a continuous local martingale, and let $f(Z) := \exp(Z - \langle Z \rangle/2)$ be its stochastic exponential. Assume that Z satisfies the *Krylov condition*:

$$\lim_{\varepsilon \downarrow 0} \varepsilon \ln \mathbb{E}\exp((1-\varepsilon)\langle Z \rangle_t/2) = 0.$$

(a) Show that the *Krylov condition* is less stringent than the *weak Novikov condition* in Exercise 19.

(b) Prove that there exists a sufficiently small $\varepsilon > 0$ such that $(f((1-\varepsilon)Z_s), s \leq t)$ satisfies the *weak Novikov condition* in Exercise 19. Hence, deduce that $(f((1-\varepsilon)Z_s), s \leq t)$ is a martingale with $\mathbb{E}f((1-\varepsilon)Z_s) = 1$ for all $s \leq t$.

(c) Use Hölder's inequality in Theorem 2.47 with $p = 1/(1-\varepsilon)$ and $q = 1/\varepsilon$ to prove that:

$$1 = \mathbb{E}f((1-\varepsilon)Z_s) \leq [\mathbb{E}f(Z_s)]^{1-\varepsilon}[\mathbb{E}\exp((1-\varepsilon)\langle Z \rangle_s/2)]^{\varepsilon}.$$

(d) Using the results above, prove that $\mathbb{E}f(Z_s) = 1$ for all $s \leq t$. Hence, deduce that $(f(Z_s), s \leq t)$ is a martingale.

21. Solve the *Schwartz log-mean reverting SDE*, which models energy prices:

$$dX_t = \alpha(\mu - \ln X_t)\, X_t\, dt + \sigma X_t\, dW_t, \quad X_0 > 0.$$

22.* Let $x \in \mathbb{R}^d$ with Euclidean norm $\|x\| =: x \geq 0$ and let W be a d-dimensional Wiener process. Define the processes $X := \|x + W\|$ and $\widetilde{W} := (\widetilde{W}_t, t \geq 0)$, where

(7.116)
$$\widetilde{W}_t := \sum_{i=1}^{d} \int_0^t \sigma_{s,i}\, dW_{s,i}, \quad t \geq 0$$

and

$$\sigma_{t,i} := \frac{x_i + W_{t,i}}{\|x + W_t\|}, \quad i = 1, \ldots, d.$$

(a) Prove that $\widetilde{W}$ is a one-dimensional Wiener process.

(b) Show that X satisfies the SDE

$$dX_t = \frac{d-1}{2X_t}\, dt + d\widetilde{W}_t, \quad X_0 = x.$$

23.* For a multidimensional SDE of the form (7.66) the Euler method generalization is to replace (7.111) with

$$Y_{k+1} := Y_k + a(Y_k)\, h + B(Y_k)\, \sqrt{h}\, Z_k$$

as an approximation to X_{kh}, where the $\{Z_k\}$ are standard multivariate normal random vectors. Implement a two-dimensional Euler algorithm to simulate the solution to the (simplified) *Duffing–Van der Pol Oscillator*:

$$dX_t = Y_t\, dt,$$
$$dY_t = \left(X_t\left(\alpha - X_t^2\right) - Y_t\right) dt + \sigma X_t\, dW_t.$$

For the parameters $\alpha = 1$ and $\sigma = 1/2$ draw a plot of X_t against t for $t \in [0, 1000]$. Also plot the trajectory of the two-dimensional process $(X_t, Y_t), t \in [0, 1000]$. Use a step size $h = 10^{-3}$ and starting point $(-2, 0)$.

24. Simulate the paths of the process Y in Example 7.58 in three different ways.

(a) Evaluate the integral $\int_0^t W_s\, ds$ via a standard integration quadrature rule.

(b) Use Euler's method to approximately simulate the process $(\int_0^t s\, dW_s, t \in [0, 1])$ and subtract this from $(tW_t, t \in [0, 1])$. How does the simulated sample path compare with that in part (a)?

(c) Simulate the Gaussian process Y via Algorithm 2.76.

25. Consider the autonomous SDE (7.77). By Itô's lemma:

$$db(X_s) = b'(X_s)\,dX_s + \frac{1}{2}b''(X_s)\,d\langle X_s, X_s\rangle$$

$$= b'(X_s)\{a(X_s)ds + b(X_s)\,dW_s\} + \frac{1}{2}b''(X_s)\,b(X_s)^2\,ds.$$

Denoting $\Delta W_t := W_{t+h} - W_t$ and $\Delta X_t := X_{t+h} - X_t$, it follows that

$$\Delta X_t = \int_t^{t+h} a(X_u)\,du + \int_t^{t+h} b(X_u)\,dW_u$$

$$= h\,a(X_t) + b(X_t)\Delta W_t + O(h\sqrt{h}) + \int_t^{t+h}\int_t^u b'(X_s)\,b(X_s)\,dW_s\,dW_u,$$

where the last term can be written as

$$b'(X_t)\,b(X_t)\frac{1}{2}((\Delta W_t)^2 - h) + O(h^2).$$

This suggests that we can replace the SDE (7.110) with the difference equation

(7.117) $$Y_{k+1} = Y_k + a(Y_k)\,h + b(Y_k)\sqrt{h}\,Z_{k+1} + \underbrace{b'(Y_k)\,b(Y_k)(Z_{k+1}^2 - 1)\frac{h}{2}}_{\text{additional term}},$$

where $Z_1, Z_2, \ldots \overset{\text{iid}}{\sim} \mathcal{N}(0, 1)$. This is *Milstein's method*. The only difference with the Euler method is the additional term involving the derivative of b.

Specify how, for the geometric Brownian motion in Example 7.112, the updating step in (7.113) is modified in Milstein's method.

26. Modify the stochastic difference equation in (7.111) to obtain an Euler approximation to the non-autonomous SDE

$$dX_t = a(X_t, t)\,dt + b(X_t, t)\,dW_t, \quad t \geq 0.$$

27. Consider approximating the strong and weak solutions of the SDE:

$$dX_t = \frac{1}{X_t}\,dt + dW_t, \quad X_0 = x \geq 0,$$

where W is a Wiener process.

(a) Simulate an approximate realization of $(X_t, t \in [0,1])$, $X_0 = x = 1$ on the grid $0, h, 2h, \ldots, 1$ with step size $h = 1/(2^{12} - 1)$ by approximating the solution of the ODE (7.106) in Example 7.105 via the Euler method ($Y_k \approx Z_{kh}$ and $Y_0 = 0$):

$$Y_k = Y_{k-1} + \frac{h}{x + W_{(k-1)h} + Y_{k-1}}, \quad k = 1, 2, \ldots.$$

In other words, simulate $(x + W_{kh} + Y_k, k \le 2^{12})$ as an approximation to $(X_{kh}, k \le 2^{12})$.

(b) Compare the approximation in part (a) with Euler's approximation $(\widetilde{X}_k, k \in \mathbb{N})$ (i.e., $X_{th} \approx \widetilde{X}_k$) to the SDE:

$$(7.118) \qquad \widetilde{X}_k = \widetilde{X}_{k-1} + \frac{h}{\widetilde{X}_{k-1}} + W_{kh} - W_{(k-1)h}, \quad k = 1, 2, \ldots.$$

(c) The explicit or vanilla Euler approximation in (7.118) fails when $X_0 = x = 0$, because the recursion requires division by 0 at $k = 1$. This failure motivates the *implicit Euler* method, in which the approximation $(\widetilde{X}_k, k \in \mathbb{N})$ with $\widetilde{X}_0 = 0$ satisfies:

$$(7.119) \qquad \widetilde{X}_k = \widetilde{X}_{k-1} + \frac{h}{\widetilde{X}_k} + W_{kh} - W_{(k-1)h}, \quad k = 1, 2, \ldots.$$

Simulate a realization of $\widetilde{X}$ on the grid $0, h, 2h, \ldots, 1$ with step size $h = 1/(2^{12} - 1)$.

(d) Let $\mathbf{W}$ be a three-dimensional Wiener process, so that its norm $\|\mathbf{W}\|$ defines a Bessel process, see Exercise 22, and the $\widetilde{W}$ defined in formula (7.116) with $d = 3$ and $\mathbf{x} = \mathbf{0}$ is a Wiener process. If we replace W with $\widetilde{W}$ in (7.119), we obtain the approximate weak solution $(\widetilde{X}_{kh}, k \in \mathbb{N}, X_0 = 0)$ to the SDE, where $\widetilde{X}$ satisfies

$$\widetilde{X}_k = \widetilde{X}_{k-1} + \frac{h}{\widetilde{X}_k} + \widetilde{W}_{kh} - \widetilde{W}_{(k-1)h}, \quad k = 1, 2, \ldots.$$

Using a realization of $\mathbf{W}$ and its corresponding Wiener process $\widetilde{W}$, simulate a sample path of the Bessel process $\|\mathbf{W}\|$ on $[0,1]$ and compare it with the corresponding sample path of $\widetilde{X}$.

SELECTED SOLUTIONS

We have included in this appendix a selection of solutions. These could be used in a tutorial setting, for example.

Various exercises in this book have been inspired by and adapted from the references mentioned in the preface. Additional sources for exercises are Feller (1970), Jacod and Protter (2004), and Williams (1991).

A.1 Chapter 1

4. Write $C = \{B_i, i \in I\}$, where I is a countable set of indexes and the $\{B_i\}$ are disjoint with union E. Let $\mathcal{E}$ be the collection of all sets of the form $\cup_{j \in J} B_j$, where $J \subseteq I$. This is a σ-algebra on E, because (a) $E = \cup_{i \in I} B_i \in \mathcal{E}$; (b) the complement of $\cup_{j \in J} B_j$ is $\cup_{k \in I \setminus J} B_k$, and so is also in $\mathcal{E}$; (c) the union of sets $\cup_{j \in J_n} B_j$, $n = 1, 2, \ldots$ is $\cup_{j \in \cup_n J_n} B_j$, which also lies in $\mathcal{E}$. Since $\mathcal{E}$ is a σ-algebra that contains C, we have $\mathcal{E} \supseteq \sigma C$. Conversely, each set from $\mathcal{E}$ lies in σC, so $\sigma C = \mathcal{E}$.

9. We check the properties of Definition 1.6: (a) $E = f^{-1}F \in f^{-1}\mathcal{F}$ since $F \in \mathcal{F}$, as $\mathcal{F}$ is a σ-algebra on F; (b) Let $A \in f^{-1}\mathcal{F}$, then there exists $B \in \mathcal{F}$ such that $A = f^{-1}B$. The complement of A in E is $E \setminus A = (f^{-1}F) \setminus (f^{-1}B) = f^{-1}(F \setminus B)$. This set belongs to $f^{-1}\mathcal{F}$, since $F \setminus B \in \mathcal{F}$, as $\mathcal{F}$ is a σ-algebra on F; (c) Let $A_1, A_2, \ldots \in f^{-1}\mathcal{F}$, then there exist $B_1, B_2, \ldots \in \mathcal{F}$ such that $A_n = f^{-1}B_n$. Taking the union, we have $\cup_n A_n = \cup_n f^{-1}B_n = f^{-1} \cup_n B_n$. Since $\mathcal{F}$ is a σ-algebra, $\bigcup_n B_n \in \mathcal{F}$ and therefore $\bigcup_n A_n \in f^{-1}\mathcal{F}$.

11. We need to show that the finite additivity and continuity imply countable additivity. Let $B_1, B_2, \ldots$ be disjoint and let $B := \cup_{i=1}^{\infty} B_i$. Also, define $A_n := \cup_{i=n}^{\infty} B_i$, $n = 1, 2, \ldots$. We have, by the finite additivity: $\mu(B) = \sum_{i=1}^{n} \mu(B_i) + \mu(A_{n+1})$. Now,

---

An Advanced Course in Probability and Stochastic Processes. D. P. Kroese and Z. I. Botev. **307**

$\sum_{i=1}^{n} \mu(B_i)$ is increasing in n and is bounded by 1, since $\mu(B) = \mu(E) - \mu(B^c) \leq 1$ by the finite additivity and the fact that $\mu(E) = 1$. Hence, $\lim_{n\to\infty} \sum_{i=1}^{n} \mu(B_i) = \sum_{i=1}^{\infty} \mu(B_i)$ exists, and $\mu(B) = \lim_{n\to\infty} \sum_{i=1}^{n} \mu(B_i) + \lim_{n\to\infty} \mu(A_{n+1}) = \sum_{i=1}^{\infty} \mu(B_i) + 0$, using the continuity property, since $A_1, A_2, \ldots$ is decreasing to $\emptyset$.

14. The complement of $(-\infty, a]$ is (a, ∞). For $a < b$, the intersection of (a, ∞) and $(-\infty, b]$ is $(a, b]$. Finally, $(a, b) = \cap_{n>N}(a, b - 1/n]$ and $\{a\} = \cap_{n>N}(a - 1/n, a + 1/n)$ for N large enough.

15. $\text{Leb}(\mathbb{Q}) = \sum_{x\in\mathbb{Q}} \text{Leb}(\{x\}) = \sum_{x\in\mathbb{Q}} 0 = 0$.

19. At step n of the construction (starting with $n = 0$) we take away the open intervals
$$D_{n,i} := \left(\frac{3i-2}{3^{n+1}}, \frac{3i-1}{3^{n+1}}\right), \quad i = 1, \ldots, 2^n.$$

The deleted intervals form the open set $D := \cup_{n=0}^{\infty} \cup_{i=1}^{2^n} D_{n,i}$, and its complement in $[0, 1]$ is the closed Cantor set C. Since D is a countable union of intervals, its Lebesgue measure is easy to determine:

$$\text{Leb}(D) = \sum_{n=0}^{\infty} \sum_{i=1}^{2^n} \text{Leb}(D_{n,i}) = \sum_{n=0}^{\infty} 2^n 3^{-(n+1)} = \frac{1}{3} \sum_{n=0}^{\infty} \left(\frac{2}{3}\right)^n = 1.$$

Hence, $\text{Leb}(C) = 1 - \text{Leb}(D) = 0$. At *any* stage n in the construction, the remaining set (that is, complement of $\cup_{i=1}^{2^n} D_{n,i}$ in $[0,1]$) consists of the union of closed intervals, and therefore has as many points as the interval $[0, 1]$. It therefore is plausible that C has as many points as $[0, 1]$. To formally prove this, we need to show that there exists a one-to-one function which maps each point in $[0, 1]$ to a point in C. This is done in Exercise 2.6.

26. (a) For fixed x, the function $f_x : y \mapsto f(x, y)$ is a *section* of f. By Exercise 12 it is $\mathcal{F}$-measurable. Since $K(x, \cdot)$ is a measure on $(F, \mathcal{F})$, the mapping $x \mapsto K(x, \cdot)f_x$ is well-defined and in $\mathcal{E}$, and the same holds for $g_x : y \mapsto g(x, y)$. By the linearity of the integral with respect to $K(x, \cdot)$, we have $T(af + bg)(x) = K(x, \cdot)(af_x + bg_x) = aK(x, \cdot)f_x + bK(x, \cdot)g_x = a(Tf)(x) + b(Tf)(x)$. In other words, T is a linear mapping.

(b) If $f_n \uparrow f$, with corresponding x-sections $(f_{n,x})$, then for each x, $(Tf_n)(x) = K(x, \cdot)f_{n,x} \uparrow K(x, \cdot)f_x = (Tf)(x)$, by the Monotone Convergence Theorem, since $K(x, \cdot)$ is a measure. That is, T is a continuous mapping.

(c) Boundedness of K ensures that Tf is bounded for every bounded $f \in \mathcal{E} \otimes \mathcal{F}$. Let
$$\mathcal{M} := \{f \in (\mathcal{E} \otimes \mathcal{F}), f \text{ is positive or bounded} : \quad Tf \in \mathcal{E}\}.$$

It contains the indicator $\mathbb{1}_{E\times F}$. Moreover, if $f, g \in \mathcal{M}$ are bounded, and $a, b \in \mathbb{R}$, then $af + bg$ are bounded and, by the linearity of T, they are in $\mathcal{E}$

again. Finally, (b) shows that for any sequence of positive functions in $\mathcal{M}$ that increases to some f, the latter also belongs to $\mathcal{M}$. Hence, $\mathcal{M}$ is a monotone class of functions. It also includes the indicators $\mathbb{1}_{A \times B}$, $A \in \mathcal{E}$, $B \in \mathcal{F}$, because $(T\mathbb{1}_{A \times B})(x) = \mathbb{1}_A(x) K(x, B)$, where both $\mathbb{1}_A$ and $K(\cdot, b)$ are in $\mathcal{E}_+$. It follows from Theorem 1.33 that $\mathcal{M}$ countains all positive (as well as all bounded) $(\mathcal{E} \otimes \mathcal{F})$-measurable functions.

A.2   Chapter 2

5. We have $F(x) = \alpha F_d(x) + (1 - \alpha) F_c(x)$, where F_d is the cdf of the constant 0 and F_c is the cdf of the Exp(1) distribution. In terms of the Dirac measure at 0 and the Lebesgue measure, the measure μ on $(\mathbb{R}, \mathcal{B})$ is given by:

$$\mu(dx) = \alpha \delta_0(dx) + (1 - \alpha) \mathbb{1}_{[0,\infty)}(x)\, e^{-x} dx.$$

6. (c) Note that q is a strictly increasing function (swap the axes in the graph of F to obtain the graph of q). Each element $x \in D_{n,i}$ is mapped by F to $u = (2i - 1)/2^{n+1}$, and this u is mapped by q to $(3i - 1)/3^{n+1}$, i.e., to the right-endpoint of $D_{n,i}$. Hence, no element of the set D is in the range of q and neither are the left-endpoints of the $\{D_{n,i}\}$. Since $F(x) \le 1$ for all $x \in [0, 1]$, there is no u such that $q(u) = 1$. Thus, the range of q does not include $D \cup C_0$. Now take any $x \in C \setminus C_0$ and consider its value $u = F(x)$. By Exercise 1.24 (b), $q(u) = x$, because $F(x + \varepsilon) > F(x)$ for every $\varepsilon > 0$. Thus, every $x \in C \setminus C_0$ lies in the range of q.

   (d) Every element in $[0, 1]$ is mapped to a unique element in the range $C \setminus C_0$ of q. Since C_0 is a countable set, C has as many elements as $[0, 1]$ (or indeed $\mathbb{R}$), even though it has Lebesgue measure 0.

7. (a) The graphs are given in Figure A.1.

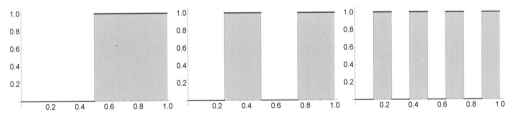

Figure A.1: The functions X_1, X_2, and X_3.

   (b) The area under the graph of each X_n is 1/2, so $\mathbb{P}(X = 1) = 1/2$, and because X_n only takes the values 0 and 1, $X_n \sim \text{Ber}(1/2)$.

(c) Take any finite set of indexes $i_1, \ldots, i_n$. For any selection $x_{i_k} \in \{0, 1\}, k = 1, \ldots, n$, we have

$$\mathbb{P}(X_{i_1} = x_{i_1}, \ldots, X_{i_n} = x_{i_n}) = 2^{-n} = \prod_{k=1}^{n} \mathbb{P}(X_{i_k} = x_{i_k}).$$

8. (a) Since X and Y are independent, the density f of (X, Y) with respect to the Lebesgue measure on $\mathbb{R}^2$ is given by the product of the individual pdfs: $f(x, y) = \mathbb{1}_{[0,1]}(x) \times \mathbb{1}_{\mathbb{R}_+}(y) e^{-y}, (x, y) \in \mathbb{R}^2$.

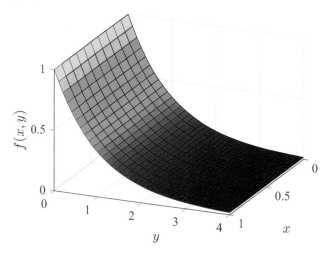

Figure A.2: The joint pdf of X and Y.

(b) It follows from direct computation that

$$\mathbb{P}((X, Y) \in [0, 1] \times [0, 1]) = \int_{[0,1] \times [0,1]} dx \, dy \, f(x, y)$$
$$= \int_0^1 dx \int_0^1 dy \, e^{-y} = 1 - e^{-1}.$$

(c) We have

$$\mathbb{P}(X + Y < 1) = \int_0^1 dx \int_0^{1-x} dy \, f(x, y) = \int_0^1 dx \, \mathbb{P}(Y < 1 - x)$$
$$= \int_0^1 dx \left(1 - e^{-(1-x)}\right) = \left[x - e^{-(1-x)}\right]_0^1 = e^{-1}.$$

9. From Example 2.42 (on multivariate normal distributions), we see that X has a multivariate normal distribution with mean vector $\mathbf{0}$ and covariance matrix $\Sigma = \mathbf{A}\mathbf{A}^\top$. Using Example 2.40 (on linear transformations), it follows that X has density

$$f_X(x) = \frac{f_Z(\mathbf{A}^{-1}x)}{|\mathbf{A}|} = \frac{(2\pi)^{-\frac{n}{2}}}{|\mathbf{A}|} e^{-\frac{1}{2}x^\top (\mathbf{A}^{-1})^\top \mathbf{A}^{-1} x} = \frac{(2\pi)^{-\frac{n}{2}}}{\sqrt{|\Sigma|}} e^{-\frac{1}{2}x^\top \Sigma^{-1} x}, \quad x \in \mathbb{R}^n.$$

This density is that of a standard normal random vector if and only if $\Sigma = I_n$. The geometric explanation is that the spherical symmetry of the standard normal distribution is preserved by any orthogonal transformation (e.g., rotation).

10. (a) Since $X \sim \text{Exp}(\lambda)$ and $Y \sim \text{Exp}(\mu)$, we have $\mathbb{P}(X > x) = e^{-\lambda x}$ and $\mathbb{P}(Y > y) = e^{-\mu y}$. Let $Z := \min(X, Y)$. Then, $Z > z$ if and only if $X > z$ and $Y > z$. Hence,

$$\mathbb{P}(Z > z) = \mathbb{P}(X > z, Y > z) = \mathbb{P}(X > z)\mathbb{P}(Y > z) = e^{-(\lambda+\mu)z}.$$

It follows that the cdf of Z is given by

$$F_Z(z) := 1 - \mathbb{P}(Z > z) = 1 - e^{-(\lambda+\mu)z},$$

which is the cdf of $\text{Exp}(\lambda + \mu)$.

(b) The desired result follows from direct computation:

$$\mathbb{P}(X < Y) = \int_0^\infty dy \int_0^y dx\, f_X(x) f_Y(y) = \int_0^\infty dy\, \mu e^{-\mu y}(1 - e^{-\lambda y})$$

$$= \mu \int_0^\infty dy\, (e^{-\mu y} - e^{-(\lambda+\mu)y}) = \mu \left[-\frac{1}{\mu}e^{-\mu y} + \frac{1}{\lambda+\mu}e^{-(\lambda+\mu)y} \right]_0^\infty$$

$$= \frac{\lambda}{\lambda+\mu}.$$

11. Since $X \sim \mathcal{U}(-\pi/2, \pi/2)$, $f_X(x) = 1/\pi$, $x \in (-\pi/2, \pi/2)$. Let $Y := \tan(X)$. We apply the transformation rule (2.39). The inverse transformation is given by $x = \arctan(y)$, with Jacobian $|dx/dy| = 1/(1 + y^2)$. Hence,

$$f_Y(y) = \frac{1}{\pi(1 + y^2)}, \quad y \in \mathbb{R},$$

which is the pdf of the Cauchy distribution.

13. Let $Z := |X|$. The distribution of Z is called the *half-normal* distribution. Its pdf is twice that of the $\mathcal{N}(0, 1)$ distribution on $\mathbb{R}_+$, so

$$f_Z(z) = \frac{2}{\sqrt{2\pi}}e^{-\frac{1}{2}z^2}, \quad z \in \mathbb{R}_+.$$

Now apply the transformation rule (2.39) to $Y := Z^2$. The inverse transformation $z = \sqrt{y}, y > 0$ has Jacobian $\left|\frac{\partial z}{\partial y}\right| = 1/(2\sqrt{y})$, so

$$f_Y(y) = f_Z(z)\frac{1}{2\sqrt{y}} = \frac{1}{\sqrt{2\pi}}e^{-\frac{1}{2}y}y^{-\frac{1}{2}} = \frac{\left(\frac{1}{2}\right)^{\frac{1}{2}}y^{-\frac{1}{2}}e^{-\frac{1}{2}y}}{\Gamma(\frac{1}{2})}, \quad y > 0,$$

which is the pdf of $\text{Gamma}(\frac{1}{2}, \frac{1}{2}) = \chi_1^2$.

19. We have

$$
\psi(t) = \int_{\mathbb{R}} dx \, \frac{1}{2} e^{-|x|} e^{itx} = \int_{-\infty}^{0} dx \, \frac{1}{2} e^{x(1+it)} + \int_{0}^{\infty} dx \, \frac{1}{2} e^{x(-1+it)}
$$

$$
= \frac{1}{2} \frac{1}{1+it} + \frac{1}{2} \frac{1}{1-it} = \frac{1}{1+t^2}.
$$

22. For $f = a \mathbb{1}_A$ of the form suggested, we have $Nf = aN(A)$, where $N(A) \sim$ Poi$(\mu(A))$. Hence, using the Laplace transform for the Poisson distribution, we have

$$
\mathbb{E} e^{-Nf} = \mathbb{E} e^{-aN(A)} = e^{-\mu(A)(1-e^{-a})} = e^{-\int \mu(dx)(1-e^{-a \mathbb{1}_A(x)})} = e^{-\mu(1-e^{-f})},
$$

where we have used the integral notation μf for the last equation. So (2.80) holds for this case. It also holds if $\mu(A) = \infty$. Next, take f a positive simple function in canonical form; thus, $f = \sum_{i=1}^{n} a_i \mathbb{1}_{A_i}$, where the $\{A_i\}$ are disjoint and the $\{a_i\}$ are in $\mathbb{R}_+$. Then, $Nf = \sum_{i=1}^{n} a_i N(A_i)$, where the $\{N(A_i)\}$ are independent, and $N(A_i) \sim$ Poi$(\mu(A_i))$, $i = 1, \ldots, n$. Therefore,

$$
\mathbb{E} e^{-Nf} = \prod_{i=1}^{n} \mathbb{E} e^{-a_i N(A_i)} = e^{-\sum_{i=1}^{n} \mu(A_i)(1-e^{-a_i})}
$$

$$
= e^{-\int \mu(dx)(1-e^{-\sum_{i=1}^{n} a_i \mathbb{1}_{A_i}(x)})} = e^{-\mu(1-e^{-f})},
$$

so that (2.80) holds again. For a general $f \in \mathcal{E}_+$, take a sequence (f_n) of positive simple functions increasing to f. By the Bounded Convergence Theorem 2.36 (applied to each outcome),

$$
\mathbb{E} e^{-Nf} = \lim_{n} \mathbb{E} e^{-Nf_n} = \lim_{n} e^{-\mu(1-e^{-f_n})} = e^{-\mu(1-e^{-f})},
$$

where the last equality follows from the Monotone Convergence Theorem applied to the measure μ and the sequence $(1 - e^{-f_n}) \uparrow 1 - e^{-f}$.

A.3 Chapter 3

4. (a) If $N \leq n$, then there is a $k \leq n$ such that $|S_k| > a$, which implies that $\max_{k \leq n} |S_k| > a$. Conversely, if $\max_{k \leq n} |S_k| > a$, then there is a $k \leq n$ such that $|S_k| > a$, which implies that $N \leq n$.

(b) For $k < n$, $S_k = \sum_{i=1}^{k} X_i$ and $S_n - S_k = \sum_{i=k+1}^{n} X_i$, so S_k only depends on $X_1, \ldots, X_k$ and $S_n - S_k$ depends only on $X_{k+1}, \ldots, X_n$. By (a), the event $\{N \leq k\}$ depends on $X_1, \ldots, X_k$ and $\{N \leq k - 1\}$ depends on $X_1, \ldots, X_k$, so that $\{N = k\}$ only depends on $X_1, \ldots, X_k$.

(c) By (b),

$$\mathbb{E}[S_k(S_n - S_k)\mathbb{1}_{\{N=k\}}] = \mathbb{E}[S_k\mathbb{1}_{\{N=k\}}] \underbrace{\mathbb{E}[S_n - S_k]}_{= 0} = 0 \quad \text{for all } k < n$$

and trivially also for $k = n$.

(d) We have

$$\begin{aligned}
\mathbb{E}S_n^2\mathbb{1}_{\{N=k\}} &\geq \mathbb{E}[(S_k^2 + 2S_k(S_n - S_k))\mathbb{1}_{\{N=k\}}]\\
&= \mathbb{E}[(S_N^2 + 2S_N(S_n - S_k))\mathbb{1}_{\{N=k\}}]\\
&\geq \mathbb{E}[a^2\mathbb{1}_{\{N=k\}}] + 2a\mathbb{E}[S_n - S_k]\mathbb{E}\mathbb{1}_{\{N=k\}} = a^2\,\mathbb{P}(N = k).
\end{aligned}$$

(e) Using (a) and (d), we have

$$\operatorname{Var} S_n = \mathbb{E}\,S_n^2 \geq \mathbb{E}\left[S_n^2 \sum_{k=1}^{n} \mathbb{1}_{\{N\leq k\}}\right] \geq a^2 \sum_{k=1}^{n} \mathbb{P}(N = k)$$

$$= a^2\,\mathbb{P}(N \leq n) = a^2\,\mathbb{P}(\max_{k\leq n} |S_k| > a).$$

10b. The characteristic function of X_n/n is given by

$$\psi_n(r) := \frac{\lambda/n}{e^{-ir/n} - (1 - \lambda/n)} = \frac{\lambda}{\lambda + n\left(e^{-ir/n} - 1\right)} = \frac{\lambda}{\lambda + (-ir + o(1))}$$

as $n \to \infty$. We see that $\psi_n(r)$ converges to $\psi(r) = \lambda/(\lambda - ir)$, which we recognize
as the characteristic function of the $\mathsf{Exp}(\lambda)$ distribution.

11. (a) The characteristic function of X_1 is

$$\phi_{X_1}(r) := \mathbb{E}\,e^{irX_1} = e^{-|r|}, \quad r \in \mathbb{R}.$$

It follows that the characteristic function of S_n is

$$\phi_{S_n}(r) := (\mathbb{E}\,e^{irX_1})^n = e^{-n|r|} = e^{-|nr|} = \mathbb{E}\,e^{nrX_1}, \quad r \in \mathbb{R}.$$

The distribution of S_n/n is thus the same as that of X_1. Hence, (S_n/n) trivially
converges in distribution to a Cauchy random variable.

(b) We have that

$$\mathbb{P}(M_n \leq x) = [\mathbb{P}(X \leq nx/\pi)]^n = \left(\frac{1}{2} + \arctan(nx/\pi)/\pi\right)^n.$$

Hence, for $x \le 0$ we get $\left(\frac{1}{2} + \arctan(nx/\pi)/\pi\right)^n \le (1/2)^n \to 0$, and for $x > 0$ we have (by L'Hôpital's rule):

$$\ln\left[\lim_n \left(\frac{1}{2} + \arctan(nx/\pi)/\pi\right)^n\right] = \lim_n \frac{\ln\left(\frac{1}{2} + \arctan(nx/\pi)/\pi\right)}{1/n}$$

$$= \lim_n \frac{1/(1+[nx/\pi]^2)}{\frac{1}{2} + \arctan(nx/\pi)/\pi} \frac{x/\pi^2}{-1/n^2}$$

$$= \lim_n \frac{-n^2 x}{\pi^2 + n^2 x^2} = -1/x.$$

Hence,

$$\lim_n \mathbb{P}(M_n \le x) = \exp(-1/x), \quad x > 0.$$

21. Since Y is integrable, we have $\mathbb{E}|Y|\mathbb{1}_{\{|Y|>b\}} \to 0$ as $b \to \infty$ by Proposition 3.30. The condition $f(x)/x \uparrow \infty$ as $x \uparrow \infty$ implies that for any $\varepsilon > 0$, we can always find a b such that $f(x)/x \ge 1/\varepsilon$ for all $x \ge b > 0$. Therefore,

$$\varepsilon^{-1}|X|\mathbb{1}_{\{|X|>b\}} \le \frac{f(|X|)}{|X|}|X|\mathbb{1}_{\{|X|>b\}} \le f(|X|)\mathbb{1}_{\{|X|>b\}}$$

implies that

$$\varepsilon^{-1}\sup_{X\in\mathcal{K}}\mathbb{E}|X|\mathbb{1}_{\{|X|>b\}} \le \sup_{X\in\mathcal{K}}\mathbb{E}f(|X|)\mathbb{1}_{\{|X|>b\}} \le \sup_{X\in\mathcal{K}}\mathbb{E}f(|X|) = c_1 < \infty.$$

Hence, for any $\varepsilon > 0$, there is a sufficiently large b such that $\sup_{X\in\mathcal{K}}\mathbb{E}|X|\mathbb{1}_{\{|X|>b\}} \le \varepsilon c_1$, which is another way to express the fact that $\sup_{X\in\mathcal{K}}\mathbb{E}|X|\mathbb{1}_{\{|X|>b\}} \to 0$ as $b \uparrow \infty$. Hence, $\mathcal{K}$ is UI by Proposition 3.30.

31. (a) This follows from:

$$|\psi_{X_n,Y_n}(\boldsymbol{r}) - \psi_{X,c}(\boldsymbol{r})| \le |\psi_{X_n,c}(\boldsymbol{r}) - \psi_{X,c}(\boldsymbol{r})| + |\psi_{X_n,Y_n}(\boldsymbol{r}) - \psi_{X_n,c}(\boldsymbol{r})|$$

$$= |e^{ir_2 c}\,\mathbb{E}(e^{ir_1 X_n} - e^{ir_1 X})| + |\mathbb{E}\,e^{i(r_1 X_n + r_2 c)}(e^{ir_2(Y_n-c)} - 1)|$$

$$\le |e^{ir_2 c}| \times |\mathbb{E}(e^{ir_1 X_n} - e^{ir_1 X})| + \mathbb{E}\,|e^{i(r_1 X_n + r_2 c)}| \times |e^{ir_2(Y_n-c)} - 1|$$

$$\le |\psi_{X_n}(r_1) - \psi_X(r_1)| + \mathbb{E}\,|e^{ir_2(Y_n-c)} - 1|.$$

(b) We use the fact that

$$|e^{ix} - 1| = \left|\int_0^x d\theta\, ie^{i\theta}\right| \le \int_0^x d\theta\, |ie^{i\theta}| = |x|, \quad x \in \mathbb{R}$$

to obtain the bound:

$$\mathbb{E}\,|e^{ir_2(Y_n-c)} - 1| = \mathbb{E}\,|e^{ir_2(Y_n-c)} - 1|\mathbb{1}_{\{|Y_n-c|>\varepsilon\}} + \mathbb{E}\,|e^{ir_2(Y_n-c)} - 1|\mathbb{1}_{\{|Y_n-c|\le\varepsilon\}}$$

$$\le 2\,\mathbb{E}\,\mathbb{1}_{\{|Y_n-c|>\varepsilon\}} + \mathbb{E}\,|r_2(Y_n-c)|\mathbb{1}_{\{|Y_n-c|\le\varepsilon\}}$$

$$\le 2\,\mathbb{P}[|Y_n - c| > \varepsilon] + |r_2|\,\varepsilon.$$

(c) The first term in the right-hand side of (3.52) goes to 0, since $X_n \xrightarrow{d} X$, and Theorem 3.24 implies that $\psi_{X_n}(r_1) \longrightarrow \psi_X(r_1)$. The second term in the right-hand side of (3.52) also goes to 0, as $Y_n \xrightarrow{d} c$ and ε is arbitrary in (b). It follows that $\mathbf{Z}_n \xrightarrow{d} \mathbf{Z}$ and, by the continuity of g, we have $g(\mathbf{Z}_n) \xrightarrow{d} g(\mathbf{Z})$; that is, $g(X_n, Y_n) \xrightarrow{d} g(X, c)$.

36. First, the symmetry of the distribution shows that $\mathbb{E}X_k = 0$ so that $\mathbb{E}\overline{X}_n = 0$. Second, a direct computation shows that

$$\mathbb{V}\mathrm{ar}\, X_k = \mathbb{E}X_k^2 = \frac{2k^2}{2k^2} + 1 - \frac{1}{k^2} = 2 - \frac{1}{k^2},$$

so that

$$\mathbb{V}\mathrm{ar}(\sqrt{n}\,\overline{X}_n) = \frac{1}{n}\sum_{k=1}^{n} \mathbb{E}X_k^2 = 2 - \frac{1}{n}\sum_{k=1}^{n} \frac{1}{k^2} \to 2.$$

Next, we show that $\sqrt{n}\,\overline{X}_n \xrightarrow{d} Z$. Since

$$\sum_k \mathbb{P}(|X_k| \neq 1) = \sum_k \frac{1}{k^2} < \infty,$$

the first part of the Borel–Cantelli Lemma 3.14 shows that $\sum_k \mathbb{1}_{\{|X_k|\neq 1\}} < \infty$ almost surely. This implies that there is an almost surely finite random integer N such that $|X_k| = 1$ for $k > N$. Suppose that $Y_1, Y_2, \ldots$ are iid with $\mathbb{P}(Y = 1) = \mathbb{P}(Y = -1) = 1/2$. Then, for sufficiently large $n > N$, we can write:

$$\sqrt{n}\,\overline{X}_n = \underbrace{\frac{\sum_{k=1}^{N}(X_k - Y_k)}{\sqrt{n}}}_{\xrightarrow{\text{a.s.}} 0} + \underbrace{\frac{\sum_{k=1}^{N} Y_k + \sum_{k>N}^{n} X_k}{\sqrt{n}}}_{=:Z_n}.$$

Since the distribution of X_k, conditional on $|X_k| = 1$, is the same as that of Y, we deduce that Z_n has the same distribution as $\frac{1}{\sqrt{n}}\sum_{k=1}^{n} Y_k$. An application of the Central Limit Theorem 3.45 to $\frac{1}{\sqrt{n}}\sum_{k=1}^{n} Y_k$ then yields that $Z_n \xrightarrow{d} Z$ or, equivalently, that $\sqrt{n}\,\overline{X}_n \xrightarrow{d} Z$.

A.4   Chapter 4

4. We have that

$$\mathbb{P}(X = x \mid X + Y = z) = \frac{\mathbb{P}(X = x, X + Y = z)}{\mathbb{P}(X + Y = z)} = \frac{\mathbb{P}(Y = z - x, X = x)}{\mathbb{P}(X + Y = z)}$$

$$= \frac{\mathbb{P}(Y = z - x) \times \mathbb{P}(X = x)}{\mathbb{P}(X + Y = z)} = \frac{\exp(-\mu)\frac{\mu^{z-x}}{(z-x)!} \times \exp(-\lambda)\frac{\lambda^x}{x!}}{\exp(-\mu - \lambda)\frac{(\mu+\lambda)^z}{z!}}$$

$$= \frac{z!}{x!(z-x)!}\frac{\mu^{z-x}\lambda^x}{(\lambda+\mu)^z} = \binom{z}{x}p^{z-x}(1-p)^x,$$

where $p := \lambda/(\lambda + \mu)$.

7. First observe that $\{N_k = j\} = \{S_{j-1} = k - 1, X_j = 1\}$. Thus,

$$\mathbb{P}(N_k = j) = \mathbb{P}(X_j = 1 \mid S_{j-1} = k - 1)\mathbb{P}(S_{j-1} = k - 1) = p\binom{j-1}{j-k}p^{k-1}(1-p)^{j-k}$$

$$= \binom{j-1}{j-k}p^k(1-p)^{j-k}, \quad j = k, k+1, \ldots.$$

This gives a negative binomial distribution starting at k.

11. The MGF of Z is given by

$$\mathbb{E}\exp(sZ) = \mathbb{E}\,\mathbb{E}[\exp(sZ) \mid X_3]$$

$$= \mathbb{E}\exp\left(\tfrac{s^2}{2(1+X_3^2)}\right) \times \exp\left(\tfrac{s^2 X_3^2}{2(1+X_3^2)}\right) = \exp(s^2/2),$$

which implies that Z is standard normal.

16. If the distribution satisfies the Markov property (a), then (b) follows trivially. Now we show that (b) implies (a).

Conditional on X_t, consider the joint distribution of X_{t+1} and X_k for some arbitrary $k < t$. By assumption, $(X_{t+1}, X_k) \mid X_t$ has a jointly normal distribution. This means that if the conditional covariance

$$\mathbb{E}[X_{t+1}X_k \mid X_t] - \mathbb{E}[X_{t+1} \mid X_t] \times \mathbb{E}[X_k \mid X_t]$$

is 0, then X_{t+1} and X_k are conditionally independent. To simplify, consider

$$\begin{aligned}
\mathbb{E}[X_{t+1}X_k \mid X_t] &= \mathbb{E}[\,\mathbb{E}[X_{t+1}X_k \mid X_j, j \le t]\, \mid X_t] \\
&= \mathbb{E}[\,X_k\,\mathbb{E}[X_{t+1} \mid X_j, j \le t]\, \mid X_t] \\
&= \mathbb{E}[\,X_k\,\mathbb{E}[X_{t+1} \mid X_t] \mid X_t] \\
&= \mathbb{E}[X_{t+1} \mid X_t] \times \mathbb{E}[X_k \mid X_t].
\end{aligned}$$

Since the conditional covariance is 0, we can deduce that X_{t+1}, X_k, given X_t, are independent. Since $k < t$ was arbitrary, this shows that given X_t, the future X_{t+1} is independent of the past $\{X_k, k < t\}$.

A.5   Chapter 5

3. (a) Integrability and adaptedness are evident. The martingale property follows from:

$$\begin{aligned}
\mathbb{E}_n Z_{n+1} &= \mathbb{E}_n[X_n + 2B_n - 1] - (n+1)(p-q) \\
&= X_n + (p-q) - (n+1)(p-q) = X_n - n(p-q) = Z_n.
\end{aligned}$$

(b) The martingale property for the stopped martingale $(Z_{n \wedge T})$ shows that $\mathbb{E} Z_{n \wedge T} = \mathbb{E} Z_0 = a$ for all n. Since T is almost surely finite, $Z_{n \wedge T} \uparrow Z_T$ and so, by the Monotone Convergence Theorem,

$$\mathbb{E} Z_T = a.$$

Since

$$\mathbb{E} Z_T = \mathbb{E} X_T - \mathbb{E} T (p - q) = b \, \mathbb{P}(X_T = b) - (p - q) \, \mathbb{E} T,$$

we have

$$\mathbb{E} T = \frac{b \, \mathbb{P}(X_T = b) - a}{p - q},$$

where $\mathbb{P}(X_T = b)$ is given in (5.34).

4. (a) Let $M_n := X_n^2 - n$ and $U_n := 2B_{n+1} - 1$ for $n \in \mathbb{N}$. Integrability and adaptedness for the process (M_n) are evident. The martingale property follows from:

$$\begin{aligned}
\mathbb{E}_n M_{n+1} &= \mathbb{E}_n X_{n+1}^2 - (n + 1) = \mathbb{E}_n (X_n + U_{n+1})^2 - (n + 1) \\
&= \mathbb{E}_n [X_n^2 + 2X_n U_{n+1} + U_{n+1}^2] - (n + 1) \\
&= X_n^2 + 2X_n \mathbb{E}_n U_{n+1} + \mathbb{E}_n U_{n+1}^2 - (n + 1) = X_n^2 + 0 + 1 - (n + 1) = M_n.
\end{aligned}$$

(b) In the same way as in Example 5.33, we may assume that T is almost surely finite. The martingale property for the stopped martingale $(M_{n \wedge T})$ combined with the finiteness of T shows that

$$\mathbb{E} M_T = M_0 = 0.$$

But also, $\mathbb{E} M_T = \mathbb{E} X_T^2 - \mathbb{E} T = a^2 - \mathbb{E} T$, so that $\mathbb{E} T = a^2$.

5. The process Y is integrable since $0 \le Y_n = 2^n (1 - X_n) \le 2^n$. As Y_n is a (measurable) function of X_n, we have $\sigma(X_0, \ldots, X_n) = \sigma(Y_0, \ldots, Y_n)$ and Y is adapted to the natural filtration. The martingale property follows from:

$$\begin{aligned}
\mathbb{E}_n Y_{n+1} &= \mathbb{E}_n 2^{n+1} (1 - X_{n+1}) = 2^{n+1} (1 - \mathbb{E}_n X_{n+1}) \\
&= 2^{n+1} (1 - (1 + X_n)/2) = 2^{n+1} (1 - X_n)/2 = Y_n.
\end{aligned}$$

10. (a) We have $(T \mid X_1 = 1) = 1$, and $(T \mid X_1 = -1)$ is distributed as $1 + T' + T''$, where T' and T'' are independent copies of T. Hence, G satisfies

$$G(z) = \mathbb{E} \, \mathbb{E}_{X_1} z^T = \frac{1}{2} z + \frac{1}{2} z G^2(z).$$

This quadratic equation has two solutions. But only the stated solution yields a valid probability generating function.

(b) We have

$$G(z) = z^{-1} \left(1 - \sum_{k=0}^{\infty} \binom{\frac{1}{2}}{k} (-1)^k z^{2k} \right) = \sum_{k=1}^{\infty} \binom{\frac{1}{2}}{k} (-1)^{k-1} z^{2k-1}.$$

From this, we conclude that $\mathbb{P}(T = 2k) = 0$ for $k = 1, 2, \ldots$ and

$$\mathbb{P}(T = 2k - 1) = \binom{\frac{1}{2}}{k} (-1)^{k-1}, \quad k = 1, 2, \ldots.$$

Consequently, using again Newton's formula, we have

$$\mathbb{P}(T < \infty) = \sum_{k=1}^{\infty} \binom{\frac{1}{2}}{k} (-1)^{k-1} = -\sum_{k=1}^{\infty} \binom{\frac{1}{2}}{k} (-1)^k = 1 - \sum_{k=0}^{\infty} \binom{\frac{1}{2}}{k} (-1)^k$$
$$= 1 - (1-1)^{1/2} = 1.$$

Also,

$$\mathbb{E}T = \sum_{k=1}^{\infty} k \binom{\frac{1}{2}}{k} (-1)^{k-1} = \lim_{z \uparrow 1} G'(z) = \lim_{z \uparrow 1} \frac{1 - \sqrt{1 - z^2}}{z^2 \sqrt{1 - z^2}} = \infty.$$

13. (a)
$$\mathbb{E}_n Y_{n+1} = \mathbb{E}_n e^{\theta M_{n+1}} \geq \exp(\theta \, \mathbb{E}_n M_{n+1}) = e^{\theta M_n},$$

where we have use the fact that $e^{\theta x}$ is a convex function in x for any $\theta \in \mathbb{R}$ and applied Jensen's inequality (Lemma 2.45).

(b) In (5.43), take $b := e^{\theta x}$ and $Y_n := e^{\theta M_n}$.

(c) By repeated conditioning (see (4.5)), we have

$$\mathbb{E} e^{\theta M_n} = \mathbb{E} \, \mathbb{E}_{n-1} \exp \left(\theta M_{n-1} + \theta (M_n - M_{n-1}) \right)$$
$$= \mathbb{E} \left[\exp(\theta M_{n-1}) \, \mathbb{E}_{n-1} \exp(\theta (M_n - M_{n-1})) \right] \leq \mathbb{E} e^{\theta M_{n-1}} e^{\frac{1}{2} \theta^2 c_n^2},$$

(d) Repeatedly applying the bound in (d), and noting that $M_0 = 0$, gives the bound

$$\mathbb{E} e^{\theta M_n} \leq \exp \left(\frac{1}{2} \theta^2 \sum_{k=1}^{n} c_n^2 \right).$$

Hence,

$$\mathbb{P} \left(\max_{k \leq n} M_k \geq x \right) \leq \exp \left(-\theta x + \frac{1}{2} \theta^2 \sum_{k=1}^{n} c_k^2 \right).$$

We now find the value θ^* which minimizes the upper bound. The exponent is quadratic in θ so $\theta^* = x / (\sum_{k=1}^{n} c_k^2)$.

15. Let r be the probability given in (5.34). Then, $(M_{n \wedge T})$ converges almost surely and in L^1 to a random variable that takes the values 0 and b with probabilities $1 - r$ and r, respectively.

23. We will show that $(M_n, n = 1, 2, \ldots)$ is a UI martingale. Theorem 5.52 then implies that it converges almost surely and in L^1 to an integrable random variable. Adaptedness is evident from the fact that M_n is measurable with respect to $\sigma(X_1, \ldots, X_n)$. For uniform integrability it suffices to show that $\sup_n \mathbb{E} M_n^2 < \infty$, using Item 3 of Proposition 3.34, with $f(x) := x^2$. Indeed, we have

$$\mathbb{E} M_n^2 = \sum_{j=1}^{n} \mathbb{E} X_j^2 j^{-2} = \sum_{j=1}^{n} j^{-2} < \sum_{j=1}^{\infty} j^{-2} < \infty.$$

Finally, the martingale property follows from:

$$\mathbb{E}_n M_{n+1} = \mathbb{E}_n \sum_{j=1}^{n+1} X_j j^{-1} = \mathbb{E}_n (n+1)^{-1} X_{n+1} + \sum_{j=1}^{n} X_j j^{-1} = M_n,$$

where we have used the fact that the $\{X_n\}$ are independent zero-mean random variables.

A.6   Chapter 6

4. Define $X := [W_{t_1}, \ldots, W_{t_d}]^\top$ and $t := [t_1, \ldots, t_d]^\top$. Let $(t^{(n)})$ be a sequence in $\mathbb{R}^d$, with all its components lying in D_n, that converges to t as $n \to \infty$. Denote by $X^{(n)}$ the random vector whose kth component is $W_{t_k}^{(n)}$, $k = 1, \ldots, d$. Since $W^{(n)}$ is a zero-mean Gaussian process with covariance $\mathbb{E} W_s^{(n)} W_t^{(n)} = s \wedge t$ for all $s, t \in D_n$, we have that $X^{(n)}$ is Gaussian with mean vector $\mathbf{0}$ and covariance matrix $\mathbf{\Sigma}_n$ with (i, j)th element $t_{i \wedge j}^{(n)}$ for $i, j \in \{1, \ldots, d\}$. Obviously, $\mathbf{\Sigma}_n$ converges elementwise to the matrix $\mathbf{\Sigma}$ with (i, j)th element $t_{i \wedge j}$. Since, $(X^{(n)})$ converges almost surely to X, it converges in distribution. The characteristic function of X is thus, by Theorem 3.24, equal to the limit of $e^{-r^\top \mathbf{\Sigma}_n r / 2}$ as $n \to \infty$, which is $e^{-r^\top \mathbf{\Sigma} r / 2}$, showing that X is multivariate Gaussian with mean vector $\mathbf{0}$ and covariance matrix $\mathbf{\Sigma}$. Consequently, (W_t) is a zero-mean Gaussian process with covariance function $\mathbb{E} W_s W_t = s \wedge t$.

9. Let T be the exit time and define $p := \mathbb{P}^x(X_T = b)$. Applying Doob's stopping Theorem 5.83 to the stopped martingale $(W_{T \wedge t})$, we have $\mathbb{E}^x W_{T \wedge t} = \mathbb{E}^x W_0 = x$ for every t. Since $a \leq W_{T \wedge t} \leq b$, $\lim_{t \to \infty} \mathbb{E}^x W_{T \wedge t} = \mathbb{E}^x \lim_{t \to \infty} W_{T \wedge t} = \mathbb{E}^x W_T$, by the Bounded Convergence Theorem 2.36. Consequently, we have

$$x = \mathbb{E}^x W_T = a \, \mathbb{P}^x(W_T = a) + b \, \mathbb{P}^x(W_t = b) = a(1 - p) + bp,$$

from which the first result follows. For the second result, define $M_t := W_t^2 - t, t \geq 0$. Obviously, $|M_t| \leq a^2 \vee b^2 + T$ for $t \in [0, T]$. Hence, if $\mathbb{E}^x T < \infty$, we can apply Proposition 5.96 to conclude that $\mathbb{E}^x M_0 = \mathbb{E}^x M_T$, leading to:

$$x^2 = \mathbb{E}^x M_0 = \mathbb{E}^x M_T = (1 - p)(a^2 - \mathbb{E}^x T) + p(b^2 - \mathbb{E}^x T),$$

which shows that

$$\mathbb{E}^x T = (1 - p)a^2 + pb^2 - x^2 = (x - a)(b - x).$$

To verify that indeed $\mathbb{E}^x T < \infty$, write

$$\mathbb{E}^x T = \int_0^\infty dt \, \mathbb{P}^x(T > t)$$

and consider $t \in [k - 1, k)$ for some $k = 1, 2, \ldots$. The event $\{T > t\} = \{W_s \in (a, b) \text{ for all } s \in [0, t]\}$ is contained in the event $\{a < W_1 < b\} \cap \{a < W_2 - W_1 < b\} \cap \cdots \cap \{a < W_k - W_{k-1} < b\}$, and so $\mathbb{P}^x(T > t) \leq \theta^k \mathbb{1}_{[k,k+1)}(t)$, where $\theta := \sup_x \mathbb{P}^x(a < W_1 < b) < 1$. Hence, $\mathbb{E}^x T \leq \theta/(1 - \theta) < \infty$.

13. Let $Y := 1/T_x$. Then, $Y \sim \text{Gamma}(1/2, x^2/2)$. Since Y has no probability mass at 0, we have $1 = \mathbb{P}(Y > 0) = \mathbb{P}(T_x < \infty)$, as had to be shown. Moreover,

$$\mathbb{E} T_x = \int_0^\infty dt \, t \, f_{T_x}(t) \geq \int_1^\infty dt \, \frac{x}{\sqrt{2\pi t}} e^{-x^2/2} = \infty.$$

22. As $T_b < T_c$,

$$\mathbb{P}(T_b < T_{-a} < T_c) = \mathbb{P}(T_b < T_{-a}, T_{-a} < T_c)$$
$$= \mathbb{P}(T_{-a} < T_c \mid T_b < T_{-a})\mathbb{P}(T_b < T_{-a}).$$

Then,

$$\mathbb{P}(T_b < T_{-a}) = \frac{a}{b + a},$$

where we used Exercise 9. The strong Markov property of the Wiener process means that $W_{t+T_b} - W_{T_b}, t \geq 0$ is a Wiener process and independent of $\mathcal{F}_{T_b}^+$. Conditioning on hitting b before $-a$, we have:

$$\mathbb{P}(T_{-a} < T_c \mid T_b < T_{-a}) = \mathbb{P}(T_{-a-b} < T_{c-b}) = \frac{c - b}{c - b + | - a - b|} = \frac{c - b}{c + a}.$$

Hence,

$$\mathbb{P}(T_b < T_{-a} < T_c) = \left(\frac{a}{b + a}\right)\left(\frac{c - b}{c + a}\right).$$

26. Recall from (5.18) that the total variation of $(N_s, s \in [0, t])$ is given by:

$$\sup_{\Pi_n} \sum_{k=0}^{n-1} |N_{s_{k+1}} - N_{s_k}|.$$

Since N_{s_k} can be interpreted as the number of arrivals up to time s_k, the process is increasing: $N_{s_{k+1}} \geq N_{s_k}$. Therefore, $|N_{s_{k+1}} - N_{s_k}| = N_{s_{k+1}} - N_{s_k}$ and the total variation equals the telescoping sum:

$$\sup_{\Pi_n} \sum_{k=0}^{n-1} (N_{s_{k+1}} - N_{s_k}) = N_t - N_0 = N_t.$$

For the quadratic variation, consider the fact that the value of $\Delta N_{s_k} := N_{s_{k+1}} - N_{s_k}$ belongs to the set $\mathbb{N}$ of natural numbers and $\Delta N_{s_k} > 0$ only if $(s_k, s_{k+1}]$ contains at least one jump/arrival. Since $\|\Pi_n\| \to 0$, and the Poisson process is right-continuous and left-limited, there is a large enough n such that $(s_k, s_{k+1}]$ contains no more than one arrival. Therefore,

$$\lim_{n \uparrow \infty} \sum_{k=0}^{n-1} (N_{s_{k+1}} - N_{s_k})^2 = \sum_{k: \Delta N_{s_k} > 0} [\Delta N_{s_k}]^2 = N_t.$$

An alternative derivation uses the results from Example 5.20. Namely, since $\int_0^t N_{s-} \, dN_s = N_t(N_t - 1)/2$ and $\int_0^t N_s \, dN_s = (N_t + 1)N_t/2$, we can write

$$\lim_{n \uparrow \infty} \sum_{k=0}^{n-1} [\Delta N_{s_k}]^2 = \lim_{n \uparrow \infty} \sum_{k=0}^{n-1} N_{s_{k+1}} \Delta N_{s_k} - \lim_{n \uparrow \infty} \sum_{k=0}^{n-1} N_{s_k} \Delta N_{s_k}$$

$$= \int_0^t N_s \, dN_s - \int_0^t N_{s-} \, dN_s = N_t.$$

A.7 Chapter 7

2. The process $(W_{C(t)})$ is zero-mean, has independent increments, and is Gaussian. The same properties hold for (Z_t). To show that the two processes have the same distribution, it thus remains to show that for every t the variance of $W_{C(t)}$ is equal to the variance of Z_t. But this is immediate from

$$\mathbb{E} W_{C(t)}^2 = \mathbb{V}\mathrm{ar} \, W_{C(t)} = C(t) = \int_0^t f^2(s) \, ds = \mathbb{E} Z_t^2.$$

11. The derivative of the function h satisfies

$$h'(t) = \left(f(t) - c \int_0^t f(s) \, ds \right) c e^{-ct} \leq g(t) \, c e^{-ct},$$

so that, by integration of $h'(s)$ from 0 to t, we find that

$$h(t) \le \int_0^t g(s)\,ce^{-cs}\,ds.$$

Multiplying both sides with e^{ct} gives

$$c\int_0^t f(s)\,ds = e^{ct}h(t) \le c\int_0^t e^{c(t-s)}g(s)\,ds,$$

which shows the result for $c \ne 0$. The case $c = 0$ is trivial.

22.   (a) The process $\widetilde{W}$ is a martingale that has quadratic variation (see Exercise 8):

$$\langle \widetilde{W} \rangle_t = \sum_{i=1}^d \int_0^t \sigma_{s,i}^2\,ds = \int_0^t \underbrace{\sum_{i=1}^d \sigma_{s,i}^2}_{=\,1}\,ds = t.$$

    Consequently, by Lévy's characterization in Example 7.61, $\widetilde{W}$ is a Wiener process.

  (b) Let $X_t := \|x + W_t\|$. By the multidimensional Itô formula in Theorem 7.51 applied to $\|x + W_t\|$, we have

$$dX_t = \sum_{i=1}^d \frac{x_i + W_{t,i}}{\|x + W_t\|}\,dW_{t,i} + \frac{d-1}{2X_t}\,dt.$$

In other words, the process X satisfies the SDE

$$dX_t = \frac{d-1}{2X_t}\,dt + d\widetilde{W}_t, \quad X_0 = x,$$

where $\widetilde{W}$ is a Wiener process.

23. The following MATLAB code generates the process $((X_t, Y_t), t \ge 0)$.

```
alpha = 1; sigma = 0.5;
a1 = @(x1,x2,t) x2;
a2 = @(x1,x2,t) x1*(alpha-x1^2)-x2;
b1 = @(x1,x2,t) 0 ;
b2 = @(x1,x2,t) sigma*x1;
n=10^6; h=10^(-3); t=h.*(0:1:n); x1=zeros(1,n+1); x2=x1;
x1(1)=-2;
x2(1)=0;
```

```
for k=1:n
    x1(k+1)=x1(k)+a1(x1(k),x2(k),t(k))*h+ ...
    b1(x1(k),x2(k),t(k))*sqrt(h)*randn;
    x2(k+1)=x2(k)+a2(x1(k),x2(k),t(k))*h+ ...
    b2(x1(k),x2(k),t(k))*sqrt(h)*randn;
end
step = 100; %plot each 100th value
figure(1),plot(t(1:step:n),x1(1:step:n),'k-')
figure(2), plot(x1(1:step:n),x2(1:step:n),'k-');
```

Figure A.3 shows two plots of interest. That left pane shows that the process oscillates between two modes.

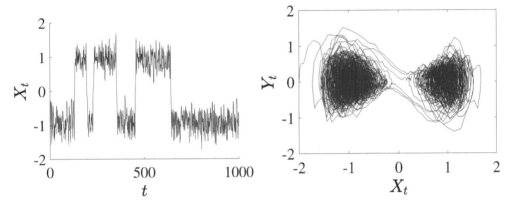

Figure A.3: Typical trajectories for the Duffing–Van der Pol Oscillator.

FUNCTION SPACES

This appendix reviews a number of topics from functional analysis, including metric, normed, and Hilbert spaces. The emphasis is on L^2 function spaces and their orthonormal bases.

The development of mathematics often involves the continual generalization of basic concepts. For example, the set of natural numbers $\mathbb{N}$ is generalized to the set of integers $\mathbb{Z}$ and rational numbers $\mathbb{Q}$, which are then further generalized to the sets $\mathbb{R}$ and $\mathbb{C}$ of real and complex numbers, which offer yet more generalizations in the form of multidimensional spaces of numbers and spaces of functions. While it may lead to a sometimes overwhelming growth of abstractions, this constant generalization can also bring about a *simplification* of ideas by identifying common patterns, leading to fundamental constructs such as metric, normed, and inner product spaces.

B.1 Metric Spaces

A *metric space* is a set of elements (or points) equipped with a *metric* function that assigns a "distance" between any two elements of the set.

Definition B.1: Metric Space

The pair (E, d), where E is a set and $d : E \times E \to \mathbb{R}_+$ is a *metric* on E, is called a *metric space* if for all $x, y, z \in E$:

1. *(Finiteness)*: $0 \le d(x, y) < \infty$; that is, d is positive and finite.
2. *(Zero)*: $d(x, y) = 0$ if and only if $x = y$.
3. *(Symmetry)*: $d(x, y) = d(y, x)$.
4. *(Triangle inequality)*: $d(x, y) \le d(x, z) + d(z, y)$.

■ **Example B.2** (**Metric Spaces**) The quintessential metric space is the space $\mathbb{R}^n$, equipped with the Euclidean metric:

$$d_n(\boldsymbol{x}, \boldsymbol{y}) := \sqrt{(x_1 - y_1)^2 + \cdots + (x_n - y_n)^2}.$$

Another important example is the space of bounded functions on the interval $[0, 1]$, equipped with the metric: $d(x, y) := \sup_{t \in [0,1]} |x(t) - y(t)|$.   ■

■ **Example B.3** (**Isometry**) A mapping $f : E \to \widetilde{E}$ between the two metric spaces (E, d) and $(\widetilde{E}, \widetilde{d})$ is called an *isometry* if f preserves distances between any $x, y \in E$:

$$\widetilde{d}(f(x), f(y)) = d(x, y).$$

For instance, suppose that $\mathbf{U} \in \mathbb{R}^{n \times m}$ is an orthonormal matrix. Then, the mapping $\boldsymbol{x} \mapsto \mathbf{U}\boldsymbol{x}$, where $\boldsymbol{x} \in \mathbb{R}^m$, is a *linear* isometry from the Euclidean space $(\mathbb{R}^m, d_m)$ to $(\mathbb{R}^n, d_n)$, because for any $\boldsymbol{x}, \boldsymbol{y} \in \mathbb{R}^m$ we have

$$d_n^2(\mathbf{U}\boldsymbol{x}, \mathbf{U}\boldsymbol{y}) = (\boldsymbol{x} - \boldsymbol{y})^\top \mathbf{U}^\top \mathbf{U}(\boldsymbol{x} - \boldsymbol{y}) = d_m^2(\boldsymbol{x}, \boldsymbol{y}).$$

■

On a metric space E we can define "open" sets to be unions (not necessarily countable) of sets of the form $\{y \in E : d(\boldsymbol{x}, \boldsymbol{y}) < r\}$, where r is a positive real number. In the case where $E = \mathbb{R}^d$ is equipped with the Euclidean metric, this set is simply an "open ball" centered at $\boldsymbol{x}$ with radius r. More generally, we can define a collection of open sets $\mathcal{T}$, as follows:

Definition B.4: Topological Space

The pair $(E, \mathcal{T})$, where E is a set and $\mathcal{T}$ is a collection of subsets of E, is called a *topological space*, provided that $\mathcal{T}$ satisfies:
1. $\emptyset \in \mathcal{T}$ and $E \in \mathcal{T}$.
2. *Any* (not necessary countable) union of sets in $\mathcal{T}$ belongs to $\mathcal{T}$.
3. Any *finite* intersection of sets in $\mathcal{T}$ belongs to $\mathcal{T}$.

The sets in $\mathcal{T}$ are called *open sets* and $\mathcal{T}$ is called a *topology* on E.

All metric spaces are topological spaces. In a topological space $(E, \mathcal{T})$ the collection of open sets $\{O_\alpha, \alpha \in \mathbb{R}\}$ is called an *open cover* of $F \subseteq E$, provided that each $O_\alpha \subset E$ and $F \subseteq \cup_\alpha O_\alpha$. A *subcover* is a subset of $\{O_\alpha, \alpha \in \mathbb{R}\}$ that still covers F.

Convergence in metric spaces is similar to convergence in $\mathbb{R}$. In particular, we say that a sequence (x_n) in a metric space (E, d) *converges* to $x \in E$, if $d(x_n, x) \to 0$ as $n \to \infty$. A convergent sequence has a unique limit and is *bounded*; that is, $\sup_n d(x_n, r) < \infty$ for some $r \in E$.

■ **Example B.5 (Metric Continuity)** Suppose that both (x_n) and (y_n) converge in E; that is, $d(x_n, x) \to 0$ and $d(y_n, y) \to 0$ for some $x, y \in E$. Then, $d(x_n, y_n) \to d(x, y)$. Indeed, by the triangle inequality we have:

$$d(x_n, y_n) \le d(x_n, x) + d(x, y) + d(y, y_n),$$
$$d(x, y) \le d(x, x_n) + d(x_n, y_n) + d(y_n, y).$$

Hence, using the symmetry $d(x_n, x) = d(x, x_n)$ and $d(y, y_n) = d(y_n, y)$, we obtain:

$$\begin{aligned} |d(x_n, y_n) - d(x, y)| &= [d(x_n, y_n) - d(x, y)] \vee [d(x, y) - d(x_n, y_n)] \\ &\le [d(x_n, x) + d(y, y_n)] \vee [d(x, x_n) + d(y_n, y)] \\ &\le d(x_n, x) + d(y_n, y) \to 0. \end{aligned}$$

■

■ **Example B.6 (Heine–Cantor Theorem)** Suppose that we have two metric spaces, (E, d) and $(\widetilde{E}, \widetilde{d})$, and a continuous function $f : E \to \widetilde{E}$. Here, the continuity of f means that for any $\varepsilon > 0$ and $x \in E$, there exists a $\delta > 0$ (depending on ε and x) such that

$$(y \in E \text{ and } d(x, y) < \delta) \implies \widetilde{d}(f(x), f(y)) < \varepsilon.$$

In other words, $\widetilde{d}(f(x), f(y)) < \varepsilon$ for all $y \in E$ satisfying $d(x, y) < \delta$.

The *Heine–Cantor theorem* asserts that if the set E is *closed and bounded*, then f is *uniformly continuous*; that is, for any $\varepsilon > 0$ there exists a $\delta > 0$ (depending solely on ε) such that

$$((x, y) \in E \times E \text{ and } d(x, y) < \delta) \implies \widetilde{d}(f(x), f(y)) < \varepsilon.$$

Without loss of much generality we next prove the Heine–Cantor theorem in the case where $E := [0, 1]$ and $d(x, y) := |x - y|$.

First, the continuity of f implies that for each $\varepsilon > 0$ and $x \in [0, 1]$, there exists a $\delta_x > 0$ such that $\widetilde{d}(f(x), f(y)) < \varepsilon/2$ for all y satisfying $|y - x| < \delta_x$. In particular, $\widetilde{d}(f(x), f(y)) < \varepsilon/2$ whenever y belongs to the open set $O_x := \{y : |y - x| < \delta_x/2\}$.

Second, the *Heine–Borel property* of $\mathbb{R}$ states that bounded closed intervals are *compact*; that is, every open cover of such a set has a finite subcover. This implies that the collection $\{O_x, x \in [0, 1]\}$, which is an open cover of $[0, 1] \subseteq \cup_{x \in [0,1]} O_x$, has a finite subcover, so that $[0, 1] \subseteq \cup_{k=1}^{n} \{y : |x_k - y| < \delta_{x_k}/2\}$ for some finite n and $x_1, \ldots, x_n \in [0, 1]$.

Finally, consider all $x, y \in [0, 1]$ satisfying $|x - y| < \delta := \min_k \delta_{x_k}/2$. For each y, define $k_y := \operatorname{argmin}_k |x_k - y|$ and apply the triangle inequality:

$$|x - x_{k_y}| \le |x - y| + |y - x_{k_y}| < \delta + \delta \le \min_k \delta_k.$$

Therefore, for all $x, y \in [0, 1]$ satisfying $|x - y| < \delta$, we have

$$\widetilde{d}(f(x), f(y)) < \widetilde{d}(f(x), f(x_{k_y})) + \widetilde{d}(f(x_{k_y}), f(y)) < \frac{\varepsilon}{2} + \frac{\varepsilon}{2} = \varepsilon.$$

∎

Given the metric space (E, d), a sequence of elements $x_1, x_2, \ldots \in E$ satisfying

(B.7) $$d(x_m, x_n) \to 0 \quad \text{as } m, n \to \infty,$$

is called a *Cauchy sequence* in the metric space (E, d).

■ **Example B.8 (Cauchy Sequences and Existence of a Limit)** Suppose that (x_n) is a sequence in (E, d) that converges to $x \in E$. By the triangle inequality, for any integers m and n, we have

$$d(x_m, x_n) \le d(x_m, x) + d(x, x_n) \to 0.$$

In other words, any sequence which converges to a limit in E is a Cauchy sequence. While every convergent sequence in E is a Cauchy sequence, the converse is not necessarily true. For example, if (E, d) is the set of real numbers on $(0, 1]$ with metric $d(x, y) := |x - y|$, then the sequence $x_n = 1/n, n = 1, 2, \ldots$ is a Cauchy sequence, but it does not have a limit in E (it has a limit in $[0, 1]$). ∎

Definition B.9: Complete Metric Space

A metric space (E, d) is said to be *complete* if every Cauchy sequence (x_n) in (E, d) converges to some $x \in E$.

That is to say, the metric space (E, d) is complete if the condition (B.7) implies that $\lim_n d(x_n, x) = 0$ for some $x \in E$. Being a Cauchy sequence in (E, d) is a necessary (but not a sufficient) condition for the existence of a limit in E. However, if the space (E, d) is complete, then being a Cauchy sequence is sufficient to guarantee a limit in E. We can deduce the completeness of the real line $\mathbb{R}$ with the metric $d(x, y) = |x - y|$ from the *Bolzano–Weierstrass theorem*; see Proposition 3.2.

We now consider a number of special metric spaces.

B.2 Normed Spaces

A set V is called a real (or complex) *vector space* if its elements satisfy the algebraic rules of addition and scalar multiplication:

If $x \in V$ and $y \in V$, then $\alpha x + \beta y \in V$ for all $\alpha, \beta \in \mathbb{R}$ (or $\mathbb{C}$).

In this section we assume that V is a vector space. Typical examples to have in mind are the Euclidean space $\mathbb{R}^n$ and the space $C[0, 1]$ of continuous function on the interval $[0, 1]$. A *normed space* is a vector space equipped with a *norm*, defined as follows:

Definition B.10: Norm

A *norm* is a function $\|\cdot\| : V \to \mathbb{R}_+$, which assigns a size to each element of V and satisfies for all $x, y \in V$ and scalar $\alpha \in \mathbb{C}$:

1. *(Zero)*: $\|x\| = 0$ if and only if $x = 0$.

2. *(Scaling)*: $\|\alpha x\| = |\alpha|\, \|x\|$.

3. *(Triangle inequality)*: $\|x + y\| \le \|x\| + \|y\|$.

A normed space with norm $\|\cdot\|$ is also a metric space, with metric

$$d(x, y) := \|x - y\|, \quad x, y \in V.$$

We say that the norm $\|\cdot\|$ *induces* the metric d and write $(V, \|\cdot\|)$ for the corresponding metric space. In principle, we can measure distances in V using metrics not derived from norms. However, only a metric induced by a norm takes full advantage of the algebraic structure of the vector space V.

Definition B.11: Banach Space

A normed space that is complete is said to be a *Banach space*.

■ **Example B.12 (Space of Polynomial Functions with Supremum Norm)** Consider the normed space $(V, \|\cdot\|)$, where V is the space of all polynomials on $[0, 1]$ and $\|x\| := \sup_{t \in [0,1]} |x(t)|$ is the *supremum norm*. Define the n-degree polynomial function:

$$x_n(t) := \sum_{k=0}^{n} x(k/n)\binom{n}{k} t^k (1 - t)^{n-k}, \quad t \in [0, 1],$$

where $x : [0, 1] \to \mathbb{R}$ is a given non-polynomial continuous function on the interval $[0, 1]$. The *Weierstrass approximation theorem*, proved in Exercise 3.33, asserts that $\lim_n \|x_n - x\| = 0$. Therefore, by the triangle inequality:

$$\|x_n - x_m\| \le \|x_n - x\| + \|x - x_m\| \to 0 \quad \text{as} \quad m, n \to \infty.$$

In other words, (x_n) is a Cauchy sequence in $(V, \|\cdot\|)$ that converges to an $x \notin V$. We conclude that, by definition, the metric space $(V, \|\cdot\|)$ is not complete. ■

The next example shows that if we enlarge the space V of polynomial functions to the space $C[0, 1]$ of continuous functions on $[0, 1]$ and retain the supremum norm, then the resulting normed space $(C[0, 1], \|\cdot\|)$ is complete.

■ **Example B.13 (Space of Continuous Functions with Supremum Norm)** The space $C[0, 1]$ of continuous functions on $[0, 1]$, equipped with the supremum norm, is a Banach space. To show that $(C[0, 1], \|\cdot\|)$ is a complete space, we need to show that $\lim_{m,n\to\infty} \|x_m - x_n\| = 0$ implies the existence of an $x \in C[0, 1]$ such that $\|x_n - x\| \to 0$.

For an arbitrary $s \in [0, 1]$, the sequence $(a_n, n \in \mathbb{N}) := (x_n(s), n \in \mathbb{N})$ is a Cauchy sequence of real numbers, because $|a_n - a_m| \le \|x_n - x_m\| \to 0$. Hence, by the completeness of the normed space $(\mathbb{R}, |\cdot|)$, we know that there exists a limit $a \in \mathbb{R}$ such that $|a - a_n| \to 0$, and we define $x(s) := a$. In other words, we have that $|x(s) - x_n(s)| \to 0$ for every $s \in [0, 1]$ or, equivalently, that $\|x - x_n\| \to 0$.

It remains to show that $x \in C[0, 1]$. For an arbitrary $\varepsilon > 0$, choose N large enough such that $\|x - x_n\| \le \varepsilon/4$ for $n \ge N$. Since $x_N \in C[0, 1]$, there exists a $\delta_\varepsilon > 0$ such that $|t - s| < \delta_\varepsilon$ implies $|x_N(s) - x_N(t)| \le \varepsilon/2$. Finally, the condition $|t - s| < \delta_\varepsilon$ and the triangle inequality yield:

$$
\begin{aligned}
(B.14) \quad |x(s) - x(t)| &\le |x(s) - x_N(s)| + |x_N(s) - x_N(t)| + |x_N(t) - x(t)| \\
&\le 2\|x - x_N\| + |x_N(s) - x_N(t)| \le \varepsilon,
\end{aligned}
$$

proving that $x \in C[0, 1]$.   ■

The following example shows that if we replace the supremum norm in Example B.13 with the L^1 norm, then the resulting space is no longer complete:

■ **Example B.15 (Space of Continuous Functions with L^1 Norm)** Let λ be the Lebesgue measure on $(\mathbb{R}, \mathcal{B})$ and define the norm on the space $C[0, 1]$ of continuous functions on $[0, 1]$ via:

$$
\|x\| := \lambda(|x| \mathbb{1}_{[0,1]}) = \int_0^1 dt\, |x(t)|.
$$

The normed space $(C[0, 1], \|\cdot\|)$ is not a Banach space. To see this, define the sequence (x_n) of continuous functions on $[0, 1]$:

$$
x_n(t) := n \left(t - \frac{1}{2} \right)^+ \times \mathbb{1}_{[0, \frac{1}{2}+\frac{1}{n}]}(t) + \mathbb{1}_{(\frac{1}{2}+\frac{1}{n}, 1]}(t), \quad t \in [0, 1].
$$

This is a Cauchy sequence: $\|x_m - x_n\| = \frac{1}{2(m \wedge n)} - \frac{1}{2(m \vee n)} \to 0$ for $m, n \to \infty$, which converges to a discontinuous function:

$$
\lim_n x_n(t) = \begin{cases} 0 & \text{if } t \in [0, 1/2], \\ 1 & \text{if } t \in (1/2, 1]. \end{cases}
$$

■

Another important example of a Banach space is the L^p space of random variables defined in Section 2.5.

> ### Theorem B.16: Completeness of L^p
>
> The space L^p is complete for every $p \in [1, \infty]$.

Proof. Let (X_n) be a Cauchy sequence in L^p; i.e., $\|X_m - X_n\|_p \to 0$ as $m, n \to \infty$. We are going to construct the limit X to which (X_n) converges in L^p norm, as follows: take a subsequence $(n_k, k \in \mathbb{N})$ such that $\|X_m - X_{n_k}\|_p < 2^{-k}$ for $m \geq n_k$, and define $Y_k := X_{n_k} - X_{n_{k-1}}$ for $k > 1$ and $Y_0 := X_{n_0}$. Then,

$$X_{n_k} = \sum_{i=0}^{k} Y_i, \quad k \in \mathbb{N}, \quad \text{where} \quad \sum_{i=0}^{\infty} \|Y_i\|_p < \infty.$$

For each $\omega \in \Omega$, define

$$Z_k(\omega) := \sum_{i=0}^{k} |Y_i(\omega)| \quad \text{and} \quad Z(\omega) := \sum_{i=0}^{\infty} |Y_i(\omega)|.$$

Since $Z_n \overset{\text{a.s.}}{\to} Z$, and hence $Z_n^p \overset{\text{a.s.}}{\to} Z^p$, we have by the Monotone Convergence Theorem that $\|Z\|_p < \infty$; that is, $Z \in L^p$. In particular, $\sup_k |X_{n_k}| \leq Z < \infty$ almost surely, which implies that $X_{n_k} \overset{\text{a.s.}}{\to} X := \sum_{i=0}^{\infty} Y_i$ as $k \uparrow \infty$. Also, $X \in L^p$, because $|X| \leq Z$ and so $\|X\|_p \leq \|Z\|_p < \infty$. Since $|X - X_{n_k}|^p \leq (2Z)^p \in L^1$, it follows that $\|X_{n_k} - X\|_p \to 0$. Finally, from the fact that (X_n) is a Cauchy sequence and the triangle inequality:

$$\|X_n - X\|_p \leq \|X_n - X_{n_k}\|_p + \|X_{n_k} - X\|_p \to 0,$$

we conclude that $X_n \overset{L^p}{\to} X$. $\qquad\square$

B.3 Inner Product Spaces

The next natural enhancement of a vector space V is to bestow it with a geometry via the introduction of an inner product, taking values in either $\mathbb{C}$ or $\mathbb{R}$. The former is used in the definition below. The resulting space V with the inner product norm is called an *inner product space*.

Definition B.17: Inner Product

An *inner product* on V is a mapping $\langle \cdot, \cdot \rangle$ from $V \times V$ to $\mathbb{C}$ that satisfies:

1. $\langle \alpha x_1 + \beta x_2, y \rangle = \alpha \langle x_1, y \rangle + \beta \langle x_2, y \rangle$ for all $\alpha, \beta \in \mathbb{C}$.

2. $\langle x, y \rangle = \overline{\langle y, x \rangle}$.

3. $\langle x, x \rangle \geq 0$.

4. $\langle x, x \rangle = 0$ if and only if $x = 0$ (the zero element).

We say that two elements x and y in V are *orthogonal* to each other with respect to an inner product if $\langle x, y \rangle = 0$. Given an inner product on V, we define a norm on V via

$$\|x\| := \sqrt{\langle x, x \rangle}.$$

The fact that $\sqrt{\langle x, x \rangle}$ is a norm and satisfies the properties in Definition B.10 follows readily from the following properties of the inner product:

Theorem B.18: Properties of Inner Product

1. *(Cauchy–Schwarz inequality)*: $|\langle x, y \rangle| \leq \|x\|\|y\|$ for any $x, y \in V$, with equality achieved if and only if $x = \alpha y$ for some constant $\alpha \in \mathbb{C}$.

2. *(Triangle inequality)*: $\|x - y\| \leq \|x\| + \|y\|$ for any $x, y \in V$.

3. *(Continuity)*: If $\|x_n - x\| \to 0$ and $\|y_n - y\| \to 0$, then $\langle x_n, y_n \rangle \to \langle x, y \rangle$.

Proof. Without loss of generality, we may assume that $\|x\|\,\|y\| > 0$; otherwise, the Cauchy–Schwarz inequality is trivial. Since $\langle x - \alpha y, x - \alpha y \rangle = \|x - \alpha y\|^2 \geq 0$, we have

$$0 \leq \langle x - \alpha y, x - \alpha y \rangle = \|x\|^2 - \overline{\alpha}\langle x, y \rangle - \alpha[\langle y, x \rangle - \overline{\alpha}\|y\|^2].$$

Substituting $\overline{\alpha} = \langle y, x \rangle / \|y\|^2$ yields the inequality:

$$0 \leq \|x + \alpha y\|^2 = \|x\|^2 - |\langle x, y \rangle|^2 / \|y\|^2,$$

from which we deduce the Cauchy–Schwarz inequality. Furthermore, $\|x\|^2 - |\langle x, y \rangle|^2 / \|y\|^2$ is 0 if and only if $\|x - \alpha y\|^2 = 0$ or, equivalently, $x = \alpha y$.

The triangle inequality follows from an application of Cauchy–Schwarz as follows:

$$\begin{aligned}
\|x - y\|^2 &= \left| \|x\|^2 + \|y\|^2 - \langle x, y \rangle - \langle y, x \rangle \right| \\
&\leq \|x\|^2 + \|y\|^2 + |\langle x, y \rangle| + |\langle y, x \rangle| \\
&\leq \|x\|^2 + \|y\|^2 + \|x\|\|y\| + \|y\|\|x\| = (\|x\| + \|y\|)^2.
\end{aligned}$$

To show the continuity statement, note that by adding and subtracting $\langle x, y_n \rangle$ and using the triangle inequality, we obtain

$$
\begin{aligned}
|\langle x_n, y_n \rangle - \langle x, y \rangle| &\leq |\langle x_n, y_n \rangle - \langle x, y_n \rangle| + |\langle x, y_n \rangle - \langle x, y \rangle| \\
&\leq |\langle x_n - x, y_n \rangle| + |\langle x, y_n - y \rangle| \\
&\leq \|x_n - x\| \times \|y_n\| + \|x\| \times \|y_n - y\| \to 0,
\end{aligned}
$$

where we used the Cauchy–Schwarz inequality twice in the last line. □

An inner product space that is *complete* is called a *Hilbert space*. One of the most fundamental Hilbert spaces is the L^2 space[1] of functions.

Definition B.19: L^2 Space

Let $(E, \mathcal{E}, \mu)$ be a measure space and define the inner product:

(B.20) $$\langle f, g \rangle := \int_E \mu(\mathrm{d}x) \, f(x) \, \overline{g(x)}.$$

The Hilbert space $L^2(E, \mathcal{E}, \mu)$ is the vector space of functions from E to $\mathbb{C}$ that satisfy $\|f\|^2 := \langle f, f \rangle < \infty$. Any pair of functions f and g that are μ-everywhere equal, that is, $\|f - g\| = 0$, are identified as one and the same.

Of particular interest is the space $L^2[0, 1]$, where E is the interval $[0, 1]$, $\mathcal{E}$ is the Borel σ-algebra thereon (i.e., $\mathcal{B}_{[0,1]}$), and μ is the Lebesgue measure restricted to $[0, 1]$ (i.e., $\mathrm{Leb}_{[0,1]}$).

A set of functions $\{u_i, i \in I\}$ is called an *orthonormal system* for a Hilbert space $L^2(E, \mathcal{E}, \mu)$ if

$$
\langle u_i, u_j \rangle = \begin{cases} 1 & \text{if } i = j, \\ 0 & \text{if } i \neq j. \end{cases}
$$

It follows then that the $\{u_i\}$ are linearly independent; that is, the only linear combination $\sum_j \alpha_j u_j(x)$ that is equal to $u_i(x)$ for all x is the one where $\alpha_i = 1$ and $\alpha_j = 0$ for $j \neq i$. Although the general theory allows for uncountable index set I, it can be proved that when E is a bounded and closed subset of $\mathbb{R}^d$, then the set I must be countable. For simplicity, we will henceforth assume that I is countable, and without loss of generality $I := \mathbb{N}$.

Let $(V, \|\cdot\|)$ be the Hilbert space $L^2(E, \mathcal{E}, \mu)$. An orthonormal system $\{u_i, i \in \mathbb{N}\}$ is called an *orthonormal basis* if there is no $f \in V$, other than the 0 function, that

---

[1]Note that Definition B.19 involves a space of complex-valued functions, in contrast to the L^2 space of square-integrable random variables in Section 2.5 (with $E = \Omega$, $\mathcal{E} = \mathcal{H}$, and $\mu = \mathbb{P}$), where we used real-valued functions.

is orthogonal to all the $\{u_i, i \in \mathbb{N}\}$. Any orthonormal system satisfies the following inequality:

Theorem B.21: Bessel Inequality

For any orthonormal system $\{u_i, i \in \mathbb{N}\}$, $\sum_{i=0}^{\infty} |\langle f, u_i \rangle|^2 \leq \|f\|^2$ for all $f \in V$.

Proof. Let $f_n := \sum_{i=0}^{n} \langle f, u_i \rangle u_i$, so that

$$\langle f_n, f \rangle = \sum_{i=0}^{n} \langle f, u_i \rangle \langle u_i, f \rangle = \sum_{i=0}^{n} |\langle f, u_i \rangle|^2$$

is real (implying that $\langle f_n, f \rangle = \langle f, f_n \rangle$) and

$$\langle f_n, f_n \rangle = \sum_{i=0}^{n} \sum_{j=0}^{n} \langle f, u_i \rangle \langle u_i, f \rangle \langle u_i, u_j \rangle = \sum_{i=0}^{n} |\langle f, u_i \rangle|^2.$$

Since $\langle f - f_n, f_n \rangle = 0$, we deduce that f_n is orthogonal to $f - f_n$. Hence, from

$$0 \leq \langle f - f_n, f - f_n \rangle = \langle f, f - f_n \rangle = \langle f, f \rangle - \langle f, f_n \rangle,$$

we obtain $\langle f_n, f_n \rangle = \sum_{i=0}^{n} |\langle f, u_i \rangle|^2 \leq \|f\|^2$. Since $\langle f_n, f_n \rangle$ is increasing and bounded from above, it follows that it converges to a limit: $\lim_{n \uparrow \infty} \|f_n\|^2 = \sum_{i=0}^{\infty} |\langle f, u_i \rangle|^2 \leq \|f\|^2$. □

Orthonormal bases satisfy the following stronger result, the consequence of which is that in a Hilbert space $(V, \| \cdot \|)$ with orthonormal basis $\{u_i, i \in \mathbb{N}\}$ every element $f \in V$ can be written as

$$f = \sum_{i=0}^{\infty} \langle f, u_i \rangle u_i,$$

in the sense that the approximation $f_n := \sum_{i=0}^{n} \langle f, u_i \rangle u_i$ converges to f in the norm on V.

Theorem B.22: Parseval Identity

It holds that $\sum_{i=0}^{\infty} |\langle f, u_i \rangle|^2 = \|f\|^2$ for any f in a Hilbert space $(V, \| \cdot \|)$ if and only if $\{u_i, i \in \mathbb{N}\}$ is an orthonormal basis for $(V, \| \cdot \|)$.

Proof. First, suppose that Parseval's identity holds for any f in the Hilbert space V. If $\{u_i, i \in \mathbb{N}\}$ is *not* an orthonormal basis, there exists a nonzero $f \in V$ such that $\langle f, u_i \rangle = 0$ for all $i \in \mathbb{N}$. This, however, contradicts Parseval's identity, because $0 < \|f\|^2 = \sum_{i=0}^{\infty} |\langle f, u_i \rangle|^2 = \sum_{i=0}^{\infty} 0 = 0$ is impossible. Hence, $\{u_i, i \in \mathbb{N}\}$ must be an orthonormal basis.

Conversely, suppose that $\{u_i, i \in \mathbb{N}\}$ is an orthonormal basis, so that $\langle g, u_i \rangle = 0$ for all $i \in \mathbb{N}$ implies that $g = 0$. Since for $n > m \to \infty$ and an arbitrary $f \in V$ we have that $\|f_m - f_n\|^2 = \sum_{i=m}^{n} |\langle f, u_i \rangle|^2 \to 0$ is a Cauchy sequence in the complete space $(V, \|\cdot\|)$, (f_m) converges (in norm) to an element of V. This element must be f, because $\langle f - f_m, u_i \rangle = \langle f, u_i \rangle - \langle f_m, u_i \rangle = 0$ for all $i \leq m$ and we can make m arbitrarily large. In other words, $f = \sum_{i=1}^{\infty} \langle f, u_i \rangle u_i$ implying Parseval's identity. $\square$

B.4 Sturm–Liouville Orthonormal Basis

Let $\mathbf{L}$ be the linear differential operator that maps a twice differentiable function u into the function $\mathbf{L}u$ defined by:

$$[\mathbf{L}u](x) := -\frac{\frac{\mathrm{d}}{\mathrm{d}x}\left[p(x)\frac{\mathrm{d}u}{\mathrm{d}x}\right] + q(x)u(x)}{w(x)},$$

where p, w are positive functions on $[0, 1]$ and p, p', w, q are all continuous.

Let $L^2([0, 1], \mathcal{B}_{[0,1]}, \mu)$ be the Hilbert space with $\mu(\mathrm{d}x) = w(x)\,\mathrm{d}x$. One of the simplest ways to construct a countable infinite-dimensional orthonormal basis $\{u_k\}$ for $L^2([0, 1], \mathcal{B}_{[0,1]}, \mu)$ is to find all the distinct *eigenvalue* and *eigenfunction* pairs $\{(\lambda_k, u_k)\}$ of the *Sturm–Liouville* ordinary differential equation with *separated* boundary conditions:

(B.23)
$$\begin{aligned}
(\mathbf{L}u_k)(x) - \lambda_k\, u_k(x) &= 0, \\
c_1\, u_k(0) + c_2\, u'_k(0) &= 0, \\
c_3\, u_k(1) + c_4\, u'_k(1) &= 0,
\end{aligned}$$

where $|c_1| + |c_2| > 0$ and $|c_3| + |c_4| > 0$. Using integration by parts and the separated boundary conditions in (B.23), it is straightforward to show that $\mathbf{L}$ is a *self-adjoint* linear operator with respect to the inner product (B.20); that is, $\langle u_k, \mathbf{L}u_j \rangle = \langle \mathbf{L}u_k, u_j \rangle$. Hence, from (B.23) we can write

$$(\lambda_k - \bar{\lambda}_j)\langle u_k, u_j \rangle = \langle \lambda_k u_k, u_j \rangle - \langle u_k, \lambda_j u_j \rangle = \langle \mathbf{L}u_k, u_j \rangle - \langle u_k, \mathbf{L}u_j \rangle = 0.$$

The last equation implies that:

1. $(\lambda_k - \bar{\lambda}_k)\|u_k\|^2 = 0$ for $k = j$ (that is, each eigenvalue λ_k is real) and

2. the eigenfunctions u_k and u_j are orthogonal to each other, provided that $\lambda_k \neq \lambda_j$.

All of the eigenvalues can be shown to be real and distinct, and so they form an increasing sequence, $\lambda_0 < \lambda_1 < \lambda_2 < \cdots$.

■ **Example B.24 (Sine Orthonormal Basis)** Choosing $w(x) = p(x) = 1, q(x) = 0$ and $c_2 = c_4 = 0$, one can verify that the eigenvalues $\lambda_k = k^2$ and eigenfunctions $\sin(k\pi x)$ satisfy (B.23). Hence, $\sqrt{2}\sin(k\pi x), k = 1, 2, 3, \ldots$ is an orthonormal basis for $L^2[0, 1]$.

An alternative half-sine basis results from choosing $w(x) = p(x) = 1, q(x) = 0$ and $c_2 = c_3 = 0$. One can verify that the eigenvalues $\lambda_k = (k + 1/2)^2$ and eigenfunctions $\sin([k + 1/2]\pi x)$ satisfy (B.23). Hence, $\sqrt{2}\sin([k + 1/2]\pi x), k = 0, 1, 2, \ldots$ is also an orthonormal basis for $L^2[0, 1]$. ■

■ **Example B.25 (Cosine Orthonormal Basis)** Choosing $w(x) = p(x) = 1, q(x) = 0$ and $c_1 = c_3 = 0$, one can verify that the eigenvalues $\lambda_k = k^2$ and eigenfunctions $\cos(k\pi x)$ satisfy (B.23). Hence, the cosine functions $u_0(x) = \cos(0\pi x)$, $u_k(x) = \sqrt{2}\cos(k\pi x), k = 1, 2, \ldots$ form an orthonormal basis for $L^2[0, 1]$. For example, consider the indicator function $\mathbb{1}_{[0,t]}$ on $[0, 1]$. The approximation

$$\mathbb{1}_{[0,t]}^{(m)}(x) = \sum_{k=0}^{m-1} \langle \mathbb{1}_{[0,t]}, u_k \rangle u_k(x) = t + 2\sum_{k=1}^{m} \frac{\sin(k\pi t)}{k\pi}\cos(k\pi x), \qquad x \in [0, 1],$$

converges to $\mathbb{1}_{[0,t]}$ in the corresponding norm: $\|\mathbb{1}_{[0,t]}^{(m)} - \mathbb{1}_{[0,t]}\| \to 0$. Therefore, from the continuity property in Theorem B.18 we can deduce the pointwise convergence of the inner product $\langle \mathbb{1}_{[0,s]}^{(m)}, \mathbb{1}_{[0,t]}^{(m)} \rangle \to \langle \mathbb{1}_{[0,s]}, \mathbb{1}_{[0,t]} \rangle$, and we can write

$$(B.26) \qquad s \wedge t = \langle \mathbb{1}_{[0,s]}, \mathbb{1}_{[0,t]} \rangle = st + \sum_{k=1}^{\infty} \frac{\sqrt{2}\sin(k\pi s)}{k\pi} \frac{\sqrt{2}\sin(k\pi t)}{k\pi}.$$

Yet another basis follows from $w(x) = p(x) = 1, q(x) = 0$ and $c_1 = c_4 = 0$, so that the eigenvalues $\lambda_k = (k + 1/2)^2$ and eigenfunctions $\cos([k + 1/2]\pi x)$ satisfy (B.23), giving the orthonormal basis $\sqrt{2}\cos(k\pi x/2)$, $k = 1, 3, 5, \ldots$ for $L^2[0, 1]$. With this basis we can write

$$(B.27) \qquad s \wedge t = \langle \mathbb{1}_{[0,s]}, \mathbb{1}_{[0,t]} \rangle = \sum_{k=1,3,\ldots} \frac{2\sqrt{2}\sin(k\pi s/2)}{k\pi} \frac{2\sqrt{2}\sin(k\pi t/2)}{k\pi}.$$

■

B.5   Hermite Orthonormal Basis

Another useful orthonormal basis can be constructed from the *Hermite polynomials*, defined via

$$\hbar_n(x) := (-1)^n \exp(x^2/2)\frac{\mathrm{d}^n}{\mathrm{d}x^n}\exp(-x^2/2), \quad n = 0, 1, 2, \ldots,$$

or explicitly via

$$\hbar_n(x) = n! \sum_{k=0}^{\lfloor n/2 \rfloor} \frac{(-1)^k}{k!(n-2k)!2^k} x^{n-2k}.$$

The definition of the Hermite polynomials implies the following properties:

$$\hbar'_n(x) = n\,\hbar_{n-1}(x),$$
$$\hbar_{n+1}(x) = x\,\hbar_n(x) - \hbar'_n(x),$$
$$\exp(xt - t^2/2) = \sum_{k=0}^{\infty} \frac{t^k}{k!}\hbar_k(x).$$

Each $\hbar_k(x)$ satisfies the Sturm–Liouville ODE (B.23) of the form:

$$\underbrace{(\exp(-x^2/2)\,u'(x))'}_{=:p(x)} + \underbrace{k}_{=:\lambda_k}\ \underbrace{\exp(-x^2/2)}_{=:w(x)}\,u(x) = 0,$$

subject to the boundary conditions: $\exp(-x^2/2)u(x) \to 0$ and $\exp(-x^2/2)u'(x) \to 0$ as $x \to \pm\infty$. Note that the boundary conditions at infinity differ from those in (B.23), but similar integration by parts computations establish the orthogonality property with respect to the inner product (B.20): $\langle \hbar_\alpha, \hbar_\beta \rangle = \alpha!\sqrt{2\pi}\,\mathbb{1}_{\{\alpha=\beta\}}$.

Since $\int_\mathbb{R} dx\, f(x)x^n \exp(-x^2) = 0$ for all $n \in \mathbb{N}$ implies that $f = 0$ almost everywhere, the Hermite polynomials form an orthogonal basis of $L^2(\mathbb{R}, \mathcal{B}, w(x)\,dx)$, with $w(x) = \exp(-x^2/2)$. Hence, any function f in this space can be approximated via $f^{(m)}(x) = \sum_{k=0}^{m-1} \frac{\langle f, \hbar_k \rangle}{k!\sqrt{2\pi}}\hbar_k(x)$ such that $\|f - f^{(m)}\| \to 0$.

■ **Example B.28 (Hermite Functions)** Using the properties of the Hermite polynomials, we can show that the *Hermite functions*:

$$\psi_n(x) := \frac{\exp(-x^2/4)}{\sqrt{n!\sqrt{2\pi}}}\,\hbar_n(x), \quad n \in \mathbb{N}$$

form an orthonormal basis of $L^2(\mathbb{R}, \mathcal{B}, \mathrm{Leb})$; i.e., $\int dx\, \psi_m(x)\psi_n(x) = \mathbb{1}_{\{m=n\}}$, and hence any function therein can be approximated with $f^{(m)}(x) := \sum_{k=0}^{m-1} c_k\,\psi_k(x)$, where $c_k := \int dx\, f(x)\psi_n(x)$. ■

B.6   Haar Orthonormal Basis

For every pair of integers $n \geq 0$ and $k = 0, 1, \ldots, 2^n - 1$, the *Haar function* $h_{n,k}$ is defined by:

$$h_{n,k}(x) := 2^{n/2} \times \begin{cases} 1 & \text{if } \frac{k}{2^n} \leq x < \frac{k+\frac{1}{2}}{2^n}, \\ -1 & \text{if } \frac{k+\frac{1}{2}}{2^n} \leq x < \frac{k+1}{2^n}, \\ 0 & \text{otherwise.} \end{cases}$$

Note that all Haar functions are 0 outside the interval $[0, 1]$. With (B.20) as the inner product for the Hilbert space $L^2[0, 1]$, we have:

$$\langle h_{n,k}, h_{m,l} \rangle = \begin{cases} 1 & \text{if } n = m, \ k = l, \\ 0 & \text{otherwise.} \end{cases}$$

In addition, $\langle 1, h_{n,k} \rangle = \int_0^1 dx \, h_{n,k}(x) = 0$, implying that the constant 1 is orthogonal to all Haar functions. The set of Haar functions with the addition of the constant function 1, that is, $\{h_{n,k}, n \geq 0, k = 0, \ldots, 2^n - 1\} \cup \{1\}$, forms an orthonormal basis of $L^2[0, 1]$. In other words, for every $f \in L^2[0, 1]$ we have the approximation

$$f^{(m)}(x) := \langle f, 1 \rangle + \sum_{n=0}^{m-1} \sum_{k=0}^{2^n-1} \langle f, h_{n,k} \rangle \, h_{n,k}(x),$$

where $\|f - f^{(m)}\| \to 0$ as $m \uparrow \infty$ and $\| \cdot \|$ is the norm on $L^2[0, 1]$.

While the expansion above uses two indexes n and k, we can define a single-index set of orthonormal functions:

$$h_0(x) := \mathbb{1}_{[0,1]}(x), \quad h_j(x) := h_{n,k}(x), \quad j = 2^n + k,$$

and thus obtain the more familiar expansion $f^{(m)}(x) = \sum_{j=0}^{m-1} \langle f, h_j \rangle \, h_j(x)$.

■ **Example B.29 (Schauder Functions)** A useful set of functions derived from the Haar basis are the tent-like or *Schauder functions*, defined for $n \geq 0, k = 0, 1, \ldots, 2^n - 1$ by

$$c_{n,k}(t) := \langle \mathbb{1}_{[0,t]}, h_{n,k} \rangle = 2^{-n/2-1} \left[1 - |2^{n+1}t - 2k - 1|\right]^+.$$

Figure B.30 shows the characteristic tent-like shape of all $c_{n,k}$ for $n = 0, 1, 2$.

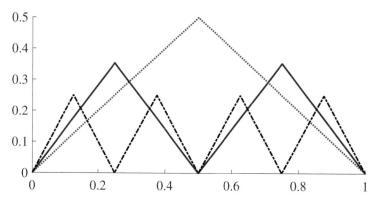

Figure B.30: Tent-like functions: $c_{0,0}$ is the blue dotted line; $c_{1,0}$ and $c_{1,1}$ are given as the red thick line; $c_{2,0}, c_{2,1}, c_{2,2}, c_{2,3}$ are depicted as the black dash-dot line.

One advantage of the Haar basis over the trigonometric basis is that each Haar function is very localized (nonzero over a vanishingly small interval), and thus easily permitting stronger types of convergence. For example, if f is continuous on $[0, 1]$, then the pointwise convergence of $f^{(m)}$ to f is uniform:

$$\sup_{x \in [0,1]} |f(x) - f^{(m)}(x)| \to 0.$$

∎

EXISTENCE OF THE LEBESGUE MEASURE

In this appendix we prove the existence of the Lebesgue measure on $((0, 1], \mathcal{B}_{(0,1]})$, following Billingsley (1995).

We first show the countable additivity of the Lebesgue pre-measure on $E := (0, 1]$. Denote by $\mathcal{I}$ the set of subintervals of E of the form $(c, d]$, and let $\mathcal{E}_0$ be the algebra of sets that are finite unions of intervals in $\mathcal{I}$.

Recall that an *algebra* $\mathcal{E}_0$ on E is a collection of subsets of E that contains E itself, and that is closed under complements and *finite* unions. Recall also that a *pre-measure* is a set function $\lambda : \mathcal{E}_0 \to [0, \infty]$, with $\lambda(\emptyset) = 0$ that satisfies the *countable additivity* property:

(C.1)
$$\lambda \left(\bigcup_{k=1}^{\infty} A_k \right) = \sum_{k=1}^{\infty} \lambda(A_k)$$

for every sequence $A_1, A_2, \ldots$ of *disjoint* sets in $\mathcal{E}_0$ with $\cup_n A_n \in \mathcal{E}_0$.

Define on $\mathcal{E}_0$ the natural "length" pre-measure λ. When λ is applied to intervals, we use the notation $| \cdot |$ instead. Thus, for disjoint intervals $\{(a_k, b_k]\}$:

$$\lambda \left(\bigcup_{k=1}^{n} (a_k, b_k] \right) = \left| \bigcup_{k=1}^{n} (a_k, b_k] \right| = \sum_{k=1}^{n} (b_k - a_k).$$

We want to show that this pre-measure is countably additive. We do this in two steps. First, we prove it for intervals, and then for general sets in $\mathcal{E}_0$. Below, $I = (a, b]$ and $I_k = (a_k, b_k]$ are bounded intervals of lengths $|I| = b - a$ and $|I_k| = b_k - a_k$.

Proposition C.2: Countable Additivity for Intervals

If $I = \cup_{k=1}^{\infty} I_k$ and the $\{I_k\}$ are disjoint, then $|I| = \sum_{k=1}^{\infty} |I_k|$.

Proof. We are going to prove the theorem by showing:

1. If $\cup_{k=1}^{\infty} I_k \subseteq I$, and the $\{I_k\}$ are *disjoint*, then

(C.3)
$$\sum_{k=1}^{\infty} |I_k| \leq |I|.$$

2. If $\cup_{k=1}^{\infty} I_k \supseteq I$ (the $\{I_k\}$ need not be disjoint), then

(C.4)
$$\sum_{k=1}^{\infty} |I_k| \geq |I|.$$

We first show (C.3) for finite sums (that is, $\sum_{k=1}^{n} |I_k| \leq |I|$), using induction. The statement is obviously true for $n = 1$. Suppose it is true for $n - 1$. Without loss of generality, we may assume a_n is the largest among $a_1, \ldots, a_n$. Then, $\cup_{k=1}^{n-1}(a_k, b_k] \subseteq (a, a_n]$, so that, by the induction hypothesis, $\sum_{k=1}^{n-1}(b_k - a_k) \leq a_n - a$, and hence:

(C.5)
$$\sum_{k=1}^{n}(b_k - a_k) \leq a_n - a + b_n - a_n \leq b - a.$$

The infinite sum case follows directly from this. Namely, from the finite sum case, we have that (C.5) holds for *every* $n \geq 1$. But this can only be the case if

$$\sum_{k=1}^{\infty}(b_k - a_k) \leq b - a.$$

To prove (C.4), we again consider first the finite sum and then the infinite sum case. We use again induction to show $|I| \leq \sum_{k=1}^{n} |I_k|$, which is true for $n = 1$. Suppose it is true for $n - 1$. Without loss of generality, we may assume $a_n < b \leq b_n$. If $a_n \leq a$, the result is obvious. If $a_n > a$, then (draw a picture) $(a, a_n] \subseteq \cup_{i=1}^{n-1}(a_k, b_k]$, so that $\sum_{k=1}^{n-1}(b_k - a_k) \geq a_n - a$ by the induction hypothesis, and hence:

$$\sum_{k=1}^{n}(b_k - a_k) \geq (a_n - a) + (b_n - a_n) \geq b - a.$$

For the infinite sum case, suppose that $(a, b] \subseteq \cup_{k=1}^{\infty}(a_k, b_k]$. If $0 < \varepsilon < b - a$, the open intervals $\{(a_k, b_k + \varepsilon 2^{-k})\}$ cover the closed interval $[a + \varepsilon, b]$. The *Heine–Borel property* of the real-number system states that bounded closed intervals are *compact*; that is, every (infinite) cover of such a set has a finite subcover; see also Example B.6. Thus,

$$[a + \varepsilon, b] \subseteq \cup_{k=1}^{n}(a_k, b_k + \varepsilon 2^{-k})$$

for some n. But then

$$(a + \varepsilon, b] \subseteq \cup_{k=1}^{n}(a_k, b_k + \varepsilon 2^{-k}).$$

Consequently, using the finite case,

$$b - (a + \varepsilon) \leq \sum_{k=1}^{n}(b_k + \varepsilon 2^{-k} - a_k) \leq \sum_{k=1}^{\infty}(b_k - a_k) + \varepsilon.$$

Since $\varepsilon > 0$ was arbitrary, we have

$$b - a \leq \sum_{k=1}^{\infty}(b_k - a_k).$$

This completes the proof for intervals. □

Next, we prove the countable additivity for arbitrary sets in $\mathcal{E}_0$. Recall that we used the notation $\lambda(A) = |A|$ if A is an interval.

Theorem C.6: Countable Additivity for Sets

Let $A \in \mathcal{E}_0$ be of the form $A = \cup_{k=1}^{\infty}A_k$, where the $\{A_k\}$ are disjoint and in $\mathcal{E}_0$. Then,

$$\lambda(A) = \sum_{k=1}^{\infty}\lambda(A_k).$$

Proof. We can write (for certain intervals $\{I_i\}$ and $\{J_{kj}\}$):

$$A = \bigcup_{i=1}^{n} I_i \quad \text{and} \quad A_k = \bigcup_{j=1}^{m_k} J_{kj}.$$

It follows that

$$A = \bigcup_{i=1}^{n}\bigcup_{k=1}^{\infty}\bigcup_{j=1}^{m_k}(I_i \cap J_{kj}).$$

Note that the $\{I_i \cap J_{kj}\}$ are disjoint and lie in $\mathcal{I}$. Moreover,

$$\underbrace{\bigcup_{i=1}^{n}(I_i \cap J_{kj})}_{A} = \bigcup_{i=1}^{n} I_i \cap J_{kj} = J_{kj},$$

as $J_{kj} \subseteq A$. Thus, we have, by the countable additivity for intervals:

$$\lambda(A) = \sum_{i=1}^{n} |I_i| = \sum_{i=1}^{n} \sum_{k=1}^{\infty} \sum_{j=1}^{m_k} |I_i \cap J_{kj}| = \sum_{k=1}^{\infty} \sum_{j=1}^{m_k} |J_{kj}| = \sum_{k=1}^{\infty} \lambda(A_k).$$

$\square$

If in Proposition C.2 the $\{A_k\}$ are not disjoint, we still have countable subadditivity:

(C.7) $$\lambda(A) \leq \sum_{k=1}^{\infty} \lambda(A_k).$$

The proof is exactly the same as in Proposition 1.42.

We next show the existence of the Lebesgue measure on $(E, \mathcal{E})$ for $E := (0, 1]$ and $\mathcal{E} := \mathcal{B}_{(0,1]}$. Recall that $\mathcal{E}_0$ is the algebra of sets that are finite unions of disjoint intervals of the form $(c, d]$, and that λ is the pre-measure on $\mathcal{E}_0$, which we have just shown to be countably additive. The proof of existence relies on the definition of two objects: $\overline{\lambda}$ and $\mathcal{M}$.

Definition C.8: Outer Measure $\overline{\lambda}$

The *outer measure* of λ is the set function $\overline{\lambda}$ that assigns to each $A \in 2^E$ (that is, for each $A \subseteq E$) the value

(C.9) $$\overline{\lambda}(A) := \inf \sum_n \lambda(A_n),$$

where the infimum is over all sequences $A_1, A_2, \ldots$ of $\mathcal{E}_0$-sets satisfying $A \subseteq \cup_n A_n$.

Note that $\overline{\lambda}$ is not actually a measure, as the set 2^E is simply too big for it to be countably additive. We can also define an *inner measure*,

$$\underline{\lambda}(A) := 1 - \overline{\lambda}(A^c), \quad A \in 2^E.$$

The outer measure $\overline{\lambda}$ has properties reminiscent of those of a measure, such as positivity $\overline{\lambda}(A) \geq 0$ and monotonicity $\overline{\lambda}(A) \leq \overline{\lambda}(B)$, when $A \subseteq B$. Also, $\overline{\lambda}(\emptyset) = 0$. The following *countable subadditivity* property is important as well:

Lemma C.10: The Outer Measure $\overline{\lambda}$ is Countably Subadditive

For any collection of sets in E (not necessarily disjoint), it holds that

(C.11) $$\overline{\lambda}\left(\cup_{n=1}^{\infty} A_n \right) \leq \sum_{n=1}^{\infty} \overline{\lambda}(A_n).$$

Proof. Take $\varepsilon > 0$ and choose $B_{nk} \in \mathcal{E}_0$ such that $A_n \subseteq \cup_k B_{nk}$ and $\sum_k \lambda(B_{nk}) < \overline{\lambda}(A_n) + \varepsilon \, 2^{-n}$. Because $\cup_n A_n \subseteq \cup_{n,k} B_{nk}$, it follows that

$$\overline{\lambda}\left(\cup_{n=1}^{\infty} A_n\right) \leq \sum_{n,k} \lambda(B_{nk}) < \sum_{n=1}^{\infty} \overline{\lambda}(A_n) + \varepsilon,$$

showing (C.11), as ε is arbitrary. □

Definition C.12: The Collection $\mathcal{M}$

The collection $\mathcal{M}$ consists of all sets $A \in 2^E$ for which

(C.13) $\overline{\lambda}(A \cap C) + \overline{\lambda}(A^c \cap C) = \overline{\lambda}(C)$ for every $C \in 2^E$.

Taking $C = E$, this implies that $\overline{\lambda}(A) = \underline{\lambda}(A)$ for every $A \in \mathcal{M}$. The following lemma shows that $\mathcal{M}$ is an algebra. Later on we will show that it is in fact a σ-algebra.

Lemma C.14: The Collection $\mathcal{M}$ is an Algebra

The collection $\mathcal{M}$ contains E and is closed under complements and finite unions. Hence, it is an algebra on E.

Proof. The collection $\mathcal{M}$ is obviously closed under complements and contains E. To show that it is closed under finite unions, we need to show that $A \cup B \in \mathcal{M}$ for any $A, B \in \mathcal{M}$. Take any $C \in 2^E$. By the countable (and hence finite) subadditivity of $\overline{\lambda}$ in (C.11), we have

(C.15) $\overline{\lambda}(C) \leq \overline{\lambda}((A \cap B) \cap C) + \overline{\lambda}((A \cap B)^c \cap C).$

Since the reverse inequality also holds:

$$\begin{aligned}
\overline{\lambda}(C) &= \overline{\lambda}(B \cap C) + \overline{\lambda}(B^c \cap C) \\
&= \overline{\lambda}(A \cap B \cap C) + \overline{\lambda}(A^c \cap B \cap C) + \overline{\lambda}(A \cap B^c \cap C) + \overline{\lambda}(A^c \cap B^c \cap C) \\
&\geq \overline{\lambda}(A \cap B \cap C) + \overline{\lambda}\big((A^c \cap B \cap C) \cup (A \cap B^c \cap C) \cup (A^c \cap B^c \cap C)\big) \\
&= \overline{\lambda}((A \cap B) \cap C) + \overline{\lambda}((A \cap B)^c \cap C),
\end{aligned}$$

we have in fact equality in (C.15), which shows that $A \cup B \in \mathcal{M}$, by Definition C.12. □

When the outer measure $\overline{\lambda}$ is restricted to $\mathcal{M}$, we have the following form of countable additivity. Take $C = E$ to get the usual countable additivity.

Lemma C.16: The Outer Measure $\overline{\lambda}$ Restricted to $\mathcal{M}$ is Countably Additive

If $\{A_k\} \in \mathcal{M}$ are disjoint, then for any $C \in 2^E$:

(C.17) $$\overline{\lambda}\big((\cup_k A_k) \cap C\big) = \sum_k \overline{\lambda}(A_k \cap C).$$

Proof. We first prove the case for a *finite* collection $A_1, \ldots, A_n$ by induction, starting with $n = 2$. In that case, if $A_1 \cup A_2 = E$, then the statement follows from the definition of $\mathcal{M}$ in (C.13). If $A_1 \cup A_2$ is smaller than E, take $C := (A_1 \cup A_2) \cap D$ and $A := A_1$ in (C.13) and use the disjointness of A_1 and A_2 to conclude that

$$\overline{\lambda}(A_1 \cap D) + \overline{\lambda}(A_2 \cap D) = \overline{\lambda}((A_1 \cup A_2) \cap D),$$

showing that the induction statement holds for $n = 2$. Suppose that (C.17) holds for $n - 1$ disjoint sets $A_1, \ldots, A_{n-1}$ in $\mathcal{M}$. Then, by the induction hypothesis and the case $n = 2$, we have for any C,

$$\overline{\lambda}\big((\cup_{k=1}^{n} A_k) \cap C\big) = \overline{\lambda}\big((\cup_{k=1}^{n-1} A_k) \cap C\big) + \overline{\lambda}\big(A_n \cap C\big) = \sum_{k=1}^{n} \overline{\lambda}(A_k \cap C),$$

completing the induction step. For an *infinite* sequence $A_1, A_2, \ldots$ of disjoint sets, we have, by the monotonicity of $\overline{\lambda}$:

$$\overline{\lambda}\big((\cup_{k=1}^{\infty} A_k) \cap C\big) \geq \overline{\lambda}\big((\cup_{k=1}^{n} A_k) \cap C\big) = \sum_{k=1}^{n} \overline{\lambda}(A_k \cap C)$$

for all n, so that

$$\overline{\lambda}\big((\cup_{k=1}^{\infty} A_k) \cap C\big) \geq \sum_{k=1}^{\infty} \overline{\lambda}(A_k \cap C).$$

The proof is completed by observing that the reverse inequality also holds, by the countable subadditivity of $\overline{\lambda}$ in (C.11). □

The following proposition contains all the ingredients needed for the existence proof:

Proposition C.18: Four Main Properties of $\overline{\lambda}$ and $\mathcal{M}$

1. $\mathcal{M}$ is a σ-algebra.
2. $\overline{\lambda}$ *restricted to* $\mathcal{M}$ is countably additive.
3. $\mathcal{E}_0 \subset \mathcal{M}$.
4. For $A \in \mathcal{E}_0$, we have

(C.19) $$\overline{\lambda}(A) = \lambda(A) = |A|.$$

Proof.

1. We have already proved that $\mathcal{M}$ is an algebra in Lemma C.14. To show that it is a σ-algebra, let $A_1, A_2, \ldots$ be in $\mathcal{M}$ and $B_1 := A_1$ and $B_n := A_n \backslash A_{n-1}, n = 2, 3, \ldots$, as illustrated in Figure 1.44 in Chapter 1. The $\{B_k\}$ are disjoint and in $\mathcal{M}$. By the countable additivity of $\overline{\lambda}$ on $\mathcal{M}$ (see Lemma C.16), we have for any $C \in 2^E$:

$$\overline{\lambda}(\cup_k B_k \cap C) + \overline{\lambda}((\cup_k B_k)^c \cap C) = \overline{\lambda}(C).$$

Thus, (C.13) holds for the set $\cup_k B_k$ and the latter therefore lies in $\mathcal{M}$. But since $\cup_{k=1}^n B_k = A_n$ and $\cup_{k=1}^\infty B_k = \cup_{k=1}^\infty A_k$, the latter also lies in $\mathcal{M}$. In other words, $\mathcal{M}$ is closed under countable unions.

2. We have already proved this in Lemma C.16.

3. We need to show that for any $A \in \mathcal{E}_0$ it holds that

$$\overline{\lambda}(A \cap C) + \overline{\lambda}(A^c \cap C) = \overline{\lambda}(C) \text{ for all } C \in 2^E.$$

Similar to the proof of Lemma C.10, take $\varepsilon > 0$ and choose sets $\{A_n\}$ in $\mathcal{E}_0$ such that $C \subseteq \cup_n A_n$ and $\sum_n \lambda(A_n) \le \overline{\lambda}(C) + \varepsilon$. This can always be done, by the way $\overline{\lambda}$ is defined. Define sets $F_n := A_n \cap A$ and $G_n = A_n \cap A^c$, $n = 1, 2, \ldots$. These sets lie in $\mathcal{M}$ as it is an algebra. Note that $\cup_n F_n$ contains $C \cap A$ and $\cup_n G_n$ contains $C \cap A^c$. Thus, by the definition of $\overline{\lambda}$ and the finite additivity of λ, we have

$$\overline{\lambda}(A \cap C) + \overline{\lambda}(A^c \cap C) \le \sum_n \lambda(F_n) + \sum_n \lambda(G_n) = \sum_n \lambda(A_n) \le \overline{\lambda}(C) + \varepsilon.$$

Since ε is arbitrary, this shows that $\overline{\lambda}(A \cap C) + \overline{\lambda}(A^c \cap C) \le \overline{\lambda}(C)$. The reverse inequality holds also, by the countable subadditivity of $\overline{\lambda}$ in Lemma C.10.

4. From the definition (C.9) of the outer measure it is clear that $\overline{\lambda}(A) \le \lambda(A)$ for all $A \in \mathcal{E}_0$. To show that also $\lambda(A) \le \overline{\lambda}(A)$, we use the countable additivity of λ proved in Theorem C.6. In particular, let $A \subseteq \cup_n A_n$, where A and $A_n, n = 1, 2, \ldots$ are in $\mathcal{E}_0$. Then,

$$(C.20) \qquad \lambda(A) \le \lambda\left(\bigcup_{n=1}^\infty (A \cap A_n)\right) \le \sum_{n=1}^\infty \lambda(A \cap A_n) \le \sum_{n=1}^\infty \lambda(A_n).$$

The second inequality follows from the countable subadditivity of the premeasure λ, which can be proved in exactly the same way as for proper measures (see Proposition 1.42). As (C.20) holds for any choice of $\{A_n\}$, it must hold that $\lambda(A) \le \inf \sum_n \lambda(A_n) = \overline{\lambda}(A)$.

□

Using Proposition C.18, the existence of the Lebesgue measure can now be proved with a few brush strokes.

Theorem C.21: Existence of the Lebesgue Measure

The outer measure $\bar{\lambda}$ restricted to $\mathcal{B}_{(0,1]}$ is a probability measure that extends the pre-measure λ.

Proof. From Proposition C.18, the outer measure $\bar{\lambda}$ restricted to $\mathcal{M}$ is a proper measure on $(E, \mathcal{M})$, as it is countably additive and $\mathcal{M}$ is a σ-algebra. Also, we have

$$\mathcal{E}_0 \subset \sigma(\mathcal{E}_0) \subset \mathcal{M} \subset 2^E.$$

So, $\bar{\lambda}$ is also a measure on $(E, \sigma(E_0))$, where $\sigma(E_0) = \mathcal{B}_{(0,1]}$. That $\bar{\lambda}$ is an extension of λ follows from (C.19). Finally, the fact that $\bar{\lambda}(E) = \lambda(E) = 1$ shows that it is a probability measure. □

Note that the proofs above hold verbatim for any probability pre-measure λ on $(E, \mathcal{E}_0)$; that is, λ is σ-additive and $\lambda(E) = 1$. Theorem 1.45 then guarantees that λ can be uniquely extended to a measure on $(E, \sigma(\mathcal{E}_0))$.

BIBLIOGRAPHY

Billingsley, P. (1995). *Probability and Measure*. John Wiley & Sons, New York, 3rd edition.

Çinlar, E. (2011). *Probability and Stochastics*. Springer, New York.

Chung, K.-L. and Williams, R. J. (1990). *Introduction to Stochastic Integration*. Birkhäuser, 2nd edition.

Feller, W. (1970). *An Introduction to Probability Theory and Its Applications*, volume 1. John Wiley & Sons, New York, 2nd edition.

Grimmett, G. R. and Stirzaker, D. R. (2001). *Probability and Random Processes*. Oxford University Press, Oxford, 3rd edition.

Jacod, J. and Protter, P. (2004). *Probability Essentials*. Springer, New York, 2nd edition.

Kallenberg, O. (2021). *Foundations of Modern Probability*. Springer, New York, 3rd edition.

Karatzas, I. and Shreve, S. E. (1998). *Brownian Motion and Stochastic Calculus*. Springer, New York, 2nd edition.

Klebaner, F. (2012). *Introduction to Stochastic Calculus with Applications*. Imperial College Press, London, 3rd edition.

Kreyszig, E. (1978). *Introduction to Functional Analysis with Applications*. John Wiley & Sons, Hoboken.

Kroese, D. P., Botev, Z. I., Taimre, T., and Vaisman, R. (2019). *Data Science and Machine Learning: Mathematical and Statistical Methods*. Chapman and Hall/CRC, Boca Raton.

Kroese, D. P., Taimre, T., and Botev, Z. I. (2011). *Handbook of Monte Carlo Methods*. John Wiley & Sons, Hoboken.

Lalley, S. P. (2012). Notes on the Itô Calculus. `http://galton.uchicago.edu/~lalley/Courses/385/ItoIntegral.pdf`.

Mörters, P. and Peres, Y. (2010). *Brownian Motion*. Cambridge University Press, Cambridge.

Williams, D. (1991). *Probability with Martingales*. Cambridge University Press, Cambridge.

INDEX

For Product Safety Concerns and Information please contact our
EU representative GPSR@taylorandfrancis.com Taylor & Francis
Verlag GmbH, Kaufingerstraße 24, 80331 München, Germany